Darwinian Physics

Second edition of the book: *El Origen de las Leyes de la Física*

Written and edited by
Claudio Lillo Canelo

Translated from Spanish to English by DeepL
(Neural Machine Translation Service)

ISBN-13: 978-3-9824701-4-6

Contac in:
darwinianphysics@gmail.com

Nothingness frightens us,
but only it can give meaning to existence,
just as darkness gives meaning to light,
and death gives meaning to life...

Dedicated to all those who are brave enough to
think for themselves,
and whose curiosity knows no bounds.

Content

Second Part
A Truly New Physics

Third Part
Answers to the Great Unknowns

INTRODUCTION

Progress in the foundations of physics has stagnated in recent decades due to a lack of theoretical progress. A large elephant in the laboratory that has become increasingly difficult to ignore. Since the completion of the Standard Model of Particles, or quantum field theory, in 1973, all the theorists' predictions have failed. No extra dimensions, no supersymmetric particles, no gravitons, no chameleon particles, no axions, no cosmic strings, no mini-black holes, no magnetic monopoles, no d-branes, no wormholes, no dark dimension, no sterile neutrinos, no WIMPs, no primordial gravitational waves to confirm cosmic inflation, no parallel universes, and so on; have been found. A long list of ideas and models that have been falsified, but instead of being discarded, they are simply modified and made more and more convoluted to fit their predictions. In this way, the cycle starts all over again, with the sole purpose of producing thousands of papers that will not be successful because they will be falsified again in the new experiments. But it does not matter, because they will serve to be recycled and thus to publish thousands of more papers in order to continue to survive in the competitive academic world that favors quantity over quality in research. This is why no scientist today can afford the luxury of studying a particular problem in depth for several years, as was normal a few decades ago, because several studies a year have to be published to put food on the table. Meanwhile, the puzzles and anomalies in the foundations of physics continue to accumulate. Without going any further, 95% of the content of the universe remains a mystery, supposedly corresponding to dark energy and dark matter.

Faced with this reality, the book will attempt to return to the fundamental problems facing the two pillars of modern physics, rather than continue to ignore them. Such problems are what is time in general relativity and what is the random collapse of the wave function in quantum mechanics. This is because for much of the 20th century, we went as far as we could by stretching these two existing theoretical frameworks in a very pragmatic way. But if want to move on, and above all if we want to unify such theories, we can no longer drag these two fundamental problems along, because they are essential for such a unification; for how can we understand what space-time is in Einstein's theory of gravity, its fundamental element, without first knowing what time is, and how can we unify it with quantum mechanics if we do not even know what happens when a simple electron is measured? And despite these obvious shortcomings, incredibly, theoretical physicists prefer to ignore them and try to solve the problems of the big bang, the expansion of the universe, or the singularities in the centers of black holes; adding extra dimensions, wormholes, and parallel universes. On the other hand, time and the problem of measurement are considered pseudoscientific issues by most physicists, since they belong to the field of philosophy, not physics. An argument that is not so wrong in principle, but who said that physics does not need philosophy? In fact, everything points to the fact that this is the real problem. Not wanting to understand what the two pillars of modern physics really mean.

When science fails to find a solution, it is usually because the problem has not been really understood, because there are gaps in our knowledge and understanding of such natural phenomena that prevent us from properly describing the problem. Therefore, the answer cannot be found because the right question has not yet been formulated. This is called a paradigm shift and has happened many times in the history of science. Without going any further, this is what Einstein had to do at the beginning of the 20th century, because in order to unify Newtonian mechanics with Maxwell's electromagnetism, he had to abandon the Newtonian principles of absolute space and time in order to be consistent with the constant speed of light in Maxwell's theory. Only in this way could the relativity paradigm be achieved. Now we have the same dilemma. We again have a theory of gravity, general relativity, which is incompatible with a theory that includes electromagnetism, quantum field mechanics. Then, like Einstein, we would have to choose one of the two in order to reinterpret the other.

Up to now, physicists have opted for the classical theory of relativity and have continued with the Laplace framework of Newton's mechanics. Since then, they have tried to reinterpret quantum mechanics according to these classical principles. This has been the modus operandi of the foundations of physics up to now, to develop a classical model and then to quantize it, to adapt it to the rules of quantum mechanics. This is why all quantum models that exist today are fundamentally classical. This is confirmed by the fact that they must respect unitarity (conservation of information). A principle that transforms an inherently random and uncertain theory, such as quantum mechanics, into a completely deterministic theory. This explains why no fundamental question about what quantum mechanics is has been answered. The reason why we do not understand such a theory. It is therefore proposed to do the opposite. To develop the first authentic quantum theory by freeing it from any classical bias and prejudice from its origins. A theory that directly addresses the problem of time and the collapse of the wave function; and whose solution allows to unify these two pillars of modern physics, because they are essentially the same problem. This is because the current formulation of quantum mechanics is atemporal, since the time operator is not defined, precisely because of the problem of measurement. Therefore, the collapse mechanism proposed by Darwinian Physics would be time itself. The element that remains to be defined in general relativity.

A presentist solution to time that is the product of the recognition of two very obvious and proven natural phenomena that have been ideologically ignored due to these classical prejudices. We are talking about becoming and quantum randomness. Now it becomes clear why it has been impossible for physicists to achieve their longed-for "theory of everything," because essential elements have been omitted in order to understand the problem correctly. And what exactly was the reason for this denial? The incompatibility of these two scientific facts with the classical dogma that has governed physics since its origins: "the absolute laws of physics." A belief that inevitably leads to eternalism - which denies the existence of the flow of time and the present - and determinism - which denies the existence of ontological randomness in quantum mechanics. Therefore, what this book proposes is a simple but radical strategy: to recognize what experiments measure, and to abandon once and for all this millenarian classical doctrine that abhors "change" and "randomness;" for how can one really understand what the space-time of general relativity is if one denies the existence of time itself? And how can you solve the measurement problem of quantum mechanics if you refuse to recognize the inherent randomness of quantum systems?

So, far from what it might seem, the book's proposal is actually much more conservative and empiricist than what is currently promoted in cutting-edge physics; because if neither space has turned out to be absolute as it is created and expands, nor is time absolute as it dilates and curves due to acceleration and gravity, or even energy is not really conserved because of that same cosmological expansion; then, why should the laws of physics - a mathematical abstraction - remain absolute for all eternity, when the universe itself is created, expands and develops? Therefore, what the book proposes is to simply assume that the laws of physics themselves emerge and evolve just as everything in the universe does. An assumption that is not only more reasonable and satisfying, but also more economical, since it avoids the dreaded assumption of infinite parallel universes. Instead, it is only necessary to recognize and integrate these two empirical facts into physics: becoming and chance.

To achieve this integration, it was necessary to search for a new theoretical framework. One so different that it was found outside of physics, in a science where there is no longer any fear of becoming or chance, because they are indispensable elements for it to have integrity and meaning. We are talking about biology. That is why the current proposal is so original, because it allows the creation of a physical mechanism capable of emulating Natural Selection. Thus, just as Darwin's theory explains how the genetic code or DNA (the most stable and fundamental regularity in biology) evolves over hundreds of thousands of years thanks to the variability provided by the randomness of genetic mutation and recombination, Darwinian Physics explains how the laws of nature (the most stable regularities in this science) evolve over time and on cosmological scales thanks to the randomness provided by quantum mechanics. It is therefore a physical mechanism that operates with time and randomness, the very elements that modern physics absurdly denies.

Thanks to this mechanism, not only will the great mystery of what is time, now, randomness and the origin of the laws of nature finally be solved (questions that today's theoretical physics would never dare to ask, because they are taboo), but it will also help to solve important problems that physicists are currently working on, such as dark energy, the invariance of the speed of light, the inside of a black hole, the problem of fine tuning, the cause of the big bang, and the holy grail of physics: quantum gravity. All this thanks to banishing an old kinematic or absolutist worldview, which conceives the universe in a static and unchangeable way; and replacing it with a much more natural dynamic worldview, where the universe and its laws evolve like everything that exists in it. A new proposal that rigorously respects relativity and quantum principles, because it is based on them, by reinterpreting such theories in a more consistent way under a Darwinian Paradigm. It is therefore a model that is consistent with the experiments and that does not need to resort to additional exotic elements, such as extra dimensions or the greatest violation of Ockham's razor imaginable: the multiverse, which is the direct consequence of the eternalism and determinism of absolute laws.

It is, therefore, an economical and even conservative solution, requiring only an axiomatic update of current physics. For this very reason, however, it will be a difficult solution to accept, especially for the most orthodox physicists, since it is so radically different from what is currently understood by physics that it will require patience to go through each of its pages. It is promised, however, that no mystical questions will be invoked, nor to the god of the physicists' holes, the multiverse. All that will be required is the official theoretical framework and the current evidence, which will be duly described and explained in the book. Therefore, it will be a simple but difficult solution, because it will be necessary to conceive these two pillars of modern physics from a completely different perspective. One that will go against the most internalized prejudices and beliefs of physicists. This will be the real challenge, to abandon old preconceptions about how nature works; for if the problem of the foundations of physics is as deep as will be argued here, then the solution will also have to be very deep.

A Darwinian solution applied to physics is rigorously and systematically explained in the book, which is divided into three parts. The first, called "Cracks in the Current Paradigm," describes and critiques the way theoretical physics is understood and practiced today. It also explores the meaning of "science" and how it has evolved and progressed throughout history. In addition, it provides the first foundations of what would become the new paradigm. In the second part, called "A Truly New Physics," the explanation of time and chance is fully revealed, and how these allow the construction of a natural mechanism that explains the emergence and evolution of the universe and its laws. And finally, in the third part, "Answers to the Great Unknowns," the most extensive part of the book, it is shown how Darwinian Physics is not only compatible with the official theories of physics, but also explains them; this is proven by offering concrete solutions to the great mysteries that the current foundations of physics have not been able to solve or simply cannot address. As you can see, there is a lot to explain and justify in these pages, so let us start this odyssey right away and see if this new proposal proves to be more credible and, above all, if it has a better future than the Laplacian approach that theoretical physicists continue to stubbornly attempt today.

First Part

Cracks in the Current Paradigm

CHAPTER I

Crises in the Foundations of Physics

The last few decades have been an excellent time for physics, thanks to spectacular discoveries and advances in solid state physics, statistical mechanics, nuclear physics, condensed matter physics, quantum optics, complexity physics, astronomy, quantum computing, and so on. A myriad of specialties in which a large number of physicists work, both theoretical and experimental, who have achieved impressive feats such as the LHC (Large Hadron Collider), which discovered the Higgs boson; large ground-based telescopes, such as the ELT (Extremely Large Telescope) in Chile, or the incredible James Webb Space Telescope, which is already rewriting the evolution of the early universe; the discovery of new materials with spectacular properties, such as the famous graphene, the development of more powerful and precise lasers, the detection of gravitational waves by LIGO (Laser Interferometry Gravitational Wave Observatory), new superconducting materials, the discovery of the CMB (Cosmic Microwave Background), dark energy, and many others. In addition, many theoretical physicists have greatly helped other disciplines by applying their knowledge and skills, for example, in engineering, computer science, economics, and even biology and neuroscience, by providing better and more diverse mathematical tools that have allowed more rigorous statistical tests and more accurate models to be developed. In fact, much of the technology we enjoy today and will enjoy in the years to come is directly or indirectly derived from the work of these physicists.

However, this great success has not been shared by theorists working in the foundations of physics. A branch of physics that, as its name suggests, cannot be derived from any other. It refers to physicists who work directly with general relativity (the theory describing gravity) and/or with the Standard Model of Particles, also known as quantum field mechanics (the theory describing electromagnetism, weak and strong nuclear interactions). The problem is that general relativity was completed by Albert Einstein in 1916, while the Standard Model of Particles was completed in 1973. Since then, the work of fundamental physicists has been to try to replace these two theories by a unification of the two. The long-awaited theory of everything, or quantum gravity. For this reason, one does not want to deny that it is a monumental goal. An obviously ambitious and difficult goal. But there is also no denying the fact that very little progress has been made in the last few decades, with great promises such as loop quantum gravity or string theory leading to great disappointments. Although the same period has been accompanied by million-dollar investments in incredible experiments and a huge increase in the number of new researchers in theoretical physics departments around the world, the program has not advanced very far despite all these available resources.

Many theoretical physicists will disagree with the above and claim that there have indeed been great advances in the foundations of physics. However, the advances they cite are not (essentially) due to fundamental theoretical physics, but mainly due to experimental physics. For example, the Higgs boson was predicted in 1964, while gravitational waves are a prediction of general relativity developed in 1916. Thus, the recent discovery of the Higgs boson is mainly due to the hundreds of physicists, engineers and technicians who helped build (and continue to maintain and operate) the LHC, the largest particle accelerator (27 kilometers in diameter) and the most complex machine in the world. The other discovery was made by the physicists and engineers who built the LIGO and Virgo gravitational wave observatories. These impressive L-shaped machines, consisting of two tunnels about 4 km long, are capable of measuring differences in the structure of space-time as small as the size of an atomic nucleus. Thanks to them, a more than 100-year-old prediction of Einstein's theory was confirmed. The same happened with the incredible discovery of dark energy in 1998. The fact that the expansion of the universe is not slowing down due to the effect of gravity. On the contrary, it is accelerating, because astronomers measured the expansion rate of the universe by observing type 1A supernovae, which act like ordinary candles by always shining with a similar intensity. An extraordinary discovery that no theorist had predicted, just like the discovery of the alleged dark matter, which was mainly due to the work of the American astronomer Vera Cooper Rubin (1928-2016) in the late 1970s, who observed that the rotation of galaxies is much faster than the theory of gravity would allow. It was an impressive discovery, but physicists did not consider it worthy of a Nobel Prize. The same happened with the oscillation of neutrinos, which confirmed that these particles have mass. Something that was also not predicted, but was discovered thanks to giant particle detectors such as the Super-Kamiokande in Japan. A reservoir of 50,000 tons of extremely pure water at a depth of one kilometer, whose interior is surrounded by hypersensitive photodetectors capable of detecting the insignificant light produced by the collisions of cosmic particles with the atoms of the water in the pond. In conclusion, these are great discoveries that are not essentially due to the predictions of theoretical physicists, but to the extraordinary work of experimental physicists and engineers. Very important work that often does not get the recognition it deserves. This is quite unfair, especially since many of these advances not only help us to unravel the mysteries of nature, but also to develop technologies that ultimately have a great impact on improving the quality of our lives, from medical applications to the introduction of new materials that improve the performance of our electronic devices, to name just a few examples.

However, the criticism of the current foundations of physics is not that they failed to predict these great discoveries, because we cannot expect that nature has no surprises in store for us. What is really troubling is that having discovered them, theoretical physicists have had very little success in understanding them. Moreover, they have had very little innovation in the process. For example, the cosmological constant or lambda (Λ) that theoretical physicists use to model dark energy was a solution discarded by Einstein himself more than a century ago, and theoretical physicists have simply recycled it. A solution that is unlikely to work, since recent observations show that dark energy may not be constant as previously thought, since measurements of the acceleration of the most recent universe are greater than the expansion rate measured in the earliest universe. In addition, its interpretation as vacuum energy differs from observations by a factor of 10^{120} (a difference of one followed by 120 zeros). The largest difference between prediction and observation in the history of science. As for the anomalies in the rotation of the galaxies, theoretical physicists decided to solve them by adding a new particle to the Standard Model. The famous dark matter. A hypothetical particle that interacts very little with light and for which no evidence has been found, even though it has been exhaustively searched for decades in many very sophisticated experiments. Meanwhile, the other option to explain the anomalies observed by Vera Rubin would be MOND (Modified Newtonian Dynamics). As the name implies, it is simply a matter of adding new parameters to the old Newtonian equation of gravity to force the model to fit the observations. In both cases, it does not explore the possibility that we may not really understand how gravity works on a larger scale, and that a deeper understanding of the principles of Einstein's general relativity is needed. Especially when we don't even know what space-time (the fundamental element of this theory) is and why it expands, or whether it is quantum or continuous. And these are only two examples, which would correspond to 95% of the content of the universe (70% dark energy and 25% dark matter). However, there are countless other problems to be added to the remaining 5% that we should understand very well, such as: the problem of strong CP symmetry, the reason for the low entropy in the Big Bang, the baryonic asymmetry, the origin of the tiny mass of neutrinos, the singularities that appear in the Big Bang and in the center of black holes. We do not even know what really happens when you measure a single electron. But the most important problem of all is the incompatibility of general relativity with the rest of the fundamental interactions (quantum field theory). Therefore, the crisis in the foundations of physics is obvious. This is recognized by more and more theoretical physicists who, since the completion of the Standard Model of Particles in 1973, have not succeeded in going further with any of their predictions. Countless new hypotheses, conjectures, and models are published every day that turn out to be false or impossible to falsify.

Given this scenario, it is reasonable to ask why this program has not progressed as far as the physicists themselves had hoped. But instead of being a bit self-critical and seriously considering the possibility that they might be on the wrong track, the vast majority of the theorists of the foundations of physics prefer to "flee forward," vigorously defending the same strategy of the last decades, excusing the lack of progress by the lack of new and more complicated mathematical tools, and/or by the need to build bigger and more expensive experiments that can confirm the same predictions, which they modify over time, making them more and more convoluted so that they remain compatible with the negative results of the experiments that falsified them. This strategy only makes the new theories and models more complicated and speculative, requiring more and more assumptions and conjectures. A picture that is far removed from the latest observations and evidence that reveal a much simpler and more elegant nature than theoretical physicists are willing to accept. For example, countless predictions have been made in the CMB over the last few decades, such as Hawkins rings, cosmic strings, mini-black holes, magnetic monopoles, primordial gravitational waves, and so on. But nothing extra has been observed. The CMB remains incredibly clean, simple, and uniform. The same goes for the prediction of countless new particles and extra dimensions at the LHC, but nothing has been discovered except the Higgs boson. It is the last missing piece of the Standard Model. It is therefore quite reasonable to think that it is time to consider other strategies. New avenues of research, a Plan B, because even if the traditional way of doing physics was incredibly fruitful in the first half of the 20th century, there is no guarantee that it will continue to work in the same way in the future. Especially with the poor results we have had in the last few decades. In fact, we do not even want to say that the current strategy should be completely abandoned, because there is still a chance that it could work. However, it must be recognized that the chances are not as good as expected. Not only because of the lack of evidence, but also because the strategy itself makes less and less sense, because it is an incredibly orthodox strategy, in the sense that it tries to explain the big recent discoveries like dark matter and dark energy, while changing as little as possible what is understood about the two pillars of modern physics. And it is not possible to make a leap of this magnitude, an authentic theory of everything, without abandoning some erroneous preconceptions of the world. Therefore, the work of the theorists of the foundations of physics is essentially to add new quantum fields to the Standard Model of Particles. It is as simple as that. However, if we want to unify the two main theories of physics, it would be logical to have to deal with them in depth, because many elementary questions remain unanswered, such as: why does light have exactly the same speed for all observers, why are we trapped in the present if it supposedly does not exist, where does the randomness in quantum mechanics come from, what does the wave function really represent, and so on. It is not surprising that modern physics is stagnating, for how can we move forward if we do not even understand well what we already have? Therefore, everything indicates that instead of speculating and assuming new elements such as: exotic particles, parallel universes, or extra dimensions; we should go back to the deposits and solve the mysteries that we have certainty that they exist.

CHAPTER II

How Science Progresses

In light of the first chapter, it is reasonable to assume that the main cause of the current crisis in the foundations of physics is an outdated strategy. To find out if this hypothesis is true, we must first understand how physics, and science in general, evolves and progresses in order to know in what situation we find ourselves today. To do this, we must turn to philosophy and the history of science, the disciplines responsible for recording and studying this process. Thus, we know from them that the epistemology used by physicists, or the way in which scientific knowledge is acquired, is mainly based on the theory of Karl Raimund Popper (1902-1994). An Austrian philosopher who developed his famous falsification principle, which stems from the logical impossibility of proving the truth of a theory or any hypothesis due to the problem of induction. This means that endless proofs are needed to ensure that a statement is completely true. The typical example of this is the statement that "all crows are black." To verify this, we would have to check every crow that exists on the planet. Every single one of them. However, to prove that such a hypothesis is false, it is enough to find a single specimen that is a different color to disprove such a claim. Therefore, proving that a theory is false (falsificationism) is infinitely easier than proving that it is true (verificationism). Thus, according to this principle, science would advance by means of a kind of theoretical darwinism, in which only the most resistant theories survive, those that have been able to survive all the experiments imposed on them without being falsified.

However, it did not take long to show that falsifiability as the sole criterion of scientificity is insufficient to fully define how science works. First, not all falsifiable hypotheses are necessarily scientific. For example, I can hypothesize that "5G antennas will turn all people within a radius of 1000 meters into colorful unicorns under a two-month exposure." Is this a falsifiable hypothesis? Yes, because it is an easily testable prediction. Just install an antenna and wait the two months to see if colored unicorns appear. So, is it necessary to fund an experiment to falsify it? Obviously not, because it makes no sense. Not only because it has no logical or theoretical support, but also because there is no empirical evidence to support such a suspicion. On the other hand, it is possible that there are hypotheses and even complete theories that are scientific, or at least considered to be scientific, but which cannot be falsified. The best example of this is the famous string theory. A theory that many physicists not only consider scientific, but believe to be the true theory of everything; however, it is still a non-falsifiable theory because everything it predicts is far beyond what we can test, according to them. Predictions like extra dimensions, strings, parallel universes, and branes. While predictions that would be falsifiable in principle, such as supersymmetry, have become so flexible that they can be modified at will to justify why experiments have failed to find them. Exactly what Popper considered unscientific. The same thing happens with the theory of cosmic inflation, which tries to solve the problems of the Big Bang theory and which many physicists think should get the Nobel Prize now, although it has not really been proven. This is because it originally predicted primordial gravitational waves in the Cosmic Microwave Background (CMB), but since they have not been found, it has been modified to produce even versions of the theory that produce primordial gravitational waves so weak that they are undetectable. In fact, there are thousands of versions of inflation to suit every taste. Therefore, the idea cannot be falsified in general either. Thus, if we were really strict with Popper's principle and considered the first predictions of the above two theories (supersymmetry at low energies and measurable primordial gravitational waves), we should have discarded them long ago. However, far from it, both theories are still very much alive today.

But the problems with Popper's falsification principle do not stop there. We have already seen that it has problems with non-scientific claims, and with claims that are supposedly scientific but cannot be falsified. In addition, however, this principle fails when applied to scientific theories that we are absolutely certain are true, given the overwhelming and solid evidence that supports them. This is the case of Darwin's Natural Selection, which Popper himself says would not be a scientific theory. We are talking about one of the best theories in all of science, without a doubt. To believe otherwise would be like believing that the earth is flat. So why did Popper think that darwinian evolution was not scientific? Because of its inability to make "falsifiable predictions" or predictions that can be rigorously tested by observation or experiment. Consider, for example, the following statement: "All biological characteristics have been naturally selected because they provide a survival advantage to the individual." This statement is false because we can observe many biological characteristics that do not serve any function in survival. On the contrary, they hinder it. The most emblematic example of this is the huge and heavy tail of the male peacock, which makes it very difficult for him to escape from predators. So, in light of the above, we could modify the statement by saying, "All biological characteristics have been selected to serve a survival or reproductive function (sexual selection)." Here we could include the tail of the male peacock and an immense variety of other biological traits, but the statement would still be false because many other traits can be found in virtually all species that serve no useful purpose, either sexual or survival, because such traits may simply be remnants of past adaptations that were conserved for no reason, or they may be anomalies that are the product of statistical deviations (genetic drift). In short, the vagaries of chance. Without going any further, in our case alone we can find several examples such as: the appendix, the palmaris longus (arm muscle for climbing trees), the wisdom tooth, the arrector pili muscle (which raised the hair so that we would appear larger like cats), the muscles in the ears, the male nipples, and so on. Many features that have vague or no function. So, any of this evidence would falsify evolution by Natural Selection. This was the problem that Popper had, the ambiguity in Darwin's theory, or his inability to make concrete statements, gave him the impression that Natural

Selection lacked scientific rigor because it was too flexible, allowing him to modify his statements as he saw fit to adapt and fit with new observations that were discovered along the way. Just the opposite of the spirit of falsification.

To understand this scientific spirit to which Karl Popper referred, let us look at some examples that he himself gave. One of them was communism, because in this theory Karl Marx predicted that the revolution of the proletariat would take place in the most industrialized or developed countries, since this social class existed there. Therefore, communism would have to begin in England, the most industrialized country of the time. In reality, however, communism arose in less developed countries such as Russia, China and Cuba. Countries where there was no proletariat because there was virtually no industry. In fact, the communist revolution really took place in the peasant class. Nevertheless, instead of falsifying their theory, communist theorists prefer to justify it by saying that it did not happen that way because of a capitalist conspiracy or something like that. Another example Popper used was psychoanalysis and the theory of dream interpretation by the Austrian neurologist and psychiatrist Sigmund Freud (1856-1939). His thesis was that all dreams are the product of wishful thinking. Although this was patently false given the existence of nightmares, for some strange reason, instead of replacing or improving his theory, Sigmund Freud continued to defend it, even going so far as to say that his patients' nightmares were caused by their desire to prove him wrong. Simply unbelievable. In conclusion, both communism and psychoanalysis reveal Karl Popper's main criticism, the attitude of never considering the real possibility of being wrong. Or, by default, the intention to confirm at all costs that one is right. This is what Popper finds unscientific, fanaticism. It produces very plastic statements that can be modified as much as necessary to fit the facts, forcing confirmation. And, on the contrary, a scientific theory is one that has the spirit or the courage to prove its falsity as quickly and easily as possible; by providing verifiable consequences that, if not found in nature, would disprove the theory. The iconic example of this spirit for Popper was the theory of general relativity, where Einstein made a very precise prediction: that the light from stars near a solar eclipse should bend before the distortion of spacetime around the Sun, so that a change in position should be observed. Exactly 1.74 arcseconds. Twice what Newton's theory predicts (0.87). So, if this exact shift of the stars was not observed, then Einstein's theory of gravity was wrong. Something we already know did not happen, since his prediction was confirmed in 1919.

Therefore, if Karl Popper were alive, he would believe that string theory and cosmic inflation are unscientific theories like communism and psychoanalysis, since they have a similar dogmatic attitude, conveniently modifying their statements because the results of experiments do not verify them. The problem, however, is that the same could be applied to Darwin's theory. A theory that we already know is not only scientific, but one of the most important in all of science. Even Karl Popper himself later had to admit this error, declaring that Natural Selection is a genuine scientific theory, although he could never explain exactly why Darwin's theory now fits his definition of science. And he could not, because it is simply not possible. Falsificationism applies only to exact sciences, such as many branches of physics, but not to all sciences. This is because the ability to make predictions that are sufficiently accurate to be rigorously tested is not a property of a science per se, of its method, but of the object it studies. Therefore, if an authentic science happens to study very complicated or even simple but chaotic or complex systems, it will unfortunately not be able to rigorously apply falsificationism. Not because it is not a science, but because in such systems it is not possible to make very precise predictions that can be rigorously tested. Natural Selection falls into this category, but so do the other branches of biology, medicine, psychology, ecology, neuroscience, paleontology, and so on. Sciences in which it is very difficult to make predictions that are falsifiable, because the margins of error of the predictions are very large, or it is very easy to make mistakes; however, this does not mean that they cease to be sciences, because the fault does not lie with the discipline, but with the nature of the system being studied. This is very evident in geology, where it has been impossible to predict earthquakes or volcanic eruptions, which would have saved thousands of lives. Or in the science that causes us the most problems in everyday life, meteorology, where its predictions about the weather often cause us a lot of inconvenience, telling us one thing and finding another. Both are true sciences in which scientists do their best, but they cannot avoid the fact that the systems they study are too complicated or chaotic. A complexity that even physics itself cannot avoid, since even the most exact science of all cannot escape uncertainty when studying complex systems, such as chaos theory itself, statistical mechanics in complex systems (here we study gases, liquids, and solids; we try to predict macroscopic behaviors based on microscopic properties), physics in non-equilibrium systems (systems where several symmetries are broken and therefore it is quite complicated to do the calculations), and many others such as the physics of fluids, plasmas, magma, and so on. In conclusion, rigorous falsification is not a characteristic of science, but of a few natural systems. That is why Natural Selection remains one of the most important scientific theories, even though it cannot be falsified like general relativity.

But even if we limit ourselves to the most exact and simple physics, falsificationism would still have problems, because it cannot be applied to framework or general theories either. It can only be applied to hypotheses or specific aspects of such paradigms. For example, the best representative of falsification for Popper, general relativity, falls into this category; because it is not falsifiable as a whole, because it is also a very flexible theoretical framework. For example, it allows a parameter, the famous cosmological constant (Λ), to be any positive, negative, or zero value. It also does not define the number of spatial or temporal dimensions, so you can do general relativity with 3, 20, or 1000 dimensions. Ideal for string theorists. The type of curvature possible is also not defined, so it can generate closed finite universes, but also flat or open infinite universes. Furthermore, it can create hypothetical objects such as wormholes, allowing theorists to play with them, as is the case with string theory through its "ER=EPR" conjecture. None of the above can falsify general relativity, since it concerns its degrees of freedom. Therefore, it does not depend on the theory itself, but on the observations. For example, if we see a closed or open universe, we will make the theory that way. Seeing it one way or the other will not determine whether the theory is wrong or not. The same thing happens, and more obviously, in the Standard Model of Particles, because here a large number of new particles can be added. Obviously, if certain conditions are met, but there is no limit to the number of quantum fields it can contain. That is why the work of theoretical particle physicists is essentially to invent new quantum fields like the inflaton, the graviton, the dark photon, other scalar fields like the Higgs, a chameleon field, dark matter particle fields, or to duplicate all existing fields by means of supersymmetry. And although none of these predictions have been verified, the Standard Model has not been falsified by them.

In fact, the inadequacy of falsificationism is particularly evident here, because if we were really rigorous, we should have already falsified the Standard Model of Particles, and not just once, but several times. For example, in quantum chromodynamics (QCD), the theory describing the strong nuclear force requires the violation of CP symmetry, just as it does for the weak nuclear force. The problem is that such a violation has been observed in the latter, but not in the strong interaction. This anomaly is called the strong CP problem, and it is due to the fact that the quarks that make up the neutron and the proton have an electric charge. Therefore, although the neutron is neutral on its scale, on a smaller scale an electric dipole moment should be seen due to the electric charges of its quarks. However, this magnetic field of the neutron was not observed. This is the origin of the famous axion, a hypothetical particle that appears from time to time in the press as a perennial dark matter candidate. But before being recycled for this function, the axion was originally postulated to clean up the strong CP problem. Hence its detergent name. The axion problem was almost immediately falsified, since its effects should have been seen in experiments by now. So, it became more and more convoluted to the point that it is now called axion-like particles, since there is not just one, but countless of them. However, the model has been modified so much that none of them fulfills its original function of solving the strong CP problem. Therefore, the idea is recycled to try to be a dark matter candidate. And this was not the only unobserved prediction in the strong interaction theory, since it also predicts gluon balls, which should also have been detected at the energy scales probed by the LHC. Meanwhile, on the weak interaction side, neutrinos were found to be massive. A Nobel Prize-winning discovery. The problem is that the Higgs field, the only mechanism the Standard Model has to give mass to particles, works through two components that massive particles have, the so-called left and right components. But all the neutrinos we observe are left-handed. In other words, they lack the right-handed component, so the Higgs mechanism cannot be responsible for giving neutrinos their tiny mass. Therefore, the Standard Model predicts massless neutrinos, and yet these particles are massive. In addition, it should not be forgotten that the Standard Model's particle model lacks a candidate for dark matter, among many other details. But despite all these problems, far from being falsified, quantum field mechanics is still very much alive. It is simply assumed that the theory is incomplete, that new constants and particles need to be added, and that it needs to be adapted to new observations and experiments. Exactly the opposite of the spirit of falsificationism.

For all these reasons, it is clear that falsificationism, as the only criterion of scientificity, is insufficient to tell us what science is and what it is not. Not only because it does not apply to all cases, but also because it generates a very naive vision of how science progresses, assuming a continuous and linear progress, denying a rather convulsive and disruptive history that science has had. A Popperian perspective that is closer to that of classical economists, who assume smooth and continuous growth, denying that there have been many, many economic booms followed by great crashes in the history of economics and finance. Thus, while it is a very practical and useful principle, as seen in much of the 20th century (especially in defining the attitude of the scientist), it is also a biased and incomplete view of how science actually progresses. For completeness, we should consider the proposal of the physicist and philosopher Thomas S. Kuhn (1922-1996), who in his book The Structure of Scientific Revolutions, published in 1962, shows that the evolution of science has not always been linear, but has sometimes advanced by leaps and bounds. A more realistic perspective, because nothing in the universe evolves forever in a continuous and linear way, but goes through stages, phases. For example, in the evolution of the universe, recombination occurred (when the plasma cooled enough to form the atoms of the universe, creating the CMB). Or, more mundanely, the abrupt changes in state that matter undergoes (plasma, gas, liquid, solid, Bose-Einstein condensate) as temperature and pressure gradually change. Or in biological evolution itself, where the notion of linear and gradual progress of classical Darwinism has also been abandoned in favor of an evolution that alternates between periods of stability and rapid evolution, as postulated by the theory of punctuated equilibrium. An updated version of Darwinism that explains exactly what is observed in the fossil record. Therefore, the alleged missing links do not disprove Natural Selection because they are based on a false type of Darwinism; because new species can appear very quickly, obviously in geological terms. For example, there was a Cambrian explosion. A period in which sex evolved and produced the sudden appearance of a great variety of complex multicellular macroscopic organisms, only about 500 million years ago. In contrast, the previous three billion years had been dominated by unicellular life such as bacteria. A much duller ecosystem than the one we have today. This evolutionary path is even more evident in human activity, where we were nomads or hunter-gatherers for about 300 thousand years, only to become sedentary and build the first civilizations about 5 thousand years ago. Or through antiquity to the Middle Ages, and from there to the Enlightenment and modernity, a period in which we have experienced four industrial revolutions, the last of which is underway with artificial intelligence and robotics. In all these cases, change has been abrupt rather than gradual. So why shouldn't science (another human activity) be disruptive?

Then, assuming a more discrete evolution of science, Kuhn identifies three stages through which science must pass. The first is normal science, where a particular scientific paradigm is developed and improved. Here, Popper's falsification principle works quite well because a generally accepted theoretical framework or paradigm already exists. Therefore, it only needs to be improved by falsification between specific competing possibilities arising from the same paradigm, as we have seen with general relativity and quantum mechanics throughout the 20th century until now. This is why Kuhn said that "Ordinary science consists mainly of a cleaning job. Experimenters carry out modified versions of experiments that have been done many times before, while theorists make small modifications to their theories that are more aesthetic than structural." For example, in the current paradigm, to explain dark energy, a fixed term (Λ) is simply added to the equations of general relativity.

A modification that does not change the essence of the theory at all. And to explain dark matter, a new particle with such properties (massive and interacting very weakly with light) is simply added to the Standard Model of Particles. Therefore, as Kuhn described it very well: "Under normal conditions, the scientific researcher is not an innovator, but a solver of puzzles. And the puzzles on which he concentrates are precisely those which he believes to be definable and solvable within the existing scientific tradition." Although this may seem to be a very negative criticism of normal science, Kuhn himself recognizes that this stage, this way of doing science, is very necessary and important. If all scientists had to start from scratch and question every fundamental and ontological assumption that underlies their paradigm on a daily basis, they would be hard pressed to achieve the level of technical sophistication necessary to make any useful breakthroughs. In other words, this lack of disbelief or too much obedience to the assumptions of the paradigm is necessary for its development, because in order to make new advances, especially at the current rate of scientific publication (the famous "publish or perish"), one cannot always start from scratch with each new study. It is necessary to make an act of faith that all previous research and experiments are valid and correctly interpreted in order to be at the forefront of science. In other words, to be hired and to have a job as a scientist today, one must almost unconsciously give up almost all freedom to question the foundations of one's knowledge.

However, although the above attitude produces great progress in much of the process of science, Kuhn also reminds us that at another stage this way of doing science can be disastrous. We are talking about when this theoretical framework or paradigm goes into crisis, a situation that has happened several times in the history of science. Without going any further, the last stage of crisis that was overcome was at the end of the 19th century, when the theoretical framework of Newtonian and Maxwellian physics could no longer accurately describe the orbit of mercury, nor the specific heat of compound gases, nor why the luminous ether was not detected, nor could it explain the radiation spectrum of a black body (the light that emanates only from the object, for example, the sun), among other experimental observations. Moreover, they could not unify these two theoretical frameworks. A situation repeated a century later with dark energy and dark matter and the other anomalies mentioned above. And again, the two main theories (general relativity and quantum field mechanics) are incompatible. Therefore, the current crisis in physics is very similar to its predecessor, which would perfectly explain the lack of progress in the foundations of physics in the last decades. And what happened after the last crisis? A revolution, a paradigm shift. Exactly what physics needs today.

But although Kuhn had a much more realistic view of the progress of science than Popper, he was wrong about how such a revolution takes place, for it is true that the history of physics is not perfectly continuous and gradual, but neither is it completely disruptive. In this sense, Kuhn was too radical in his approach, denying any kind of continuity and progress, which is obviously false. Obviously, 20th-century physics is much better than 19th-century physics (which was better than medieval physics). Moreover, modern physics does not completely deny ancient physics, but assimilates it. For example, Kepler's laws did not disappear, but emerged from Newton's laws, which in turn can be derived from Einstein's general theory of relativity. In fact, Newtonian mechanics is still used, for example, to send satellites into space, people to the International Space Station, and robotic probes to Mars, even though we have had general relativity for more than a century. This is because Einstein's theory is only necessary when gravity is very strong and/or when we are traveling at very high speeds, close to the speed of light. However, when gravity is less intense and velocities are low, Newtonian mechanics is an excellent approximation. In fact, Newton's theory became a special case of Einstein's theory. This is an excellent example of how science progresses, because it shows that the new theory cannot negate the old one. On the contrary, it must absorb it, integrate it, transform it into an approximation of a more fundamental and general theory. In other words, the new theory must explain everything that the old theory explained, and more, in a completely coherent way.

This semi-disruptive evolution, by replacing the wrong parts of the previous physics while preserving its right parts, is best described by the Hungarian economist, philosopher, and mathematician of Jewish origin, Imre Lakatos (1922-1974). In his proposal, paradigms or theoretical frameworks are called Scientific Research Programs (SRPs), and within them there is a hard core where the principles, assumptions, or ontological postulates assumed to be evident in the paradigm or SRP would be located. These ideas are assumed not because of empirical evidence, but for convenience, as they provide internal consistency to the SRP itself. In other words, they would be the basic premises or axioms from which the postulates of the framework theory are derived. Then, because of their obvious importance in supporting the entire logical apparatus of the SRP, a protective belt is built around this hard core, consisting of a series of secondary or auxiliary hypotheses that can be eliminated, replaced, or appended with new ones. This is where Popper's falsification principle would be applied and executed. Then a normal science would be developed, which in the case of Lakatos would be applied by means of a positive heuristic that guides the development of the SRP, determining the problems and the investigations to be carried out. At the same time, a negative heuristic is established, which seeks to discourage or avoid attacking the dogmas of the paradigm, punishing, ridiculing or isolating those who dare to criticize the postulates of the hard core. Thus, these two heuristics generate the so-called progressive stage in SRP. The Lakatosian equivalent of Kuhn's normal science stage. The stage in which the paradigm or SRP successfully evolves, strengthening or hardening the hard core more and more. In this sense, the negative heuristic would not be so negative for the same reason as Kuhn's normal science stage, since one cannot constantly question the basic principles or postulates, especially if they are validated as the conclusions derived from them are confirmed by experiments. In other words, it is the very success of the SRP that reaffirms confidence in the hard-core ontological postulates. Thus, there is always controversy and (healthy) skepticism at the beginning of a paradigm or an SRP, but as it succeeds in proving its predictions, the critics either disappear of their own accord or are considered less and less. So, the negative heuristics would be natural, a product of the SRP's own success. However, precisely for this reason, these heuristics begin to become a big problem when the theoretical framework or SRP stops working as usual, with fewer and fewer successes and more and more anomalies. This is when we enter the crisis stage, or as Lakatos calls it, the degenerative stage. A stage where predictions are no longer experimentally confirmed and the SRP stagnates. This is where the scientific revolution or the birth of a new SRP would be needed.

In Lakatos' approach, which is much less radical than Kuhn's, it allows us to better deduce how this paradigm shift would occur, which would be the result of altering or modifying the hard core that sustains the old SRP. In other words, when one or more of the ontological or fundamental postulates are eliminated, added, or replaced. It is not possible to change the entire hard core by renewing all the fundamental premises, which is exactly what Thomas S. Kuhn would be suggesting, because there would be no continuity, no progress, but a completely new science, totally disconnected from the previous science. And therefore not a better science. On the contrary, it would omit or deny the successes proven by the experiments of the previous scientific paradigm. That would be very unscientific. Not everything can be wrong in an SRP, otherwise there would not have been a normal scientific phase or progressive stage. Thus, an authentic new SRP must rescue and maintain the postulates or ontological premises of the old hard core that have been validated by the previous progressive stage, and identify and replace the few postulates that prevent the explanation of the anomalies that appear. This is how science progresses and improves in the long run, by making small adjustments to the hard core. Adjustments that, although small, produce great revolutions in the cosmovision we have of nature, because we have to abandon principles that were thought to be obvious and elementary. And at the same time, thanks to the fact that a large part of the hard core is preserved, a continuity is generated between the different paradigms. This makes it possible to generate coherence in the progress of science over time, ensuring that progress is always made towards a better paradigm. This is what makes science science.

In summary, we can say that science would be the product of a delicate balance between two very powerful forces. Rationalism (the power of reason), which believes that the only way to truth is through deductive logic, and Empiricism (the power of sense or experience), which believes that the only way to truth is through rigorous observation. The former is represented by theory or the logical body, which in physics has mathematics as its highest representative. Empiricism, on the other hand, is represented by evidence and by meticulous and systematic measurements carried out according to the principle of reproducibility. In the case of physics, this distinction can be clearly seen, since theoretical physicists are essentially rationalists, while experimental physicists are essentially empiricists. At the stage of normal or progressive science, rationalism coexists harmoniously with empiricism. There is no conflict. On the contrary, they reinforce each other, since experiments and empirical observations confirm and drive the development of theory; while theory, in turn, allows the design of new experiments in search of more evidence to further confirm the theory by confirming its predictions. This creates a strong positive heuristic that makes scientists increasingly confident in the ontological postulates of the hard core. On the other hand, in the crisis or degenerative stage, this confidence turns against them, since the long and successful previous stage has made scientists have a conviction, a very strong faith in the paradigm and therefore in the premises of the hard core. This is where empiricism comes into conflict with rationalism, as various anomalies begin to appear and accumulate that the theory is unable to explain. But since scientists are trained only to do normal science, they are not even aware that a paradigm shift is an option. So, the last thing they think or want to believe is that their beloved paradigm is wrong, so a strong negative heuristic is generated to desperately defend the hard-core postulates by ignoring or downplaying the various anomalies as they appear. And if science is still science, in the sense that empiricism must always have the last word, then a scientific revolution should take place, since the highest authority in science is the experiments and not any particular reason. This is the great difference that places science above philosophy and politics. Therefore, the stagnation of the paradigm or the SRP would force scientists to rethink and question the postulates of the hard core, until finally the ontological postulate that prevents explaining such anomalies would be found, which would allow to create a new paradigm with an updated hard core, ready to generate a new stage of normal science, and thus repeat the whole cycle again.

In conclusion, to overcome a crisis stage, each of the postulates or ontological principles of the hard core of the paradigm must be re-evaluated. A critical analysis to determine which might be the premise that would fail. As the French philosopher, mathematician, and physicist René Descartes (1596-1650) once said: "In order to arrive at the truth, it is necessary, once in a lifetime, to get rid of all the received ideas and to rebuild our whole system of knowledge from scratch." It is a very difficult, meticulous and delicate work, because any mistake in the modification of the hard core can be disastrous. It is like trying to cure a genetic disease. It is not enough to modify the DNA by eliminating the gene that causes the disease or by adding the healthy gene, because it is also necessary to avoid creating new diseases by disturbing the rest of the genes that were fine. For example, it is not enough to cure multiple sclerosis by genetic modification if the lungs eventually become deformed and the person is unable to breathe. We must make sure that the medicine is not worse than the disease, by making the proper analysis, impartially evaluating all the postulates, to find the premise that needs to be replaced, explaining why the paradigm or SRP has become stagnant, carrying an assumption that no longer fits the degree of precision or rigor that has been achieved from nature. In this way, the mere change of a single ontological postulate of the hard core is enough to provoke an authentic scientific revolution, deriving from it an authentic new paradigm. Recall that scientific theories are deductive in nature; therefore, a modification in the set of axioms will significantly change the conclusions. For example, the last paradigm shift in physics occurred when Albert Einstein eliminated the fundamental dogma or postulate that space and time are absolute and immutable entities, proposed by Newton himself, and replaced it with the axiom that "the speed of light is the maximum possible and the same for all observers." With this small change, Einstein derived his entire theory of special relativity, in which space and time are the protagonists, able to contract and expand. The next example is the birth of the other pillar of modern physics, also helped by Einstein, by changing the postulate of infinitesimal continuity of electromagnetic waves to discrete behavior: photons. A small change that led to quantum mechanics. Two examples that show how important it is to protect the hard core in order to maintain the SRP, because any attack, no matter how small, on the heart of the paradigm can kill it, allowing another radically different paradigm to emerge and take its place.

CHAPTER III

HOW TO REACH THE NEW PARADIGM

As we saw in the previous chapter, the progress of science is not so obvious, because we must be very careful in the crisis stage. Unfortunately, the progress of science is taken for granted by the vast majority of scientists due to their ignorance of the history and philosophy of science and their specialization in the current paradigm where only normal science is taught. This has led to the generation of a dogmatic rationalism, which makes scientists fanatical or very uncritical of their current paradigm, because after a long and successful stage of normal science, it has turned confidence in the theory into blind faith. This is why it has been so difficult to overcome the stages of crisis in science, especially in physics, which explains the great resistance there has been to abandoning beliefs such as: the earth is the center of the universe, space and time are independent and absolute entities, light needs a medium to propagate (the famous luminiferous ether), conservation of matter, and so on, along with many other ontological assumptions that seemed quite obvious, elegant, and beautiful at the time, but which ultimately turned out to be wrong.

In fact, so strong is the devotion to such ontological assumptions that in every paradigm great physicists appear predicting the end of physics, being convinced that nothing can be better, that they are in the ultimate SRP, the perfect paradigm. For example, the best physicist of the 19th century, the Englishman William Thomson, better known by his honorary title Lord Kelvin (1824-1907), addressed the British Association for the Advancement of Science in 1900 with these words: "There is nothing new left to discover in physics. All that remains is to make more and more accurate measurements." He could not have been more wrong, since just a few years later a German physicist named Max Karl Ernst Ludwig Planck (1858-1947) revolutionized everything, taking the first step towards quantum mechanics. The same man who, in 1875, when he began to study at the University of Munich, was advised by his physics professor, Philipp von Jolly, not to devote himself to that science, since there was nothing left to discover in that discipline. And if you think that these are mistakes of the past, that modern physics cannot be so naive, on April 29, 1980, the famous English cosmological physicist Stephen Hawking (1942-2018) dedicated his inaugural lecture as Lucasian Professor at the University of Cambridge to this question: "Is the end of theoretical physics in sight"? And his answer was yes, because of his enthusiasm for the then fashionable supergravity theory and superstring theory. A widespread opinion among physicists at that time, supported by the great success of the electroweak theory, which unified two of the four fundamental interactions. Therefore, it was believed that it was a matter of time to obtain GUT (Grand Unification Theory), which would integrate the strong nuclear interaction. And with a little more time, it was believed that the theory of everything, which adds gravity, would be obtained. A great enthusiasm that has almost completely disappeared today, due to the great disappointment that has resulted from all these attempts.

Therefore, the lack of knowledge of how science evolves has meant that practically every scientific revolution in history has required the fortuitous appearance of a true rebel. Someone who is not afraid to contradict authority. People who have exiled themselves from the orthodox stream of science and then, against all odds, managed to convince their colleagues. Unfortunately, after much hostility and even, on many occasions, several years after they have passed away. And these are only the cases we know of, the few that have succeeded, but we completely ignore the many who swam against the current and, despite having been right, drowned in complete indifference. A great resistance on the part of the scientific community has even led to suicide, as was the case of the Austrian physicist, father of statistical mechanics and modern thermodynamics, Ludwig Eduard Boltzmann (1844-1906) who, although he already suffered from some mental disorders, it is undeniable that the indifference and even the constant attacks on his work have contributed significantly to his unfortunate decision to take his own life. Criticism was aimed at his thermodynamic theory, which assumed the existence of atoms. A hypothesis that at that time many considered pseudo-scientific because it was believed to be impossible to prove. He was ridiculed on several occasions for that reason. Fortunately, there are still a few rebels stubborn enough to withstand such hostility. One of them was Albert Einstein, who not only had to face the strong conformism of his time, but also had to endure the direct attack by his professors, who came to hate him for his constant questioning and criticism of the obsolete physics they taught. A hatred that reached the point of sabotaging his job possibilities, preventing him from finding a job as a physicist. For this reason, Einstein had to work in a patent office in Switzerland. In fact, he seriously considered giving up physics and selling insurance. Fortunately, Einstein had enough conviction, tenacity, and luck to carry on; because, most probably, without him we would not have the degree of development that physics has reached, losing in the process much of the technology we enjoy today.

However, it is neither responsible nor wise to depend in every crisis on the chance appearance of a genius hard-headed enough to withstand all the resistance of the scientific community. Especially, if the present situation is in no better condition than the last crisis in promoting the scientific revolution. In fact, it is much worse, because we are educated so intensively that it does not even occur to us to revolutionize science, because the entire academic system is designed to do only normal science, with the sole objective of developing the paradigm as much as possible and getting the maximum practical benefit from it (generating technology).Therefore, today we have an intense positive and negative heuristics that shield the hard core through a protective belt, which generates an almost blind obedience to the dogmas or fundamental postulates of the reigning paradigm; one of its consequences being the culture of publish or perish that promotes short and fast research and studies, privileging quantity over quality. A modus operandi that discourages penetrating the paradigm to reach its core, because there is no time available to rethink the worldview and resolve fundamental questions. All because of the need to publish soon, since it is evaluated by number of publications and citations. This means that scientists have to research topics that are in fashion so that their publications are accepted by scientific journals and thus be in demand. An essential criterion, since without it they simply cannot find a job, and cannot survive in the incredibly competitive market of science. This causes researchers to remain permanently on the surface of the paradigm. And not even in a large area, let alone the entire surface, because they can only investigate a very specific area of it. An academic phenomenon that we will call extreme specialization.

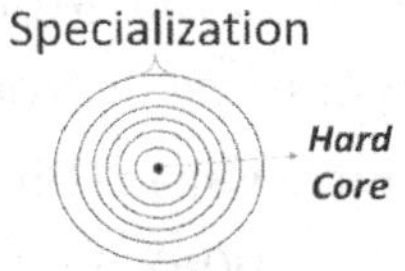

As can be seen in the figure, the sole purpose of current academic training is to expand the scope of the paradigm, little by little, by adding new layers to it. A fantastic academic strategy in the normal science stage, but which becomes a double-edged sword in the crisis stage. A stage where the paradigm has to be replaced, not developed. And extreme specialization makes it very difficult to do just that, because it not only distances scientists from the place where the problem really is, the hard core; but also distances them from where we could find the possible solution, in other disciplines and sciences. This is because to revolutionize you have to be very creative and innovative. And a great help in generating this is multidisciplinarity, because studying other disciplines and sciences is the best way to get new ideas that can be successfully applied in other areas of research. A skill that is almost impossible to develop today due to extreme specialization. Hence the famous phrase of the Spanish philosopher José Ortega y Gasset (1883 - 1955):

"We know more and more about less, until we know everything about nothing." A trend that makes scientists more and more ignorant of how much they do not know, drastically reducing the creative process by having a more and more biased and myopic knowledge. This generates an increasingly rigid way of thinking, which makes it very difficult to think outside the box, since we are practically educated to believe that everything that exists is inside that box.

A good example of the worrying consequences of extreme specialization is what is known as Nobel Prize disease, the Nobel effect, Nobel syndrome or Nobelitis. A pathology that has been suffered from Pierre Curie (Physics, 1903) to Santiago Ramón y Cajal (Medicine, 1906), including Richard Smalley (Chemistry, 1996) and Luc Montagnier (Medicine, 2008), to name but a few. This is a term that has often been used to refer to the fact that being very intelligent in a small area does not necessarily mean being very intelligent in other areas of knowledge as well. For example, Linus Pauling (1901-1994), an American scientist and winner of two Nobel Prizes (Chemistry in 1954 and Peace in 1962), was a pioneer of modern chemistry for his discoveries about the nature of chemical bonds and the molecular structure of matter through the application of quantum mechanics. However, Pauling did not hesitate to assert that high doses of vitamin C could be effective in curing diseases such as cancer and the common flu. His studies contained multiple errors, and such effectiveness was never demonstrated. Or James Watson (1928 - present), American scientist who won the Nobel Prize for Medicine in 1962 (which he shared with Maurice Wilkins and Francis Crick) for his discovery of the double helix structure of DNA. A finding considered one of the most important in modern science. However, the controversial biologist maintained that black people were less intelligent than white people. A difference in IQ that he said was due to genetic factors. He also said that exposure to sunlight in regions near the equator increases the sex drive and that fat people are less ambitious. Or Luc Montagnier (1932-2022), a French virologist who won the Nobel Prize for Medicine in 2008 for having succeeded in isolating the human immunodeficiency virus (HIV) for the first time. A year after being honored with this award, he claimed without proof that water can recall electromagnetic waves supposedly emitted by the DNA of viruses and bacteria. He also recommended fermented papaya against Parkinson's disease and criticized vaccines against Covid-19, which he accused without foundation of being the cause of the emergence of new variants of the virus. Another case is the Norwegian-born American physicist Ivar Giaever (1929-present), who shared the Nobel Prize in Physics with Leo Esaki and Brian Josephson in 1973, for his discoveries on tunneling phenomena in solids. On several occasions, the researcher expressed his skepticism towards global warming, which he said was not a problem and even called it a "new religion." Or Frederick Banting, the discoverer of insulin in the early years of the last century, who barely cured some patients with diabetes, thought he could cure cancer as well. He tried, but since he knew so little about the subject, he obviously did not succeed.

Examples like the above abound, which reveal the danger of extreme specialization, because it makes you very ignorant about the rest of scientific knowledge outside your specialty. But because you are recognized in your field, you gain confidence, falling into the Dunning-Kruger effect on steroids.

A cognitive bias that makes people think they are smart on issues where they know very little. The same ignorance that causes them to lack the skills that would allow them to correctly measure their real capabilities in that area, so they end up overestimating themselves. As Charles Darwin once said: "Ignorance more frequently begets confidence than does knowledge." But the real danger of extreme specialization is not to make a fool of yourself when you give an opinion outside your specialty, but what happens when the paradigm is in the crisis stage, as it becomes a major obstacle to make a scientific revolution. This is because extreme specialization makes researchers become very obedient and loyal to the paradigm they should overthrow, being all that is known. Thus, they are deprived of the sources of knowledge and inspiration that would provide the keys to overcome it, which are just outside their specialty. This is confirmed by the fact that most of the innovations (whether scientific, technological, or business) have been the result of having a broad and diverse knowledge that allowed combining two or more different disciplines, which were thought to be unconnected. It is because of this that multidisciplinary groups are promoted to solve difficult problems in a disruptive way, because creativity arises when different ideas are combined, when knowledge that was thought to lack any relationship is connected, but that, from a deeper or broader perspective, one discovers that they are compatible or coherent.

But the paradigm's efforts to remain on the throne are not limited only to the institutional or structural aspects of the academic system; it relies on heuristics at the most informal levels as well. An excellent example of this is a mantra that the vast majority of theoretical physicists repeat religiously, and which is promoted by the most prestigious and respected physicists. The idea that "the best guide to the foundations of physics is mathematical beauty." An idea that, despite being evidently unscientific, is so popular that it is practically a must for theoretical physicists. This sociological phenomenon is well documented by German theoretical physicist and quantum gravity specialist and researcher at the Frankfurt Institute for Advanced Studies Sabine Hossenfelder in her book Lost in Math: How Beauty Leads Physics Astray, published in 2018. Here she reveals how theoretical physicists are led astray by some very strong principles of how nature has to be, in aesthetic terms. Obviously subjective assumptions that very few question. And which, despite their irrationality, strongly guide the course of research in physics. Some examples given in that book are phrases like: "The physicist should strive for mathematical beauty in his attempt to express the fundamental laws of nature," or worse: "It is more important to have beauty in the equations than to make them fit the experiments." Both phrases said by the British physicist and mathematician Paul Dirac (1920-1984), Nobel laureate and a legend of 20th century physics. Or: "The more symmetry, the more beautiful" by Anthony Zee, professor of physics at the University of California; or: "I see beauty as a path to knowledge, or even the only knowledge worth knowing" by astrophysicist Edward Arthut Milne (1896-1950); or: "My work has always been to unite truth with beauty, but when I have had to choose between the one and the other, I usually choose beauty," by mathematician and physicist Hermann Weyl (1885-1955); or: " In the foundations of physics, a beautiful and elegant theory is more likely to be correct than a less beautiful one" by Murray Gell-Mann, American physicist and winner of the Nobel Prize in 1969; or also: "In this expedition, I do not think there is no more promising guide than beauty itself" by Frank Wilczek, another American physicist and winner of the Nobel Prize in 2004. And so, we could continue to fill the whole book with similar phrases from very prestigious physicists, who are role models for many new physicists.

But, however attractive, romantic, poetic, or corny this love of mathematical beauty preached by physicists may seem, the problem is just that, it is not a scientific criterion. To begin with, it is evidently subjective, since it leaves out, in a completely arbitrary way, other possible mathematical aesthetics. For example, chaos theory, with its hypnotic unrepeatable patterns, its nonlinearity, and its infinite fractal complexity, may be incredibly beautiful, even sublime, to many (myself included); but according to the current theoretical physicists' canons of beauty, this kind of mathematics is repugnant. In fact, that was how they referred to it when chaos theory was just being developed, receiving indifference and even hostility from much of physics, who labeled it as not doing science. So, clearly physicists are as human as any other person, perhaps too human, being carried away too much by their prejudices and subjectivity at the moment of doing science, because beauty in physics has changed many times. Something that the great German physicist, father of quantum mechanics, Max Planck, experienced firsthand when he said: "A new scientific truth does not triumph by convincing its opponents and making them see the light, but rather because its opponents eventually die, and a new generation grows up that is familiar with it." These sayings were his response to the end of the beauty of classical physics. An era of physics where everything was certain and continuous. Whereas quantum physics, in comparison, was extremely ugly for physicists because it was discrete and probabilistic. Even one of its most important founding fathers, the Austrian-Irish Nobel Prize-winning physicist Erwin Schrödinger (1887-1961) went so far as to say that he regretted having had anything to do with quantum mechanics. A sentiment he shared with other physics greats such as Louis de Broglie and Albert Einstein. Both Nobel Prize winners for participating in the birth of quantum mechanics as well, but who also could not accept what their creation became. So, clearly aesthetic tastes in physics have changed, because horrible quantum physics is incredibly beautiful to physicists now, apparently. So, we cannot use "beauty" as a criterion of scientificity because it is simply unscientific, being subjective. And, despite this, today's theoretical physicists continue to use aesthetics not as a mere inspiration, but as the maxim in physics. This is clearly unprofessional, and even dangerous for the progress of this science.

This obsession for a very particular mathematical beauty is reinforced and even provoked by another type of specialization, but this time of a methodological nature, the monopoly of mathematics in theoretical physics. This is because mathematics has ceased to be a tool at the service of the profession, to become itself the profession. Now the physicist is more mathematician than physicist. And please do not misunderstand me, no merit is being taken away from mathematics. Quite the contrary, it is being recognized that it is of vital importance not only in physics, but for all sciences. In fact, mathematics really made science, providing the rigor and precision that characterize it. The only thing to remember is that it is a tool. Most probably, the most powerful tool ever invented by human beings, but a tool, after all. And like any tool, its usefulness depends on how it is used. A hammer can be very useful for building a house, but it can also be used to hit yourself in the head. So, what is the usefulness of mathematics in physics? Well, in the hands of physicists it has become an incredibly rigorous and precise logical tool, specialized in answering "how?" questions. For example: how do systems evolve in time, how do objects fall, how do electrons move? Questions whose function is essentially descriptive. In other words, mathematics serves to build models. However, what the vast majority of theoretical physicists always forget is what a Polish-American philosopher Alfred Korzybski warned: "The map is not the territory." This means that a model is simply a graphic or symbolic representation of reality. A simplification. And no matter how accurate and precise it is with observations, it does not necessarily mean that reality is exactly that way. Therefore, mathematics alone is not sufficient to tell us what is really real, the ultimate purpose of science. However, mathematics is the only tool that theoretical physicists use to study nature. In fact, they take great pride in it. For this reason, the approach of today's theoretical physics is excessively pragmatic. They are not interested in whether their model is realistic or not. They are only concerned that it works, that it serves to deliver predictions that can be contracted with experiments, causing them to constantly confuse a good effective representation with the way nature itself works. A mistake very similar to that made by economists, simply because their equations are a good approximation of what happens in the markets. Except for the big difference that the predictions of physicists are much more reliable than those of economists, obviously.

But why exactly should theoretical physicists be wary of taking mathematics too seriously? Because of what has just been said, that it is only a tool. Specifically, a language. Again, the most precise and rigorous language mankind has ever invented, but a language, nonetheless. And just as there are thousands of spoken languages, so too can there be many different mathematics. In fact, in principle, infinite. It is therefore a mistake to consider mathematics as the only way to truth because there are infinite mathematical paths and truths. For example, the expression "2 + 2 = 5" has historically been considered as an analogy or metaphor for what is obviously false or wrong. However, in mathematics "2 + 2" can be "5." In fact, it can be any number, because in mathematics you must define what you are adding and define well what it means to add. For example, if you are adding temperatures, it will not yield the result we normally think of. If I add a glass of water at 20°C with a glass of the same content at 20°C as well, the result will not be 40°C. It will still be 20°C.

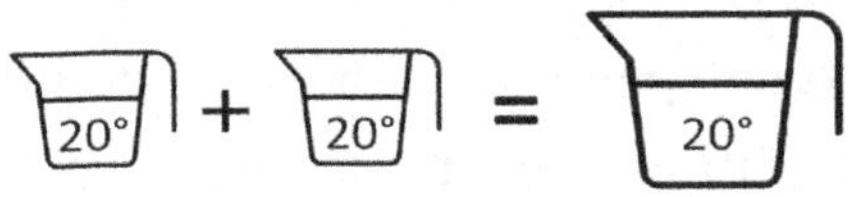

This is because temperature is the average of the kinetic energy of the molecules, and averages do not add up in the same way that we add apples. Another example of how important it is to know what is being added is the speed of light. We will cover this in much more detail in chapter XIII, but, as a preview, one cannot add up the speed of light with the speed of the source or where the light is coming from. For example, if one goes inside a train at 50 km/h and throws a ball at 10 km/h, the speed of the ball with respect to a person watching from outside the train would be 60 km/h (50 km/h + 10 km/h =

60 km/h). Nothing unusual so far, since the speed of the ball is added to the speed of the train. However, if instead of throwing a ball, the person traveling inside the train were to turn on a flashlight, light would come out of it at a speed "c" or 299 792 km/s (assuming it is in a vacuum. This is just a thought experiment). The strange thing is that a person standing outside the train will not see the light moving at the speed "c + 50 km/h," but only at "c." This is true. Even if the train travels at 90% of "c," the light will not leave the flashlight at "1.9 c." Or if the person were to point in the opposite direction that the train is traveling, the speed of light would not be "0.1 c." In other words, the speed of light is neither added nor subtracted, it is always constant. This is the most important principle of Einstein's relativity, something that has been proven and confirmed by experiments countless times.

So, we have already realized that in the real world 2 + 2 is not always 4, since it depends on what is being added. But this is also true depending on how we define the operation "sum," or what addition rule we use. For example, we can create an ordered group that is composed of 4 elements: 0, 1, 2, 3. Thus, "1 + 1 = 2," the usual. No surprise here. But if we add "2 + 2" this time the answer will be "0." It will seem strange to you, but it is supernormal in set mathematics. In fact, this operation is called "modular arithmetic." And although it may seem far-fetched, you use it every day. For example, if we take a set of 12 elements, we basically have a clock. Thus, "11 + 8 = 7" and "9 + 6 = 3," because you cannot exceed the number 12 in the clock. These examples are called a cyclic group, and clocks are an excellent case of that, because after striking 12 o'clock we redial the 1, 2, 3....

In conclusion, in mathematics 2 + 2 can perfectly well be 5 or any other number imaginable, since it depends on what is being added and what addition rule is being used. And this is completely valid in mathematics; in fact, mathematicians are working very hard to show that the sum of the infinite powers of the number two yields -1. That is $2^0 + 2^1 + 2^2 + 2^3 + 2^4 + 2^5 + 2^6 + \ldots$ or $1 + 2 + 4 + 8 + 16 + 32 + 64 + \ldots = -1$. In fact, they are pretty sure that this is so. And they achieve this by redefining the concept of distance, which they have called the 2-adic metric. In fact, mathematicians (and physicists too) believe that this holds not only for the number 2, but for every prime number p (p-adic metric), where each of these metrics gives rise to a new type of numbers that are neither real nor complex, and which have become the central notion in modern number theory.

After all of the above, it has become clear that mathematics is a language, because just like spoken or written language, there are many types of mathematics as well. And although it is the most exact and rigorous language of all, it is not perfect. A fact ironically demonstrated by mathematics itself. It all started when mathematicians wanted to prove that their discipline was not a language, or at least not an ordinary or vulgar language, but a perfect one. A language that was exact, rigorous, and free of contradictions; for ordinary language can be paradoxical at times. For example, we can write the sentence: "This statement is false." A sentence that makes no sense, because if the statement were true, then it would be wrong to say that it is false, and if the statement were false, then it would be true to be right. These types of paradoxes are called self-referential, since they assert something about themselves, which can generate contradictions. Mathematicians believed that mathematics was free of this defect of vulgar language, that in mathematics one could always rigorously prove whether a mathematical statement is true or false. The greatest representative of this current, one of the most important mathematicians of recent times, was David Hilbert (1862 - 1943). He postulated that mathematics was complete (that all necessary axioms can be obtained), consistent (that such axioms do not contradict each other) and decidable (that in a finite number of steps a mathematical statement can be proved to be true or false). He therefore encouraged his colleagues to prove each of these three characteristics to prove that mathematics is perfect. However, there was someone who was not so sure about that. That someone was none other than another great Austrian mathematician and logician named Kurt Gödel (1906-1978). To find out he had a brilliant idea, to translate a self-referential sentence, as we saw above, into mathematical equations through a code. In this way, a complex mathematical idea could be expressed in a single number, allowing mathematics to talk about itself. Through this method, Gödel was able to write: "This statement cannot be proved" as an equation, thus creating the first self-referential mathematical statement. Thus, if the statement is false, it means that the statement has a proof, but if a mathematical statement has a proof, then it must be true, and its truth says that it cannot be proved. This contradiction means that Gödel's statement cannot be false, because if there is a proof it would be validating the statement, but if there is not, it would be validating the statement as well. Therefore, it means that we now have a true mathematical equation that states that it cannot be proven. This revelation is at the heart of the incompleteness theorem developed by Gödel in 1931, which introduces a whole new class of mathematical statements, for statements are still true or false, but true statements can be provable or unprovable within a given set of axioms.

To understand what this means, we must know that a theorem is a logical proof that seeks to demonstrate whether a statement is true or not. This is done through a body or logical structures that are deduced from a set of assumptions or fundamental premises that we call axioms, which are statements considered self-evident and therefore do not require proof (i.e., a theorem), because if these premises could be justified from other premises, then they would cease to be axioms and would become conclusions of another theorem. All mathematically based systems, from the most complex proof to the simplest arithmetic, are built from axioms. Thus, if a statement about a certain number is true, mathematicians can confirm it through an axiomatic proof. A procedure that has been used since ancient Greece. Thus, mathematicians were convinced that mathematics lacked contradictions, that it was a perfect language. Therefore, the ultimate dream was to discover all the axioms from which all mathematics could be deduced and extracted. In other words, to axiomatize the whole discipline, to reduce all that vast mathematical universe to a set of a few axioms. The source code from which all mathematical truths can be extracted.

However, Gödel destroyed this dream, because in all axiomatic systems there are undemonstrable true statements. Therefore, it is impossible to create a complete system using mathematics, because there will always be true statements that we cannot prove. All this is because, if the axioms in such a formulation do not contradict each other, then there would be statements deducible from them that could neither be proved nor disproved. In other words, there would always be statements deducible from a given coherent set of axioms whose validity could not be proved, so that, if one wants to consider such a statement as true, it would have to be entered as a new axiom. This would produce a new set of axioms from which new propositions would be derived, from which again propositions would be derived whose validity could not be proved, so if you wanted to take these propositions back as true, you would have to add them as new axioms, which would produce a new set of axioms from which new propositions would be derived... and so on and so forth. No matter how many axioms you add, there will always be true statements that cannot be proved or demonstrated. This result put an end to the mathematicians' aspiration that all mathematical statements would someday be proved or disproved. It proved, ironically, that any formal mathematical theory cannot be simultaneously consistent and complete, bringing uncertainty to the heart of the most exact and rigorous discipline of all.

And if this result was not enough to bury this yearning for perfection, the third characteristic postulated by Hilbert, the last remaining hope, was also proven false, this time by another great genius named Alan Turing (1912 - 1954). A British mathematician, logician, theoretical computer scientist, cryptographer, philosopher, and theoretical biologist who proved that mathematics cannot be decidable either. A demonstration that is known as the halting problem, and that he discovered when he was developing his famous theory known as: the Turing machine. The theoretical foundation that allowed the construction of all our computational technology. So, thanks to Turing we now know that there is no guarantee that there are mathematical truths that can be demonstrated in a finite number of operations. In other words, a computer, no matter how fast and powerful it is, can be computing for thousands or millions of years, and even then, we could not know whether or not it will prove a certain mathematical statement, because it could be forever trying. In conclusion, thanks to Gödel and Turing we know that mathematics cannot guarantee that it is complete, consistent, or decidable; so not even mathematics, the most exact and precise language that can exist, turned out to be a perfect language.

And despite this irrefutable mathematical truth, theoretical physicists continue to trust them blindly, having a practically religious faith that mathematics is all they need. Their only way to the truth. However, the most important lesson these two geniuses leave us with is that we cannot blindly trust any set of axioms or fundamental premises, because no matter how obvious they may seem to us, they may turn out to be simply erroneous biases or incomplete truths about how the world works. Something that not even a solid and rigorous development of the deductive body of such assumptions can guarantee us that they are not, as we saw in the incompleteness theorem. A mathematical truth that is unquestionable and that is also directly transportable to physics, since the equivalent of axioms in mathematics are the ontological assumptions of the hard core that sustains the SRP of physics. Therefore, if even rigorous mathematics, the great goddess of theoretical physics, is not able to ensure that its axioms are complete truths, then how can theoretical physicists be so sure that the ontological postulates of their current paradigm are all true? This is the great teaching of the incompleteness theorem, that we must always doubt those ideas that we justly believe to be obvious and self-evident, because the theorems (our logical deductions or models) will be as true as their axioms (fundamental premises). And there is no way to prove that those axioms are really true. A reasonable and healthy doubt that becomes even more necessary and even indispensable when the paradigm is going through a stage of crisis as it is happening nowadays.

However, many physicists will disagree with the above by saying that it is not really a problem, because it does not matter what ontological assumptions we have, or if there are different mathematical ones, or even if they cannot be complete and consistent; because physics has the experiment. In other words, mathematicians are not interested in the real world. They are not limited by it. In fact, mathematicians create their own worlds, where their theorems will be true for all eternity; because whatever axioms they invent will be true in one of those Platonic worlds. A physicist, on the other hand, must be concerned that the mathematics and fundamental premises he is using are representing something of the real world, an authentic aspect of nature. Thus, the physicist, no matter how mathematical he has become, needs to monitor, and verify his equations through observations, to ensure that those mathematical truths are physical truths as well. Then, if the logical conclusions of the theory (the predictions) do not agree with the experiments, the theory is dead, period. Therefore, it is empiricism that determines whether the ontological premises of the model or theory are correct or not, not us. This is the famous scientific method, the idea that by integrating experiment and rigorous observations one is guaranteeing that the mathematical model is an accurate description of nature. That is, modern science would essentially work by taking the available data to design a mathematical model that describes it, and then taking new data to confirm the model. And when anomalous data is found in the new experiments, it is simply replaced by a new one that better describes the updated observations. This is the naive vision of continuous progress that many physicists imagine. An arms race between experiment and theory, where new experiments destroy established theories through Popper's falsification principle. This encourages the creation of new theories, which in turn encourages the construction of new experiments, and so on. However, we already know that this only works at the level of normal science, that is, within a paradigm, because in order to design new experiments and interpret their results, it is necessary to have a worldview. Precisely what would be the problem in a crisis stage. Therefore, the constant accumulation of anomalies could help us to realize that we are in the crisis stage, a sign that the paradigm is becoming obsolete; however, it would not necessarily help us to inspire the creation of a new and better paradigm.

We need only review the history of physics to discover that experiment or observations alone are incapable of revealing to us what nature is really like. A classic example of this is the argument between the geocentric model of Ptolemy and the heliocentric model of Copernicus. The former ruled unchallenged from the 2nd century to the 16th century CE, that is, for more than a millennium, defending the popular and beautiful (for the astronomers of that time) belief that the Earth was the center of the universe. A model that, as erroneous as it was, was surprisingly better adjusted to the observations of the planets than the model of Copernicus, who correctly placed the Sun at the center. The explanation for the superiority of Ptolemy's model lay in allowing the inclusion of as many epicycles (small circular orbits within the circular orbits of the planets) as necessary. In fact, in the best Ptolemaic models, between 40 and 60 epicycles per planet were incorporated. They were unknowingly making a complex Fourier series, where anything imaginable can be drawn by adding more and more epicycles.

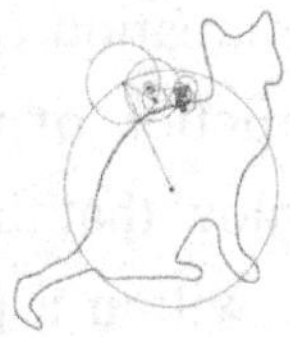

This freedom allowed this convoluted description to better fit the astronomical observations. Recall that the Copernican model, although it got the central position of the Sun right, still retained the erroneous (but equally beautiful to them) dogma that the orbits had to be perfect circles, just like Ptolemy's epicycles. Therefore, it did not ideally fit the trajectories of the planets. This was corrected only with the heliocentric model of Kepler, who finally abandoned the dogma of circular orbits by implementing elliptical orbits. A model that was strongly criticized by the astronomers of that time, finding the ellipses horrible in comparison to the beautiful circle.

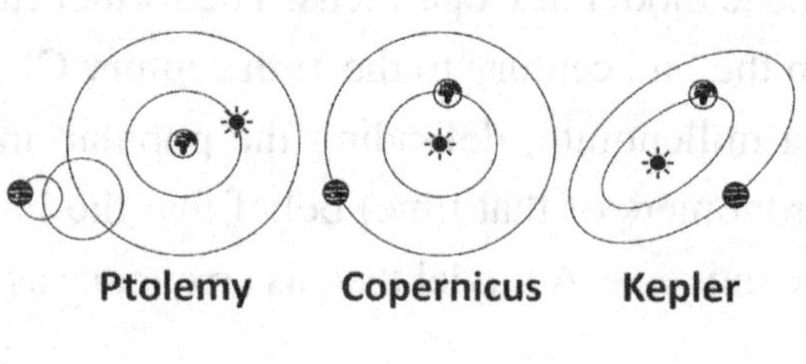

However, despite the alleged failure or lack of empirical success of the Copernican model compared to the Ptolemaic model, we all agree that the Copernican model is much more realistic or valid than the Ptolemaic model, even if the latter is more practical or useful in predicting the position of the planets. Therefore, the fact that a model fits better to empirical observations does not necessarily make it a reliable description of nature. Moreover, we can say this now, with all the highly accurate astronomical information we have today; however, at the time, one could not know this in the face of the available evidence. Therefore, it was as valid to defend the geocentric model as the heliocentric one, since seeking precision (what Ptolemy's model did) was as valid as seeking simplicity (what Copernicus' model did). A millennial dilemma that is still as present today as it was then, because no matter how developed our technology is, nor how advanced our experiments are, this is an intrinsic problem of science. For example, in 2020, governments around the world had to make a very important decision: how to deal with the Covid-19 pandemic. There were basically two strategies: suppression (imposing strong quarantines, social distance, and masks to limit virus transmission as much as possible) and mitigation (maintaining normality as much as possible, concentrating efforts on the health system so that it does not collapse, treating only the most vulnerable, while betting on mass immunity). Many countries chose the latter strategy, including England, because of the obvious economic and political benefits, being the least unpopular. However, in March of that year, the head of the Covid-19 response team at Imperial College London published a study in which a model appeared showing that the suppression strategy would save hundreds or thousands more lives than the mitigation strategy.

The first author was Neil Ferguson, professor of mathematical biology at Imperial College, and leader of the Imperial College Covid-19 response team. Faced with this study, many governments of different countries decided to change their strategy to suppression, including England, Germany, USA, and France. However, there was one exception, Sweden. Swedish scientists strongly criticized the English study, the main critic being Johan Giesecke, professor at the Karolinska Institute, Stockholm, and WHO epidemiologist, who declared that the conclusion of the Imperial College study was wrong. Regardless of what we can say now, having much more information, the truth is that at that time it was a completely valid debate, although I believe that one side was much more criticized than the other. However, the question arises, how did such respected scientists have completely opposite conclusions if they were based on the same data at the time? This is a confusion that many common people have, but also many scientists, to believe that science is completely objective, being based on experiments and verifiable observations. This phenomenon is called underdetermination in philosophy of science and refers to the situation where two or more theories are compatible in the face of the same evidence. In fact, this phenomenon is supported by a mathematical theorem called interpolation. To understand how such a principle works, let us look at the following graphical example.

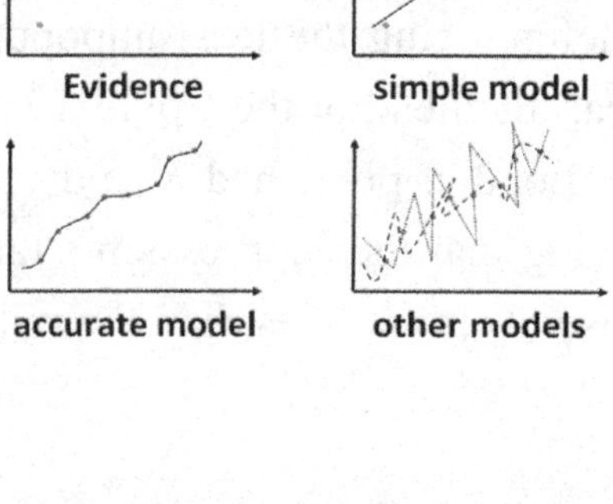

In the first graph we have the data, the evidence. With them we can build a simple model, a straight line that does not go through all the data, but needs very few parameters, and makes relatively accurate predictions. However, someone may complain and say that the model is too simple, because it is not very accurate with the available data. Therefore, a more rigorous theory can be constructed, which does explain all the data. However, it requires more parameters, which we will have to define what they mean, what they refer to in the real world. This would be the third graph, the one on the bottom left. However, as you can see in the last graph, we can actually build infinite models or theories with many more parameters that fit the data incredibly well. This is what the interpolation theorem refers to, that given a set of data there are infinitely many curves that can fit it. This is a problem that physicists know very well since much of their work is to add parameters to their models to make them fit new observations. A problem already identified by a famous Hungarian mathematician and physicist and naturalized American citizen named John Von Neumann (1903-1957) who said: "With four parameters I can adjust an elephant, and with five I can make it move its trunk." Something very similar to the epicycles of the complex Fourier series that we saw earlier.

So how do scientists decide which model is the most appropriate or valid if there are potentially infinite models compatible with the available scientific evidence? This is what Thomas Kuhn also investigated, and he identified three criteria that scientists use to prefer one model over another. These are precision or accuracy, simplicity, and broad scope. So, for example, the dilemma that existed between English and Swedish scientists with respect to Covid-19 was summarized in the fact that the former favored simplicity, while the latter criticized it precisely for being too simple. This makes sense, since a pandemic in a civilization such as the present one is obviously complex, with cultural, economic, political, technological, commercial, and so on. variables. And as you may have noticed, it is essentially the same dilemma that occurred hundreds of years ago between the model of Ptolemy and Copernicus, where the former prioritized accuracy, while the latter prioritized simplicity. So, although we have a set of very precise and reliable data, thanks to the scientific method, still, from a very fundamental point of view, the choice of the model that represents them will be determined by subjective criteria, because there will always be potentially infinite possible models that are able to explain the same set of data.

But if you are still not convinced of the above, believing that it is not a serious problem in science, or at least not for physics, being the most exact and rigorous science of all; then, let us look at the comparison between Newtonian mechanics and analytical or Lagrangian mechanics in classical physics. Both models, mathematically speaking, are indistinguishable because they give the same result. However, they are clearly not the same, because they describe very different ways of how nature works. Newtonian mechanics assumes a single trajectory determined by forces, while analytic mechanics assumes a choice between infinite trajectories according to the principle of least action. Thus, in one we have a single story, while in the other we have infinite possibilities, of which one always materializes. But if you are still not convinced because it is still a very old example, then let us look at an example in one of the pillars of modern physics: quantum mechanics. Here we have Heisenberg's matrix version, which many prefer because it is totally abstract and therefore more faithful to the spirit of quantum mechanics according to them, since it makes no reference to anything real or known to us in nature. However, the most popular is Schrödinger's wave version, which is the most intuitive for physicists, since they know the mathematics of waves much better, due to Maxwell's electromagnetism and the harmonic oscillator. But in addition to these two we can add Feynman's path integral formulation, based on analytical mechanics and where particles take infinite trajectories to go from point A to point B. Here again the three theories are mathematically equivalent and explain the same experimental data, although they describe radically different ways of how nature would operate.

So, according to the above, which is the correct one? Which model is more realistic? Which is the most faithful description of quantum systems? Abstract matrices, probability waves or infinite parallel worlds? This question, in the world of mathematics, would make as much sense as having to choose between the "2+2" model and the "8/2" model to represent the number 4. This is the problem of mathematics, which is not really a problem, because mathematics specializes in just that, in describing, in modeling. That is its purpose, to describe the world in a practical way for a certain purpose. Not to say that the world is necessarily so. For that reason, we still use the model of electricity as a fluid of electrons flowing through copper wires, even though we have known for decades that electricity does not work that way. But even though we know that the view of electricity as a fluid is unrealistic, we still use it in electronics because it is practical. The same goes for Newton's current use of gravity for space missions, and his idea that objects with mass attract; even though we have known for over a century that gravity is not a force, but the product of the curvature of space-time. So, does this mean that mathematics is no good? Quite the opposite, it means that mathematics is too good at its job, because if we want a model that fits the observations, but at the same time satisfies certain desires or whims of how we want nature to be, then mathematics can give you what you want. Obviously, it's not that simple. That's why you have to be a physicist. Someone who is very good at mathematics. But if you master this tool very well, and you know what you're looking for, then you can always find some mathematical gadget that meets your needs, because the world of mathematics is so vast (in fact, infinite), that you're very likely to find a tailor-made solution. And that is extremely useful and practical, but also dangerous. If we want the Earth to be at the center of the universe and the orbits to be beautifully circular, there we have it. If we also want it to conform to empirical observations, no problem, more epicycles. So, that mathematics is so good at giving us what we want may turn out to be a curse, because maybe what we want doesn't fit what nature is like. In fact, it is most likely, based on the history of science, since we have always been wrong about how the world is. And all indications are that it will be that way forever. So, we cannot blindly trust mathematics, even if it is supported by empirical experiments and observations.

This care with mathematics is even more subtle, because not only must we be careful with which mathematical model to choose (interpolation theorem), but we must even be careful with a mathematical model already enshrined, which we know to be valid. This is because the theories must fulfill certain physicality conditions as well. What does this mean? Models can have several solutions, which are perfectly valid in mathematical terms; however, some of these may lack any physical sense. Take for example one of the most important theories today, special relativity. Many have heard that this theory prohibits anything with mass from reaching the speed of light, let alone exceeding it. However, the truth is that this mathematical model does allow solutions for hyperluminal objects. In fact, for such objects traveling faster than light is the simplest thing to do. They are called tachyons. However, although it does not technically violate Einstein's theory, very few physicists believe that these hypothetical particles exist, because they violate many other principles shared by other accepted theories. For example, tachyons at their minimum energy level would have infinite velocity. In other words, the easiest thing for them to do is to travel instantaneously anywhere in the universe, completely violating the principle of locality and causality. Whereas, slowing down to approach the speed of light would require infinite energy. In other words, they work completely backwards to the world we observe, since they need to gain energy to travel slower, but to go faster they need to lose it. In addition, it generates instability in quantum fields, and they would have negative energy, which would make as much sense as negative distance. And if this were not enough, the tachyon should have imaginary mass, what does that mean? Imaginary energy too. The same goes for Einstein's other theory, general relativity, since its mathematics allows for solutions called Einstein-Rosen bridges, more popularly known as wormholes. These are solutions that would need negative matter or energy again. Antigravitational objects that instead of going down would go up if we let them fall to the ground, and that are necessary to keep the wormhole open and thus not collapse into a black hole.

Then, despite obtaining valid solutions in the theories that are mathematically consistent, they can be completely physically invalid, since they do not respect the conditions of realism or applicability; which are simply the properties or parameters that we observe and measure in nature, and from which the physical model must be located and respected. Therefore, they would be exotic solutions that would be discarded by experiments; however, they would not be falsifying the theory, because they are suppressed solutions. This is the difference between a purely mathematical model and a physical one. The real world. So far so sensible. The problem is that physicists are making more and more mathematical sense than physical, forgetting that mathematics is only a tool, so they allow themselves to be tempted or seduced by some of these exotic solutions of their models, taking more and more mathematical liberties at the expense of realism. Then, they transform these solutions into predictions or extensions of the models, putting them in the only place where they can place them, in the places where experiments have not yet arrived. That is, at very early times in the universe or at very high energy scales. In this way, theoretical physicists take advantage of this empirical vacuum to play with extra spatial and temporal dimensions, negative energy, imaginary time, AdS geometries, retrocausality, parallel universes, supersymmetries, variable speed of light, axions; in short, a large part of the cutting-edge research in the foundations of physics. A work that becomes more and more speculative and extravagant as the conditions of physicality or realism become more and more diluted. That is why the title of German physicist Sabine Hossenfelder's first book was so appropriate: lost in math.

In conclusion, because of the incompleteness and interpolation theorems, together with the realism or physicality conditions, we cannot blindly trust mathematics, nor can we blindly trust experiments, because neither can assure us that we are not going down the wrong path. Therefore, it is necessary to add another tool to science to further ensure that we are not fooling ourselves by wanting to confirm our own preconceptions of how the world should be through an effective theory. And who better to do this job than the very one responsible for discovering these flaws in the first place: philosophy. But before justifying the above, we must make it clear that we are not referring to a particular philosophical school or current, but the discipline itself: its method of rigorous, systematic and above all critical thinking, which provides the ability to question everything, especially what we assume as obvious. This clarification is important, because it is necessary to separate the defense of philosophy as a discipline from the defense of a particular type of philosophy. Especially if the current state of philosophy leaves much to be desired. For example, today the postmodernist position dominates. A very anthropocentric thought, bordering on narcissism, which denies the existence of a unique and objective truth, and even of rationality itself. This is because it postulates that everything is a social construction, so that reality can be deconstructed, contradicting history, psychology, biology and even physics. Therefore, it is an anti-scientific philosophical current, denying that it is possible to systematically approach the truth, by means of an increasingly better paradigm, through reason and evidence. In other words, postmodernism applies a radical relativism, by denying the existence of facts, by declaring that there are only interpretations or accounts. Something that is not wrong in principle, but that leads to absurdity by putting scientific arguments on the same level as political, religious, traditional, spiritual, and so on. arguments. Which is obviously ridiculous, since there are arguments that are much more valid than others, the most reliable being the scientific ones. This is what this book defends, the original spirit of philosophy, its search for truth through wisdom or the love of knowledge.

Then the closest thing to what this book defends would be the philosophy of science, but not only in its historical or epistemic character, but also ontologically, as a relevant actor in the process of generating new scientific knowledge about the world. But even within the current philosophy of science we should be careful, because if we take the philosophers who are most related to the themes of the book, that is, those who study the philosophy of physics, we see that most of them are neither very relevant nor very critical, on the contrary, they are rather complacent and conformist to the current paradigm. On many occasions they are transformed into true lackeys, dedicating themselves mainly to defend and justify the positions of physicists. Something that goes totally against the rebellious (but rigorous) spirit that a philosopher should have. This attitude could be explained, perhaps, due to the inferiority complex of philosophy in front of the most rigorous and exact branch of science, so that, for fear of appearing unserious, too metaphysical, they prefer to have positions more similar to those of physicists so that they have more opportunities to be considered and taken into account. But in addition to the above, it is undeniable that the defense of the absolutist, eternalist and legalistic ideas of physicists (precisely what will be criticized later in the book) is shared by the vast majority of philosophers, also those of physics; because they are all immersed in the same Western paradigm, that of Plato and his world of ideas. Therefore, the vast majority of philosophers of physics defend the idea that time is an illusion, as do theoretical physicists; because both share in a certain sense that fascination for that immutable perfect mathematical world that is more real than the earthly, mutable and imperfect world we inhabit.

So, if the goal is to change the paradigm and really contribute to physics, it will be better to distance ourselves from the more traditional positions of the philosophy of physics; and get closer to the original spirit of philosophy. A discipline that we will see will be essential to continue with scientific progress, because to obtain a new paradigm we must make an update of the principles or axioms of the hard core. The most fundamental epistemic elements that neither mathematics nor experiments can address. Whereas the only one that has power in this area and is capable of rigorous axiomatic analysis is philosophy. Therefore, this apparent incompatibility between science and philosophy that many scientists preach is not real; on the contrary, they need each other, because, although scientific philosophy requires mathematics and experiments to work, only the former can go where the latter two cannot, towards the fundamentals. As Józef Maria Bocheński (1902-1995), a prominent Polish philosopher, said in his book Philosophy: An Introduction (ch. 2): "Philosophy is the science of foundations. Where the other sciences stop, where they do not ask questions and take a thousand things for granted, the philosopher begins to ask questions. The sciences know; he (the philosopher) asks what it is to know. Others establish laws; he (the philosopher) asks what law is...

Thus, philosophy is also a radical science because it goes to the root in a deeper way than any other science. Where the others are satisfied, philosophy continues to ask and investigate." So, if philosophy is so compatible and necessary for science, who is to blame for the fact that it is so despised in the scientific world, the word philosophy being used several times in a pejorative way? Again, the culprit would be extreme specialization, because this long separation between the two has made them irreconcilable. On the one hand, philosophy has become increasingly postmodernist precisely because science was what allowed it to keep its feet on the ground through evidence and experiments. On the other hand, this distancing has prevented scientists from appreciating the true virtues of philosophy. There are not even courses on the history of science, much less philosophy of science, in scientific careers. Or if there are, the universities that teach them are very few, and secondly, they are not very important subjects. Therefore, students of physics or science in general do not pay much attention to them. This generates great deficiencies, not only with respect to ontology, but more worryingly, with respect to ethics and epistemology. Hence, many scientists have problems with the technological applications of their research, but even with defining terms that are essential in science, such as: what is scientific knowledge, and how to acquire it, what is a model, a theory, a hypothesis, a conjecture, a principle, a law, falsification, and so on. Ambiguities that generate much confusion in scientific discussions and reflections, but also problems in the scientific work itself, due to the methodological application and interpretation of experimental results.

Thus, the lack of philosophical skills in science is worrisome, even in the field where their need is most obvious, such as ethics and epistemology. But this need is especially urgent in the case of physics, because philosophy fulfills a completely different function than mathematics. While mathematics has an essentially descriptive approach, philosophy has an essentially explanatory character. But far from being contradictory functions, they are totally complementary and necessary; because philosophy seeks to understand, to relate the fact or phenomenon to all the rest of knowledge, in a consistent and systematic way. In other words, it seeks meaning. That is why philosophy specializes in answering questions such as "why?" and, for the same reason, it has been one of the reasons why philosophy has been a despised discipline in the academic world in recent times, since, as has been repeated on several occasions, the academic system is specialized in doing normal science. So constantly asking why these principles or axioms can be annoying for scientists who just want to develop the SRP as early as possible and advance their careers. It is easier to describe a particular part of the world than to understand it.

However, to be fair, this distancing from philosophy is not the responsibility of physicists and scientists alone, for philosophy is also partly to blame. As stated above, philosophy specializes in answering "why?" questions. But how can one correctly answer why nature behaves the way it does, if one does not first have a good description of that behavior? In fact, this was the main problem of natural philosophy before Newton, that many wrong things were assumed, such as, for example, that heavier things fall faster than lighter things, as Aristotle (384-322 BCE) believed. No philosopher after him bothered to confirm such an assumption, because it was enough for them that it sounded logical or rational to convince them that it must be true. It took the great Italian astronomer, engineer, mathematician and physicist Galileo Galilei (1564-1642), two thousand years later, to test this assumption by experiment. He finally proved that it was in fact a false assumption, since all objects, regardless of their weight, fall at the same speed if we eliminate air friction, of course. Something moreover easily demonstrable, since you have to drop at the same time something heavy like an apple together with a sheet of paper, but this time crumpled into a ball (to minimize the friction with the air). There you will realize that both reach the ground practically at the same time. Therefore, philosophy was despicable at the beginning of modern science because one must first have a much more rigorous description of nature, and then ask why it is so. Therefore, the first objective of science was to make more rigorous measurements of nature, through increasingly precise experiments and instruments of observation. And second, to take those measurements and relate them logically to find patterns and build models that symbolically represent that reality in order to make falsifiable predictions. A job that mathematics has done extraordinarily well.

So, it is recognized that in the beginning it was more important to make measurements and models than to think philosophically about them. But now we have reached the other extreme. We have impressive experimental and observational technology, and we have mathematical models that describe nature from its smallest scales (molecules, atoms, and subatomic particles) to its largest scales (planets, stars, and galaxies) with astonishing accuracy. Therefore, the time has come to understand such models, to ask ourselves why nature behaves the way it does, what these mathematical models want to tell us about the universe. In other words, physicists must start calculating less and philosophizing more. There is no better proof of this than quantum mechanics, since all the calculations, mathematics, experiments, and observations have already been done here. The only thing that physicists have not done in this century of development of this theory is to make sense of all this information. Something that is surprising to have to say, since it seems obvious, because if we want to go further, how can we achieve it if we do not understand in depth the theories we already have?

Again, a perfect example of this need is again the last revolution in physics. As we already know, there was an inconsistency between Newton's model of gravity and Maxwell's model of electromagnetism. This inconsistency was due to the fact that they were based on mutually contradictory ontological assumptions. For example, the fundamental assumption in Maxwell's equations is that the speed of light is finite and constant. In contrast, the fundamental assumption of Newton's gravity is that this force propagated instantaneously, at infinite speed. This contradiction becomes apparent when we consider, as a thought experiment, what would happen if the sun disappeared in an instant. If this were to happen, we would see that the sun would have disappeared only 8 minutes later. The time it takes light to travel the distance from the Sun to the Earth. On the Newtonian gravity side, on the other hand, we would immediately feel that the Sun disappeared, since the Earth would stop orbiting around it, and start traveling straight out of the solar system. It would be like spinning a ball on a string, and in the middle of the spin we let go of the string. The ball would shoot away from us, as in the Olympic sport of hammer throwing. Well, the same thing would happen to the Earth, so we would apparently freeze to death if we immediately moved away from the center of the solar system before knowing the reason, because 8 minutes later we would just notice that the sun had disappeared. This theoretical incompatibility was solved by Albert Einstein through his theory of relativity (special and general), by banishing Newton's instantaneous assumption and adapting it to the assumptions of Maxwell's theory. Only after this metaphysical work did he continue the mathematical development, arriving at the conclusion that gravity propagates at the same speed as that of light, through gravitational waves. Thus, overcoming the crisis of physics at the end of the 19th century was not solved by a mathematical trick, but by means, fundamentally, of an axiomatic correction. Through a deep understanding, from the hard cores, to understand the meaning of such theories and to solve their incoherence among them.

Interestingly, the current crisis is quite similar to its predecessor, because once again we have an incompatibility between a theory of gravity, this time Einstein's general relativity, and a theory that includes electromagnetism, quantum field mechanics. And instead of physicists devoting themselves to a deep understanding of such theories to find out which ontological assumption(s) might be wrong, they devote themselves only to calculations. This null effort in understanding is so evident that there are many very important open problems in this respect, because physicists are not interested in reflecting on what is time, what is energy, what is information, what is a scientific theory, what is light, what is space-time, what are the laws of physics, what is a quantum field, and so on. Very important elements, which are even defined differently in each pillar of physics. Therefore, it is not surprising that physicists have not succeeded in unifying such theories, if they have not even understood very well what the problem is about. Again, the best example of this is quantum mechanics, a theory that works remarkably well. In fact, the most accurate theory that has ever existed. Yet no one knows why. As the legendary American physicist and Nobel laureate in physics Richard Feynman (1918-1988) put it: "nobody understands quantum mechanics." This lack of understanding is due to the extreme specialization and monopoly of mathematics, which have turned the physicist into a simple calculator, whereas physics is not only about designing models, describing nature, but also about understanding it. Therefore, to recover this essential function in physics, it is necessary to reincorporate an ancient tool. One that paradoxically was the one that allowed the birth not only of physics but of all science: philosophy. This is rather ironic, because every physicist is already a philosopher, even if he does not know it or does not want to be one, since he needs, yes or yes, a few reasonable and well-founded metaphysical assumptions to carry out his work. Some hypotheses, even if they are vague, of what nature and its most basic elements such as energy and time are. For example, physicists should be realists. A philosophical stance that assumes that there is an external reality out there, independent of us, even though it cannot be proven, because everything we experience could be a dream of a brain floating in a jar. But even though it is a possibility impossible to rule out, Ockham's Razor is applied assuming the simplest thing, that this world we experience does exist. That is why nature is studied as objectively as possible. However, due to this lack of philosophical skills, there are more and more idealistic voices in physics, who say that reality is a product of consciousness, or that we are inside a simulation (some even venture to predict that the possibility is 50%, as the famous American astrophysicist Neil deGrasse Tyson once stated). Meanwhile, the vast majority of physicists do not even know what their position is, because they ignore that they do philosophy. And, for that very reason, they end up doing bad philosophy, which also results in bad physics. A defect that was not so noticeable in the normal science stage, because the main purpose was to develop the paradigm, not to understand it. The famous shut up and calculate. However, this lack of philosophical skills is becoming more and more noticeable in the current stage of crisis, where it is necessary to deepen, question and criticize the assumptions that are taken for obvious and unquestionable.

Then, as said in the previous chapter, in order to revolutionize it, it is necessary to modify the hard core of the SRP. Something for which, as we have already seen, mathematics is of little use, since its specialty is not to study the axioms, but to deduce the conclusions (solutions) derived from these axioms. In other words, calculus. And as we saw in Gödel's incompleteness theorem, it is not even capable of assuring us of the completeness and consistency of such axioms. Therefore, to study the axioms or ontological premises of the hard core requires a qualitative analysis, not a quantitative one, in order to determine which assumptions are most likely to be incorrect. And who better to do this work than philosophy, the discipline of rigorous and critical thinking. A tool specialized in systematically examining ideas that no one dares to question, or even knows are questionable. And just as in the case of mathematics, if this tool is used well, it can be a great source of scientific creativity, promoting truly innovative solutions that generate new solutions that open up new avenues of research. This is something that today's physics, which is essentially mathematical and loyal to the current paradigm, desperately needs. In fact, the vast majority of theoretical physicists are completely unaware that there is a hard core, since they assume that their ontological premises are the only possible ones, that all previous paradigms have been wrong, that they were ignorant or naive physicists; but that they are not, because they are in the correct paradigm. That is why philosophy is so necessary in science, especially at this time, because philosophy allows us to fight against the blindness of specialization, thanks to its focus on criticizing intellectual authority, overturning dogmas, and questioning the unquestionable. In short, philosophy fosters the healthy doubt that is essential to science. And even if philosophy does not produce clear answers, this does not matter, because its great value lies in producing good questions, because to get the right answers you must first have the right questions, and that is the role of philosophy. Therefore, far from being a rival or a nuisance to mathematics and experiments, philosophy would be a valuable ally. In fact, we can summarize this strong relationship between science and philosophy by saying: "Science without philosophy is short-sighted, while philosophy without science is pure intellectual masturbation."

At this point, one could appeal to the above by saying that the last revolution or paradigm shift in physics was accomplished without the help of philosophy, because it was due primarily to the work of a physicist: Albert Einstein. But although it hurts the ego of many physicists, Einstein was more than a great mathematician (which he was), he was an incredible philosopher. Albert Einstein studied philosophy when he was preparing to enter college. In fact, he loved it so much that he even seriously considered changing his major and studying philosophy. But, as his father (a strict engineer who even believed that theoretical physics was nonsense) was not going to like it, he continued with his original plan. However, that philosophical way of thinking, something innate in him, he applied it intensely and constantly to physics. In fact, this was what made Einstein so unique among his colleagues. So much so that other physicists began to hate him for it. His professors ended up hating him for constantly challenging their ideas in class. For constantly asking "why? As mentioned before, the resentment was so great that Einstein could not find a job after graduation because no professor would give him a recommendation. Not only that, but they went so far as to sabotage jobs for him. Ironically, the part of him they hated so much was the key to his success. His philosophical side. His ability to deeply analyze the problems of physics, breaking them down to their most basic and essential principles, questioning even the most fundamental, even the most obvious and obvious to his colleagues. That is why Einstein was a rebel by nature, because his philosophical attitude of questioning everything allowed him to go against the scientific authority of his time. And in fact, against any authority. Hence, he was one of the few who criticized the Nazis from the beginning, even writing a manifesto. Unfortunately, he failed because of the little support he received from his colleagues. In fact, one of his phrases is: "Unthinking respect for authority is the greatest enemy of truth." Therefore, this philosophical rebelliousness of questioning everything, especially what should not be questioned, allowed him to discover which of the ontological assumptions of the hard core of the previous paradigm was wrong. Specifically, the Newtonian dogma that space and time were absolute, thus eliminating the need for a luminous ether. Thus he did what no physicist had dared to question because of his conformism and obedience to authority.

In fact, a common belief among physicists is that Einstein did not accept quantum mechanics because he was too old. But this is completely wrong. First of all, he was one of the founders of the theory when he was only 26 years old. An age that is hard to call old. But more importantly, Einstein did not dislike quantum mechanics because he did not understand it, but because he understood it too well, so well that he realized that it broke completely with the classical paradigm. Unlike many physicists today who love quantum mechanics precisely because they don't understand it. So, as Einstein proves, the intervention of a scientific philosopher by profession is not essential, although such a possibility is not denied either. But if he is a physicist, he must have a strong philosophical spirit, such as Albert Einstein had. One strong enough to survive the whole academic process specialized in destroying the philosophical spirit, plus all the attacks that will come later from the scientific community. But whether it is a physicist, a philosopher, or both, it is essential to have a more critical and reflective mentality, together with a scientific community that not only tolerates it, but also promotes it. All this without losing scientific rigor, because they are not mutually exclusive. This is the only way to ensure that science always progresses and does not get stuck in conformism and dogmatism. Or worse, it degenerates.

Chapter IV

The Current Paradigm

Before even proposing a new paradigm, it is necessary to analyze in depth the current SRP. A paradigm that has been called in many ways, for example: classical or Laplacean; however, here we will add another name and call it the Parmenian paradigm, in honor of the pre-Socratic philosopher called Parmenides of Elea (515-470 BCE, approximately). The reason for this name will become much clearer in the next chapter, when its origin and meaning will be discussed in detail. In the meantime, in this chapter we will limit ourselves to describing in general terms what the current paradigm consists of. A conceptual framework that, as its name implies, goes back much further to ancient Greece. And we are not going to resort to the ancient Greek philosophers as a clichéd rhetorical resource of physics popularization books, which has been abused on innumerable occasions to fill in and/or pretend intellectual depth. This time it will not be so, because this invocation of ancient Greece is essential, because it is the origin of the most important dogma of the current paradigm: the axiom of the absoluteness of the laws of physics. A fundamental premise that is not only present in all previous physical paradigms, but also at the very beginning of Western philosophical and scientific thought. This is because it is derived from the mother of all axioms, the principle of non-contradiction. Named by Aristotle as the first principle, since there is no previous principle from which it is derived, because it is self-evident, and because it is presupposed in any other demonstration. In other words, it is indemonstrable, because it is needed in order to demonstrate anything, for it is the most basic principle of all. And what does this principle, so obvious to Aristotle, say? That "A" cannot be "A" and "not-A" in the same sense of "A" and simultaneously. Or in a more simplified way: "A ≠ not-A." This is the paradigm of principles. For many, the fundamental principle of our knowledge. This is why it is most often used in philosophical argumentation and, of course, in mathematics, since the most common way to refute a hypothesis (whether in mathematics or in logic) is to show that it leads us to a contradiction, for example, that "x = a" and "x = b." From this follow two other principles that are also essential. The principle of identity, which says that the same is the same, or that every identity is identical with itself, or what Parmenides said: what is, is; and what is not, is not. Or in a simpler way: "A = A." On the other hand, the second principle derived from non-contradiction is the principle of the excluded third, which says that between two contradictory judgments there is no possibility of an intermediate one. That is, "A" or "not-A," no other alternative. This is the cornerstone of all current scientific thought. A principle that is based on the belief that it is ontologically impossible that something can be and not be at the same time. Or epistemologically speaking, that there cannot be a statement that is true and false at the same time. We will see later that, unlike what Aristotle believed, it is not only possible to conceive of a different logic, but also that it would be truly fundamental.

But back to the current paradigm. On the basis of this Aristotelian or classical logic, for more than two thousand years a philosophy and a science have been built based on absolutism, in the sense that, ontologically speaking, there are only certainties. A very radical way of thinking, since there would be no intermediate points or nuances. Things are or are not, period. This obviously excludes randomness, uncertainty, nothingness, and even change and time, since they are considered logical impossibilities. Thus, it is not surprising that the foundation of modern physics, the dogma of absolute or conservative laws of physics, has just such attributes. Laws with infinite validity and applicability, being valid throughout the universe (and now in the multiverse); and also eternal, existing at all times, even before the Big Bang. In other words, the laws of physics are omnipresent and omnipotent. A great similarity with the characteristics of classical theism that describes the Abrahamic god (the god of Christians, Muslims and Jews) as an immutable, impassible and timeless or eternalist entity. And despite such characteristics, it can affect and cause our mutable and temporal world. A power very similar to that of the laws of physics. Therefore, it should not be surprising that physicists speak of mathematical beauty when they refer to their immutable and eternal laws, because it is the same aesthetics that theists use to describe their perfect God, since the mutable and uncertain is something ugly, because it is imperfect. That is why God, who is perfect, already knows everything, because the future already exists, it is there, in front of his eyes, immovable, again like the laws of physics. Therefore, the concepts of God and laws are completely interchangeable, since they share the same characteristics.

A similarity that is not accidental, because from the beginning they have been intimately related. As already mentioned, in ancient Greece there was already a very special devotion to the absolute. A quality of being that was related to the immutable and eternal. Let us remember that "A =A." And what better than God to represent that absolute, the perfect. This is seen very clearly in the world of ideas of the father of Western philosophy, Aristocles of Athens, better known as Plato (427-347 BCE), and in the immobile motor of his most famous disciple, Aristotle. Both cases correspond to absolute deities, being immutable and eternal. In the Middle Ages, where philosophy was considered a servant of theology, Christians took the Greeks as a frame of reference, where they saw in Platonic and Aristotelian ideas a perfect support to describe their absolute and eternal God. So, it is not surprising that the father of modern physics, the Englishman Isaac Newton (1642-1727), a devout believer who spent more of his life searching for God's secret messages in the Bible than doing physics, transferred those same attributes to his laws of motion. This is not blasphemy, but totally true, because at that time, practically all natural philosophers were theologians as well, like the other great physicist and mathematician of that time, the German Gottfried Wilhelm Leibniz (1646-1716), who had passionate discussions with Newton about the role of God in nature.

So, like Newton and Leibniz, many natural philosophers were deists; that is, they believed that science was a way to know God, because by understanding creation one could understand the Creator. Therefore, clearly in the times when modern physics was born there was no distinction between science and religion, so we should not be surprised that the laws acquired the same absolutist qualities, since they were divine, being dictated by God. An absolutism that was reinforced by the very definition of laws that Newton used, as mathematical relations. A mathematics that was not only based on Aristotelian logic, but was intimately related to the divine as well, through the Pythagoreans, who believed that God was number, as many physicists today apparently do. In fact, it was the Pythagorean ideas that greatly inspired Plato to create his world of ideas. An absolute world more real than our earthly world, where mathematical entities also inhabited. So, it was not difficult to conclude that, if equations are static and eternal, the laws of physics were too. And since mathematics is practically the only tool used by theoretical physicists today, as we saw in the previous chapter, this belief is not questioned. This is the cornerstone of determinism. The belief that the laws of physics are universal and eternal. The cornerstone of modern physics, which is based on the belief that being is immutable and eternal, just because some influential thinkers in the past considered it beautiful and sublime. While change, chance and nothingness were considered repugnant and logically impossible.

However, the fact that a belief is rooted in religious ideas or ancient traditions does not necessarily make it invalid. Therefore, we will see below some scientific and philosophical arguments why this way of understanding physics is beginning to fail. The modern version of this science originated with the Italian astronomer, engineer, philosopher, mathematician, and physicist Galileo Galilei (1564-1642), who introduced systematic experimentation and precise observation using mathematics. In other words, he invented the scientific method. At least in the West, because the Arabs were already conducting experiments centuries earlier. This is because the Arabs, instead of burning the books of the ancient Greeks as was done in the European Middle Ages, devoted themselves to rescuing them. In fact, they financed expensive expeditions to search for books, not only in Europe, but also in Asia, to translate them and learn from them. A rather ironic fact, since something quite similar happens today, but in reverse, unfortunately. But at least at that time, the Muslim religion was not always the great enemy of science. On the contrary, thanks to the fact that the Arabs rescued and continued with the philosophical and scientific studies of the ancient Greeks, much of that knowledge was recovered and returned to Europe, which allowed for a Renaissance in the West. For example, one of these important Arab philosophers and scientists was Alhacen (born in Iraq in 965 CE), who is considered the grandfather of the scientific method, thanks to his experiments and important contributions in the field of optics. In fact, he is also considered the father of this branch of physics, thanks to his work with lenses, mirrors, and camera obscurae, which allowed him to study the reflection and refraction of light. He was the one who refuted the idea established by Plato and Ptolemy that the eyes had an active action on sight, as if they were a pair of lanterns; through a rather clever reasoning, realizing that, in order to be able to instantly see very distant objects (such as the moon), the light coming out of the eyes would have to be too fast (according to current data, we should see the moon three seconds later approximately, since that is how long it takes for the light to go to and return from the moon). He therefore reasoned that it is the light from objects that reaches our eyes. He also wrote the first comprehensive treatise on lenses, describing the image formed on the human retina by the crystalline lens. Many of his texts were later read and studied by many European scientists such as Leonardo Da Vinci, Christiaan Huygens, Descartes, Johannes Kepler and also by Galileo Galilei.

However, independent of the official birth of the scientific method, science, and especially physics, matured in the hands of Isaac Newton, when he discovered that to predict the behavior of any physical system, only two ingredients are needed: the initial conditions (e.g., location and velocity) and the laws of motion (the mathematical equations describing the evolution of the physical system). Newton, influenced by his Catholic-Greek heritage, the immutable nature of mathematics, simplicity, and (why not) common sense, concluded that these two ingredients are absolute. Which for him obviously included space, time and the whole universe. However, who formalized the Parmenian paradigm as such was the French astronomer, physicist, and mathematician Pierre-Simon Laplace (1749 - 1827), when he recognized in Newton's laws a very special symmetry: the conservation of information. This means that the information is the same at any time. No matter if it is a very remote instant or belonging to a very distant future, any moment can be used to reconstruct the whole history of a system, all its past and all its future. Eternity would be revealed to you with perfect certainty if only a single, fleeting instant is known. An extraordinary conclusion, but based on reasonable assumptions for that time. However, after three centuries of scientific progress, is it still so? The absolutist view of Newton and Laplace began to crack at the beginning of the 20th century, with the advent of Einstein's theory of relativity. Here it was demonstrated (and later proved) that time and space are not absolute, but relative, variable, producing such impressive phenomena as Lorentz contraction and time dilation. Space and time turned out to be so flexible, even for Einstein, that they form a fabric that is capable of bending, twisting and even vibrating, generating gravitational waves. Later, against Einstein's own wishes, the Soviet mathematician, meteorologist and scientist Aleksandr Fridman (1888-1925); and the Belgian priest, astronomer, mathematician and physicist Georges Lemaître (1894-1966) demonstrated using the very mathematics of general relativity that the universe is not static and eternal, because it was expanding. Something that was later supported by observations made by the lawyer and astronomer Edwin Hubble (1889-1956), using the Hooker telescope, the largest at that time. A fact that was definitively proven thanks to the discovery of the cosmic microwave background or CMB in 1965, giving the ultimate confirmation to the Big Bang theory. But the strongest blow to absolutism in physics was the arrival of the second revolution, quantum mechanics, which destroyed the concept of absolute initial conditions by means of the uncertainty principle of the German physicist and Nobel Prize winner Werner Karl Heisenberg (1901-1976), which prohibits two conjugate or related variables, such as position and momentum (velocity), from being known simultaneously with absolute precision. A principle that establishes a natural limit, not a technical one, in the knowledge of the initial conditions of fundamental systems. In conclusion, after three centuries we have discovered that neither the universe, nor space, nor time, nor initial conditions have turned out to be absolute as previously believed. Then, why are physicists so convinced that the laws of motion are, if not even the universe turned out to be absolute?

This is where the dogmatic nature of this belief begins to manifest itself, because the vast majority of physicists have never questioned it, even though it is an impossible premise to prove, as the Scottish philosopher David Hume (1711-1776) revealed in the 18th century, because to believe in absolute laws is to believe in an absolute causality. A principle based on induction. An approach that science knows very well is not a valid demonstrative proof, as we saw in Chapter II when Popper's falsification principle was explained. But absolute laws are just that, a statement obtained by induction over time. Therefore, just as seeing 1,000,000 white swans was not definitive proof that all swans are white, seeing the same sequence of events 1,000,000,000 times is not definitive demonstrative proof that such a law exists, because nothing guarantees that event number 1,000,001 will not be different. As a historical anecdote, there was a saying in Europe: "black swan" to refer to something impossible. However, in 1697, it turned out that they encountered the impossible, black swans. A new species was discovered in Australia. An event that came as a great surprise in Europe and inspired David Hume in his critique of induction, realizing that the only way to confirm absolute laws would be to observe it eternally. Something evidently impossible, so that the laws will never cease to be a conjecture since no one will be 100 billion years in the future to ensure it. A time horizon that seems to us literally eternal, enough to confirm such a law, but that would still be an insignificant time in comparison with eternity. Therefore, for Hume laws are nothing more than convenient ways of summarizing what happens. A way to order and describe the world in a purely pragmatic way, because of our impossibility to know how nature really works.

In this sense, physicists may have become Russell's turkeys. A story created by the famous British philosopher, mathematician, and Nobel laureate Bertrand Russell (1872-1970) to illustrate the problem and danger of induction. In this story, on his first day on the farm, a turkey discovers that he eats at 9 o'clock in the morning. Something he found odd. The next day again a human comes and gives him food at 9 a.m., and the day after that, and the day after that, and the day after that. In this way, after months, the turkey acquires enough confidence and decides to postulate a universal law, that every day, no matter if it rains or is sunny, if it is hot or cold, a guy appears with food at 9 o'clock in the morning. A law that unfortunately was violated when Christmas Eve arrived, and instead of receiving food, the turkey became food. With this account Bertrand Russell, as before with David Hume, is alerting us to the risk of assuming as absolute truths inferences that are drawn inductively. A mistake that could be perfectly being made by physicists, in the same way as the same mistake was made in another science, biology, where it was assumed that species were immutable, absolute; just because it took time horizons unimaginable to us (hundreds of thousands or even millions of years) to see the change and evolution in animals. That is why it was assumed as dogma that species cannot change, being the creations of an absolute god. How can creations be mutable and imperfect if their creator is eternal and perfect? However, we now know that they do change. Then, what assures us that, in hundreds of billions of years (a tiny interval of time in comparison with all the history that the universe will have) the laws will still be the same?

But one does not have to resort only to philosophy and logic to seriously question this assumption, for in physics itself one can find evidence, both theoretical and experimental, that seriously questions the absoluteness of the laws of physics. For example, the Big Bang theory says nothing about what caused that bang, and the theory itself has several problems (discussed in much more detail in chapter XV) that would supposedly be solved by cosmic inflation. However, this theory is unproven, and it also has its own serious problems. In addition, it is not understood how galaxies formed so early, since the oldest ones look very similar to the present ones. Nor is it known how the supermassive black holes at the center of such galaxies, which can reach the mass of billions of suns, were generated. A size that makes it impossible to have been reached by the merger of stellar black holes, therefore, the mechanism that produced them is unknown. In addition, it has already been confirmed that the tension with the cosmological constant is real, and that everything indicates that the expansion of the universe would be accelerating even more with time. Something that would make the universe younger, generating even more tension with the early formation of stars, galaxies and supermassive black holes mentioned above. Therefore, perhaps all these problems are signs that the evolution of the universe is more complex and dynamic than a simple explosion and expansion based on absolute laws.

However, another sign that is not a possibility, but a certainty is the violation of the conservation of energy. The principle that would maintain the absolutism of the laws through time reversal symmetry (this will be explained in more detail in the third part of the book). A violation that already occurred in general relativity, but which has been accentuated by the discovery of dark energy or the accelerated expansion of the universe. So, the amount of energy in the universe varies in time. At present, it is estimated that the amount of dark energy corresponds to almost 70% of the total energy, but before, when the universe was smaller, the percentage was much lower. Whereas, in the future the percentage of dark energy will increase even more. Especially, if it is not constant. Therefore, the amount of energy is quite flexible. But the violation of the principle of conservation of energy does not happen only by this exotic way, but also by a much more familiar channel, the light itself. Because of this very expansion of space, electromagnetic waves traveling through it are stretched. A phenomenon called cosmological redshift of light. So, if the wavelength of light is stretched, its frequency (v) decreases. Consequently, as its energy depends on the latter ($E = h\,v$), its energy decreases as well. Therefore, energy can not only be created, as in the case of dark energy, but it can also be destroyed, as in the case of light. So, when it comes to large scales of time and space, the principle of conservation of energy ceases to be valid. And if someone is cleverly thinking that the creation and destruction of energy would be compensating to keep the amount of energy fixed, it must be remembered that 70% of the total energy of the universe is dark energy. This percentage is growing faster and faster. While the percentage of light energy that is being destroyed is really negligible in comparison, because electromagnetic radiation corresponds to only 0.01% of the content of the universe. Therefore, the two phenomena are far from canceling each other out.

However, these cosmological observations are not the only empirical evidence that we can find in modern physics that casts serious doubt on the absoluteness of the laws, because if the laws of absolute physics are supported by the principle of conservation of energy, the latter in turn is supported by symmetries. Something that was discovered by one of the most brilliant mathematicians of the 20th century, the German Emmy Noether (1882-1935), who proved in a theorem in 1915 that: any symmetry entails the conservation of certain physical quantities. In other words, certain properties of a physical system remain invariant under certain transformations if there are certain symmetries. The classic example is to rotate or turn a circle, because this leaves the figure invariant. We would not be able to tell whether the circle was rotated or not, since the figure before and after the action is identical. Therefore, the circle has continuous rotational symmetry. The same happens with the square, but it would be a discrete rotational symmetry, since the figure remains invariant before rotations of a multiple of 90°: 180°, 270°, 360°, 450°, and so on.

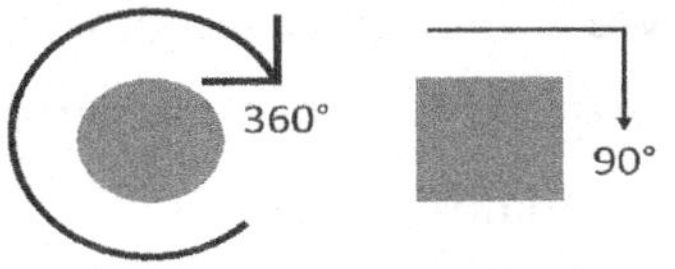

Thus, Noether mathematically proved what had only been dogmatically assumed, since the laws of physics would be absolute not because they are the will of an absolute God, but because they are in fact the manifestation of symmetries that would exist in nature. For example, the law of conservation of linear momentum (mass x velocity) would be the product of symmetry under spatial translation; while the law of conservation of angular momentum would be the product of symmetry under rotations (as we saw in the previous examples). This is quite impressive, because Noether says that the reason that any physical system always works the same way (the same laws) is because nature is symmetric. For example, if we design an experiment correctly, it should give the same result no matter if we move it to a laboratory in Brazil, Germany, China or on Mars, or if we rotate the whole laboratory 180°. These transformations should not influence the results if we keep the same conditions of the experiment. That is why Noether's theorem is so important for modern physics, because it finally provides a justification for its divine laws of physics, which would be absolute because the symmetries of nature would be absolute. That is why the laws do not change, do not vary, because nature is perfectly symmetrical. But Noether not only gave physicists a justification for their absolute laws, she also gave them a tool to discover new laws, new equations of physics, by assuming the existence of new symmetries in nature. Thanks to this, the Standard Model of Particles was constructed, by assuming new symmetries. Here we can already understand the fascination of physicists for symmetries, because not only do they greatly facilitate their work by making their calculations simpler (which is very true), but also because symmetries represent their favorite assumption, the absoluteness of laws. Their invariance and immutability.

Then the assumption of absolute laws of physics would be justified by the fact that nature is essentially symmetrical. From which natural symmetry, then, would the law of conservation of energy be derived? The answer would be from time-translation symmetry, which states that the energy of a system is conserved, or remains invariant, even as it moves through time. If this is true, then energy is conserved, because the laws of physics are the same today as they were a thousand years ago, or in the next billion years, because time is symmetric. That is, there is no difference between past and future. In this way, physicists found the perfect justification for their absolute laws, through symmetry. The small problem is that the universe is not symmetric according to the theory of general relativity, because it is dynamic. A fact that even Einstein himself tried to deny. However, the very mathematics of his theory dictated that the universe cannot be static, it must contract or expand. And we already know, from astronomical observations, that the universe expands. Therefore, time translation symmetry is clearly violated on a cosmological scale, because the universe was very different 10 billion years ago, being much smaller and denser than today. And in the future, it will be much larger and dilute. Thus, if the entire universe changes in time, the energy changes with it. Just what happens with dark energy and the redshift of light. This removes the support for absolute laws because the universe as a whole is not symmetric in time.

But the argument of time symmetry to maintain the absoluteness of the laws of physics encounters problems not only in general relativity, but also in the other pillar of physics, in the Standard Model of Particles. In this theoretical framework we find not only the local symmetries that describe its fundamental interactions, the so-called gauge symmetries, but also a very special general symmetry that is essential for maintaining the logical consistency of the whole model: the CPT symmetry. We will get a bit more technical here, however, we will return to this topic in much more detail in the third part of the book, so that this point will be clearer. However, for now, we should only know that it is an exotic symmetry product of combining charge symmetry, parity symmetry and time-reversal symmetry. The first consists of the invariance of the laws of physics in the face of exchanges of particle charges. For example, we could not distinguish between a hydrogen antiatom (antimatter) that has a positively charged electron (positron) and a negatively charged proton (antiproton), from a traditional hydrogen or matter atom, because both atoms would behave in the same way.

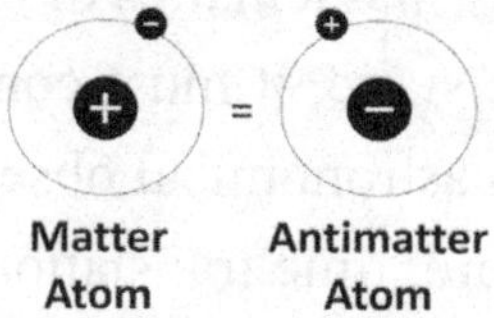

The second symmetry corresponds to the parity symmetry, which would be the indistinguishability of physical systems in the face of a specular inversion. In other words, the system functions in the same way as its mirror image.

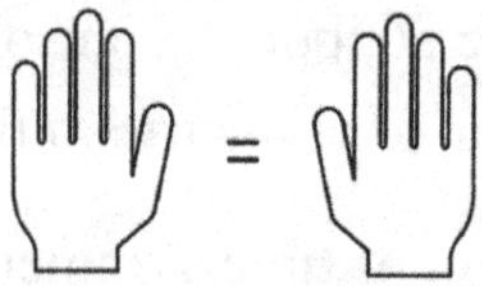

And the last one is the time reversal symmetry which consists in the indistinguishability of a system moving normally forward in time with the same system, but moving backward in time. In other words, the foundations of physics does not change if the time variable t (-t) is multiplied by -1. There would only be a preferential time direction in thermodynamic or macroscopic systems, due to the increase in entropy, which is a product of the interaction of many elements such as a gas formed by a large number of molecules (this will be explained in more detail in chapter XI). But fundamental systems cannot distinguish a direction of time. For example, a molecule can move in any direction in air. Therefore, if we recorded such a motion on video, we would not be able to distinguish whether the recording is moving normally forward or backward in time, because it is normal to see a molecule or a speck of dust moving up or down or going to the right or to the left. However, the recording of a gas made up of millions of molecules we could distinguish the direction of time, because we know that a gas will never contract spontaneously. It always expands. Well, the irreversible behavior of the gas is thermodynamic, being a compound phenomenon. On the other hand, the behavior of the fundamental elements, such as atoms and particles, are reversible in time.

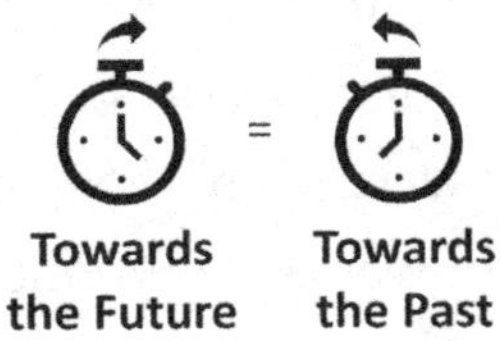

Each of these symmetries was thought to be true, absolute. However, to the great surprise of all physicists, the violation of each of these symmetries has been experimentally demonstrated separately, due to the weak nuclear interaction. It all began in 1956 with the publication of a scientific paper stating that parity symmetry in the weak nuclear interaction had never been verified. On Christmas Day of that year, a Chinese-born American physicist, Professor Chien-Shiung Wu (1912-1997) of Columbia University, decided to cancel her vacation in order to be the first to verify such a symmetry. She did this with her husband, who was also a physicist. They took a group of cobalt-60 atoms and lowered their temperature to just three thousandths of a degree above absolute zero. They then applied a strong magnetic field to align them. Since these atoms are radioactive, they release electrons via the weak interaction, and because of their alignment, the electrons should come out both left and right with the same probability. 50% on each side. However, what Professor Wu discovered was that they were very preferentially emitted in one direction more than the other, violating parity symmetry as much as is physically possible. Physicists were shocked. Why does nature prefer one direction over the other, as if it were left-handed or right-handed? Such was the impact of the experiment that the legendary Austrian-born Swiss-American physicist and Nobel Prize winner Wolfgang Ernst Pauli (1900-1958) declared: "Such a result is nonsense, the experiment is obviously wrong." But when the results were independently replicated by other experiments, physicists had to accept the truth. The Nobel Prize was awarded in 1957, the same year the first result was published; however, very embarrassingly, Professor Wu was excluded from the recognition. Perhaps because she was a woman and/or Chinese. It is not known. But leaving this embarrassing event aside, this result was a great blow to the foundations of physics, so, in order to save the paradigm and the internal consistency of quantum field theory, physicists desperately decided to invent new symmetries by combining them. Thus, they said that it was okay for the parity symmetry to be violated, because the true symmetry of nature was charge and parity symmetry together, i.e., CP symmetry. However, again, in 1964 it was discovered that some particles violated this new supposed symmetry. And again, the Nobel Prize was awarded for it. In panic, physicists resorted to the last available strategy to save their paradigm. They combined the three symmetries to create the CPT symmetry (charge + parity + time reversal). So far, no experiment has been found that observes the violation of this new symmetry. So, the internal consistency of the Standard Model should be secure. However, this seems cheating on the part of physicists, because it is difficult to imagine how it could be possible that this conveniently invented new symmetry could be violated, because it implies that each of the three symmetries is broken. In other words, for the CPT symmetry to be violated would mean that one of the three symmetries must be true, something that was already proven not to be true for each of them. Therefore, the conservation of the CPT symmetry, far from what physicists want to believe, shows that nature is very asymmetric. Above all, because there is no good justification for combining symmetries, especially when they are of such a different nature (charge and spacetime symmetries).

But beyond the legitimacy or not of the CPT symmetry, what is important for this topic is that it implies the violation of each symmetry separately, including time reversal symmetry. This has been proven by the decay of particles called B mesons, which evolve differently when time moves forward or backward. This is because they are particles composed of two quarks oscillating inside another particle. However, the time of the oscillation is different in each direction, making it possible to distinguish the direction of time. This shows that nature is fundamentally asymmetric in time, because it behaves differently when you go into the future than when you go into the past.

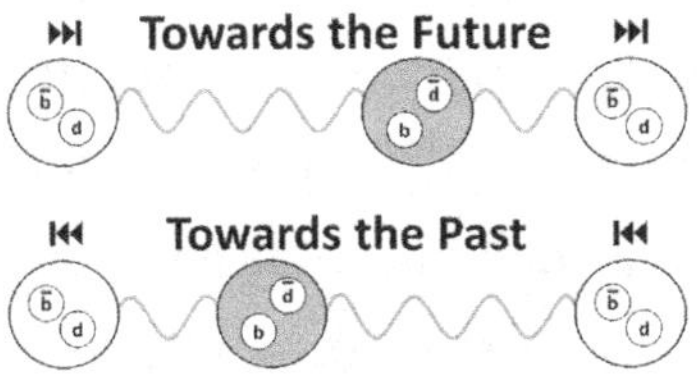

This was unthinkable in modern physics because it was believed that the asymmetry of time was an exclusively thermodynamic phenomenon, macroscopic but not fundamental. However, we have already discovered that this is not the case. A violation of time symmetry is, moreover, a mandatory condition for compliance with the conveniently invented CPT symmetry. This is very serious, because in the current Parmenian paradigm it is incompatible with a violation of time reversal symmetry, since it is completely inconsistent with the absolute laws of physics, which are perfectly time reversible. A deterministic mathematical equation cannot distinguish a temporal direction because they do not change with the flow of time, precisely because they are absolute. This happens with Newton's, Maxwell's, Einstein's equations and even in Schrödinger's equation, because they are invariant, static, timeless, perfectly symmetric in time. You can solve the equations and get the same results whether the time parameter is positive or negative. It does not matter. Therefore, the idealized, absolute, immutable view of current physics clashes with reality, due to the evidence of the violation of time translation symmetry in general relativity and the violation of time reversal symmetry in the Standard Model of Particles. Therefore, the time argument does not serve to justify the absoluteness of the laws of physics.

Faced with the collapse of the conservation of energy argument and the symmetry of time argument, physicists had to search desperately for another argument to support the dogma of absolute laws. And they supposedly found it in quantum mechanics, through unitarity. However, this only confirms that we are still in the Parmenian or classical paradigm, because far from being a very innovative principle, it is the same principle that Laplace recognized and defended in Newtonian mechanics in the 17th century: the conservation of information. Only this principle has been forcibly adapted or updated to quantum mechanics, by means of its physical law: the Schrödinger equation, which determines the evolution of the wave function (Ψ). A mathematical object that describes the probability distribution of the properties of a quantum system such as its position, momentum, spin, and so on. And like any equation or law of physics, the Schrödinger equation is immutable, which preserves information and ultimately justifies the symmetry of time. This is because if all instants are equal or have the same importance by having exactly the same information, then, it does not matter which instant we go to in time (time translation symmetry), nor if we go to the past or to the future (time reversal symmetry). The laws will always be the same. Then, if all the information of an instant is known, the information of any other instant can be known, since it is the same. So far so good for the Laplacean or classical paradigm, except for the small detail that, empirically speaking, in quantum mechanics information is not conserved, due to the problem of measurement. The conservation of information means the conservation of the wave function (Ψ); however, we know that when a measurement is made in quantum mechanics the Ψ collapses. And it does so in a truly random manner. An indisputable scientific fact that destroys the absoluteness or symmetry of time.

To see why unitarity is not satisfied, let us use the analogy of dice. If we know that there are six dice available to be thrown, and we are told that the result of one of them was, say, two; then we could not know which of the six dice gave such a value, because it could be any of them, since they all have the same probability of giving that result (1/6 or 16.666%). This, which is a technical or practical truth in the case of dice, is a fundamental truth in the case of quantum mechanics, because its randomness is ontological. This means that quantum results are genuinely irreversible, since there is no way of knowing which Ψ produced such a measurement, in view of the fact that it collapsed randomly. Therefore, that information was destroyed when that particular Ψ collapsed. In fact, we have never seen a Ψ. Only its collapse, which is measured. For example, with respect to the position of an electron, it is always there or here when we measure. On the other hand, the Ψ represents the electron in a superposition of places, scattered as if it were a wave. Precisely what we have never observed because the electron is always there but not there. Always in a specific place.

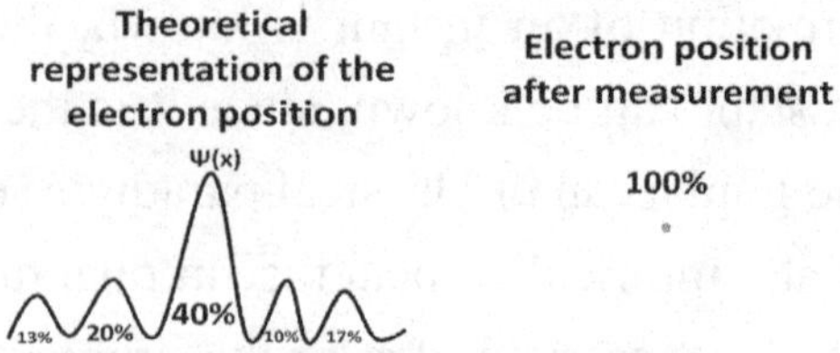

So in quantum mechanics, empirically, information is not conserved because a lot of Ψ disappears when it collapses, taking that information with it. And without that information we cannot go back in time. And, of course, we cannot go forward into the future with complete certainty, because Ψ collapses randomly; just as we do not know what result will come up when we roll a dice. So, empirically, quantum mechanics kills unitarity, and with it the absoluteness or reversibility of the quantum law, precisely because of this random collapse. Let us remember that randomness is the opposite of absoluteness:

uncertainty or lack of information. And despite this irrefutable scientific fact, physicists continue as if nothing had happened. That is why most physicists avoid talking about the implications or interpretations of quantum mechanics, even going so far as to deny that there is any problem with measurement.

As we can see, we are far from assuring the absoluteness of the laws. A question that can be extended beyond their philosophical and experimental justifications, by taking it to the field of their absurd consequences. For example, one of the most absurd consequences of the absoluteness of laws is that time ceases to exist. If information and energy were truly conserved, this would mean that time is perfectly symmetric, fulfilling both translational and time-reversal symmetry, making our universe totally deterministic. This means that, the becoming of things, all history, can be literally reduced to the laws of physics, because in them would be contained all the information needed to know the evolution of any physical system. Therefore, if we perfectly knew the initial conditions of a mere instant, and we had the correct laws, then we could know any other instant. We could go back indefinitely into the past, or we could know the future completely. We would be what is known in physics as Laplace's demon. A hypothetical entity that, having the laws and initial conditions of a single moment, the entire history of the universe would be revealed before its eyes. All because any instant is indistinguishable from the others, because they have the same amount of information, therefore, by knowing only one instant you can know all eternity. The problem is that, if there is no special or different instant, then, there is no instant that can serve as "now," since it is not possible to identify an instant that can fulfill the important function of separating time into past and future. And if there is no present, there is only eternity. All the future and the past existing simultaneously without separation from each other. A single temporal block that evidently eliminates the generative character of time, its flowing and becoming, being immutable, static. The dream of Parmenides.

This is a consequence of the unitarity of quantum mechanics, but time also disappears in general relativity interpreted in Parmenian terms, even though time can be deformed and curved here. This is because, in the present paradigm, Einstein's time is interpreted as a dimension equivalent to the spatial dimensions. Thus, if left and right exist simultaneously in space, then past and future also exist simultaneously, making all instants real, just as all points in space exist simultaneously. A vision of the universe as a tetra-dimensional block, where time no longer flows because it is petrified, because before and after are fixed as here and there in space. Therefore, both pillars of physics (quantum mechanics and general relativity) become absolute; not because these theories necessarily are (we will see that they can be interpreted in a better way), but because they are forcibly interpreted to be so through the current paradigm. Generating a very questionable absolutist worldview, because if the past and the future are as real as the present, why are we imprisoned in the now? If the present is only an illusion, as Albert Einstein said, how is this flow generated, what is the mechanism (physical or neurological) that puts us in this instant and not in any other? Nobody knows.

Another serious consequence of assuming that the laws are absolute is that it makes it impossible to explain the Big Bang. The origin of the universe. This is because the laws, along with all the energy and information, had to be created spontaneously at the first moment of the universe. Something that is difficult to justify in an eternalistic worldview like the present one. In fact, the Parmenian paradigm is incapable of answering questions about origins. A problem known in philosophy as the first cause or the cause without a cause. A type of problem that current physics cannot address, because it necessarily requires initial conditions and laws of motion to work. Therefore, it is paradoxical to try to answer the origin of everything, which includes only these two ingredients. So, the only thing that the Parmenian paradigm can do is to postpone the answer to the origin, extending the laws and initial conditions further back than the Big Bang. This is the real reason why physicists are so fanatical about the multiverse today, because it is an inevitable consequence of trying to answer what caused the Big Bang, assuming that the laws are eternal or absolute. That is why physicists invent brane universes, eternal cosmic inflation, many worlds, bouncing universes, symmetric universes, among many others; because they are desperate attempts to explain the Big Bang assuming that the laws are absolute. The big problem with this is that they provoke a regression ad infinitum, because maybe they explain the Big Bang with a multiverse, but where did that multiverse come from, from a multi-multiverse, and that one, from a multi-multi-multiverse...? In the end, the answer to the origin of everything is only postponed through an infinitely greater whole. So, this faith in absolute laws is again approaching its religious origin. Not only because of its blind faith in the absolute (God and laws), but also because of its reasoning, because believing in infinite parallel universes requires a leap of faith as great as believing in a god who created everything.

To recapitulate, modern physics was built on the foundations of Parmenides' thought, as was Platonism. A dualistic world view that separates what exists between a perfect and eternal world, where ideas and mathematics dwell, and the imperfect, finite, and transient earthly world. A philosophy later adopted by Christianity and known as Neoplatonism. Here Plato's idea that the body is the prison of the absolute and immutable soul is literally copied. Thus, many Christians torture and abuse their own bodies as penance, imitating the suffering of Jesus on the cross, because the pleasures of the body are sinful, the product of the devil. Strong beliefs that could not possibly have permeated when this new science was being built, because quite simply, the early physicists were also devout Christians. So, it should not be surprising that this contempt for the earthly in Platonism and Catholicism was also inherited by modern physics, in the form of the denial of becoming, emergence, and chance. Natural phenomena that contradict their Platonic ideas of absolute mathematical laws and infinite parallel universes in which all past and future exists in eternity, just like heaven for the Christians and ideas and equations for the Platonists and Pythagoreans.

In conclusion, when we consider all of the above. The questionable origin, the weak justifications, and its large and far-fetched consequences; it becomes clear that the ontological assumption of "absolute laws of physics" is not so absolute after all. Therefore, it would be a good candidate to be removed and replaced in the hard core of the current SRP. Especially when there is a different way of thinking, a different logic, which happens to be more fundamental, based on the principles of quantum mechanics. Exactly what we will see in the following chapters. Therefore, this small modification will have dramatic consequences in the foundations of physics, since it is one of the oldest and, therefore, most important dogmas of modern physics. An assumption that has survived all its scientific revolutions, for it has evidently been quite useful, especially during much of the last century. Therefore, as the great astronomer and popularizer Carl Sagan (1934-1996) said: "Extraordinary claims require extraordinary evidence." Therefore, the rest of the book will be devoted to arguing why eliminating this absolutist dogma will allow not only to overcome the current crisis in physics, but also to give a great boost to science in general. An idea that will seem very radical to the vast majority of physicists, only because they forget or do not want to recognize that the absolute law is a rather radical assumption as well. In fact, absolutist. So, as shocking as this new proposal may seem at first, the truth is that in comparison it will turn out to be much more conservative and natural than what is currently offered by state-of-the-art theoretical physics.

CHAPTER V

NIHILOPHOBIA

We have already seen that the dogma of the absoluteness of laws would be a serious candidate for elimination in the next revolution of physics, because of its weak justification, the empirical evidence that casts doubt on it, and its absurd consequences. But before proposing what a non-absolutist physics would be like, in this chapter we will go deeper into the origin of this dogma, because it will be very important to understand the great resistance that there will be in the community of physicists to renounce it. A resistance that is due, as we saw in the previous chapter, to the fact that absolutism is at the core not only of physics, but of all science and philosophy. Therefore, it is not only a question of theoretical physicists, but of a large part of all Western thought, which is essentially based on a logic that was believed to be the only possible one: Aristotelian or classical logic. A radical way of thinking that was apparently more evident in its beginnings, since it brought many problems to the first Western philosophers, who were the first to abandon mythologies and traditions, to use only reason to explain nature. These pioneers were the pre-Socratic philosophers, who designed monistic solutions, in the sense that they proposed a single arché or fundamental origin to explain the multiplicity of phenomena that exist in the world. For example, Thales of Miletus (625-546 BCE, approximately) proposed that the arché was water. In other words, at the origin of the universe there was only water, and from it the rest of the elements such as earth, air and fire were obtained. Or Anaximander of Miletus (610-545 BCE, approximately), a disciple of Thales, who proposed that the arché could not be something concrete, since it had to generate a great variety of things. Therefore, he postulated that the origin of everything was a new element. A chaotic, amorphous and indeterminate one that he called apeiron ("without limits" in Greek). An element that is not anything because it is all things in potency. Or Anaximenes of Miletus (around 590-525 BCE), who proposed air as arché; or Pythagoras of Samos (around 569-475 BCE), who postulated number as arché (like many physicists today, especially the Swedish-American cosmologist and writer Max Tegmark, who preaches that the most fundamental thing is mathematics); or Heraclitus of Ephesus (540-380 BCE), who postulated fire as arché. In all these explanations there is a transmutation of the elements. A change from one thing to another. From air to fire, from fire to earth, from water to fire, and so on. However, there was a pre-Socratic who was not at all happy with these solutions, Parmenides of Elea (515-470 BCE, approximately). In fact, he found it an impossibility, because for him it makes no sense for one thing to cease to be in order to become something else, because it simply violates the principle of principles, that of identity or that of non-contradiction. This because the monists were implicitly postulating the possibility that "not-A" could become "A," which is a logical contradiction for Parmenides, since it would be equivalent to saying that "not-A = A," eventually. Therefore, it is not surprising that Parmenides denied the existence of "change," not only of the transmutation of the elements, but of all kinds of transformation, because it was simply illogical for him. Therefore, all that we see through the senses, the multiplicity of things and their variability and mutability, is only an illusion; because there can only be "being": unique, immutable, and spherical. Yes, exactly, because for the Greeks the perfect was not only the eternal and uncreated, but also round. This is how God revealed it to them, according to Parmenides in his poem.

This was the solution achieved by Parmenides through absolutist or classical logic. The greatest difference in history between experience and reason, which denies any kind of multiplicity or separation, and also any kind of mutability or becoming. Simply because the particular thought of Parmenides was incapable of conceiving the logical possibility of change. This is the origin of absolutism in physics and its worship of the invariant and the eternal (symmetries). All because it did not enter the heads of the first Western thinkers how change could be possible, because it would imply the intervention of nothingness. And this, for Parmenides of Elea, was simply unacceptable, for what would change be if not a transition from "non-being" (nothingness) to "being"? Something impossible for Parmenides, because ex nihilo nihil fit (from nothing, nothing comes out). Therefore, from "non-being" or "nothing" cannot come "being;" and "being" cannot become "nothing." Ergo, change must be an illusion, because "what is, cannot cease to be," and "what is not, cannot become." To understand the scope of his reasoning, we must distinguish between two types of change: translational change (motion), and transformational change (emergence of novelty). Both are impossible for Parmenides, because the former consists in being (existing) at a point in space, then being nothing at that point, and then being at another point in space. In other words, not-being here and then being there. A dynamic of translation of the same being through space or a change in position or locality. This problem of change was deepened by the disciple of Parmenides, Zeno of Elea (490-430 b. e. c), known for his famous dichotomous paradoxes or the problem of cutting in two as: the paradox of Achilles and the tortoise, or the paradox of the arrow. All these are based on the same principle. For example, to travel any distance, you must first travel half of that distance. And to travel that half, you must first travel half of that half; and before that, half of that half of that half. And so on, since every real number is always divisible by two. And since it takes a finite amount of time to travel each half (even if it is shorter and shorter), it takes an infinite amount of time to travel any distance, since there will always be an infinite number of halves. Ergo, motion must be an illusion, as its master Parmenides said.

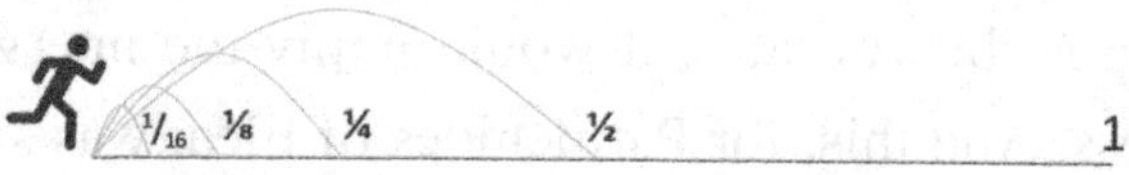

It should not have taken an infinite amount of time, but in any case, it was a long time, about two thousand years, to solve this problem. Moreover, it took two geniuses since Leibniz and Newton solved it in parallel by inventing the calculus. Thanks to this new mathematical tool, we can reveal the logical error made by Zeno, who believed that an infinite sum of numbers always equals infinity. Using Zeno's own example, if his distance is one kilometer, and he advances at a speed of one kilometer per hour; then, when he travels half of the distance ½ hour will have elapsed, and when he travels half of the next half kilometer ¼ hour will have elapsed, and half of the next half ⅛ hour will have elapsed, and so on. Therefore, the summation of the elapsed time would be ½ h + ¼ h + ⅛ h + 1⁄16 h + 1⁄32 h +... However, this infinite sum of digits does not diverge to infinity as Zeno believed, because each new addition is smaller. Instead, the summation converges to a finite number, to the hour. Just what it would take Zeno to walk one kilometer at a speed of 1 km/h.

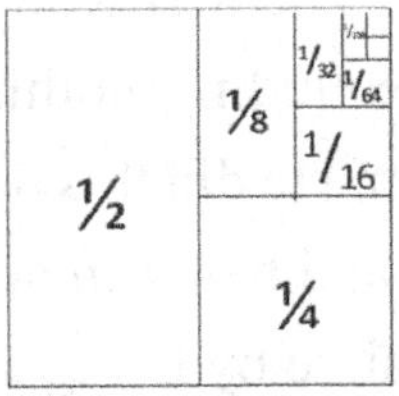

Therefore, an infinite sum of terms can produce a finite result if the sum converges or has a certain limit. This is exactly what you see in the figure above, where the fractions fill the area of the square infinitely, but never exceed the limits of the square. This is essentially calculus. The basis of the derivative and integral. In this way, the Parmenian paradigm demonstrated that translational motion or change is logically possible and not an illusion, thus resolving the first paradox of change postulated by Parmenides of Elea. However, the second type, transformational change or the emergence of novelty, has proven to be more complicated. This is because it is not a simple movement of "being" through space, but a transformation of "being" itself in time, where one "is" something now, ceases to be or "is nothing" in relation to that thing, in order to "be" then something else that did not exist before. A change that implies a transition from "not being" (nothing) to "being." An impossible phenomenon according to Parmenides, due to his principle of ex nihilo nihil fit. Something that also makes a lot of sense in current physics, since it would be a temporal translation, so if the universe is temporally symmetric, there should be no change about a thing moving from the past to the future or vice versa. It must remain invariant in time, at least in principle. This was illustrated by Parmenides as follows: a seed is now a "non-tree" or "nothing tree," but eventually it will be a "tree" if we plant it. So how can something that is not (nothing) come to be, if what is and what is not is not? The first who tried to solve this problem was Aristotle, by excluding nothingness from the whole process of transformation. He did this by inventing two types of being: being in potency and being in action. In this way, transformation would be possible by being a transition between two types of being, from being in potency to being in action. Using the same example as above, a seed "is" in "action," but it also "is" a tree in "potency." So, the seed can become a tree without nothingness intervening in the process, because it is a change or transition between two types of being.

A solution that would not have pleased Parmenides at all, but he would have accepted it anyway, since it saved his fundamental premise, absolutism, by allowing change without the intervention of nothingness. Therefore, science adopted this Aristotelian solution, especially physics, for example in the case of energy, where there is no creation or destruction of energy, but a transformation from one type of energy to another. Specifically, from potential energy (the work or motion that could be or will be), the physical analog of being in potency; and from kinetic energy (the effective work or motion, that which is being done), the physical analog of being in action. Two types of energy that are absolute, since the total energy (the sum of the two) is always the same, invariant, by means of the law of conservation of energy. In this way, potential energy is converted into kinetic energy, and vice versa, the sum of the two always remaining constant (if we do not consider heat or friction, which would be another type of energy. In such a case, the sum of the three types of energy remains invariant). However, even if we disregard the fact that we already know that energy is not conserved on a cosmological scale, Aristotle's solution is not entirely satisfactory, since it still leaves unsolved the great mystery of how the act of potency is generated, how exactly this change is produced, because we must not forget that transformational change implies a transformation of being, a transmutation of a thing or the emergence of a new thing (the creation of novelty). For if translational change or motion is a change of being, that is, a change of being through space, then transformational change or emergence is a change of being itself, that is, a change in time. Aristotle answered these doubts by ascribing purpose or will to things; for example, the seed "wills" or seeks to become a tree, the stone falls because its "wills" to be on the ground, because that is its purpose or nature. Even if all things really had a purpose, it would not really answer why it does so. So, it has the same explanatory power as saying that things do what they do because of some law or divine command. It simply describes what happens without providing a mechanism that allows for such change.

With the advent of modern science, this problem could be better addressed, especially in the non-physical sciences (chemistry, biology, and sociology), and some aspects of emergence could be explained thanks to the systemic approach. In it, the interaction between a set of components through long-lasting connections can generate new global properties that its components lack. For example, a set of neurons connected by a complex network of synapses generates the property of mind or consciousness, something that each individual neuron obviously lacks. Or a group of people and their relationships generates a structure that creates properties corresponding to a company, government, church or family; allowing it to achieve and accomplish things that would be out of reach for its members separately. Or the strong bonds between cells produce the unique characteristics possessed by a dog, for example. A set of attributes that each cell of the dog evidently lacks. In all these cases, the new properties of the system are the product of the relationships between its constituents. However, although this approach explains many things, it is not perfectly understood how those links generate such emergent properties, nor is it clear from when. What is the magic number of connections between individuals that generates such novelty. For example, does the mind arise from two connected neurons, 10, 69 or a million? Nor is it known how and why these components self-organize to group and link in such a way. Where do these new links that generate such complex relationships arise from? In conclusion, the exact mechanism of emergence is unknown. But not only in the contingent sense; for example, how the cell division of a single fertilized egg can so spectacularly self-organize to form the embryo of a human being or of any multicellular living being. An extraordinary, practically miraculous phenomenon, but one that occurs millions of times every day in our species alone. However, something even more difficult to explain is how this extraordinary process emerged in the first place: why did a group of different types of atoms tend to generate such complex organic molecules; why did these complex molcecules group together into even more complex structures to form the first living cell; why did these single-celled life forms group together to form the first multicellular life forms; and why did consciousness and intelligence emerge from a multicellular life form? This type of emergence is called disruptive, and is the most difficult to explain, since it is the emergence of primordial novelty or the appearance of the first individual of its type or kind. On the other hand, the second type is progressive emergence, also known as relative, which would be easier to explain (although it is still a very difficult problem), because it corresponds to the emergence of novelty as a simple improvement or modification of something similar previously existing. Therefore, if all novelty is the product of a mechanism of assembly and creation of links, this means that disruptive emergence is the appearance of that mechanism; while progressive emergence would be the modification (adaptation or improvement) of that assembly, by varying a little the already existing links. Applied to a technological example, emergence or progressive innovation would be the small improvements of a pre-existing technology, such as the latest cell phone, television or computer that goes on sale every year. On the other hand, emergence or disruptive innovation would be equivalent to the invention of the first cell phone, the first television or the first computer. A phenomenon that is more difficult to explain, since it is not a simple variation, but the creation of an entire assembly mechanism. The appearance of a new system that in essence did not exist before, because it is the first of its kind; but that can improve and become more stable through progressive emergence.

The science that has most addressed and advanced this type of problem of change is biology since it has been able to explain the emergence of biological systems (living beings) through Darwin's Natural Selection. One of the most respected theories in all of science, it has been extraordinarily successful in explaining the progressive evolution of new phenotypic characteristics of living organisms thanks to small genetic mutations that occur between generations and are conserved by Natural Selection. Although it is not denied that much more research is needed regarding disruptive emergence, such as the appearance of new categories of species like the first eukaryotic cell, the first multicellular living being, the first reptile, the first mammal, the first human being, and so on. One of the most accepted theories is that they would be the product of a great accumulation of progressive mutations suppressed in the genetic code, which at one moment are activated at the same time and expressed together to generate such a disruptive novelty. However, much research remains to be done on the details of this process. But despite these advances, the main problem with this theory is that it is unique to the biological world. Thus, it cannot address, for example, the mother of biological disruptive emergences: the appearance of LUCA (the last universal common ancestor) or the first form of life on Earth. What chemical mechanism could have produced it? The same is true for the other boundary of biology, the origin of human consciousness. So, there is no backward integration with chemistry (the origin of life); nor is there forward integration with neurology and psychology (the origin of consciousness). Therefore, Darwin's Natural Selection works fantastically in the biological realm, but has no power outside of it. Fortunately, recently there has been a small progress in the integration with neurology, psychology, and sociology, through evolutionary psychology and memetic selection (an equivalence between ideas and genes, due to the competition in the cultural ecosystem, postulated by the famous theoretical biologist Richard Dawkins); but still very tentative, due to the difficulty of developing this multidisciplinary research path in extreme specialization. However, where there has been even less progress, almost none, is in backward integration, due to the great incompatibility between biology and physics. The reason for this is that physics denies emergence as a fundamental or real phenomenon, precisely because it contradicts its Parmenian or absolutist principle. Let us remember that in the SRP of physics everything is conserved, energy and information; therefore, nothing authentically new can be created, because absolutely everything has already been planned by God, sorry, by the laws of physics. Thus, while biology recognizes emergence as a perfectly natural or real phenomenon, by accepting that genuine novelties (new species) can arise spontaneously; physics denies the existence of real novelty, since it goes against the determinism of its laws. As California Institute of Technology astrophysicist and best-selling author Sean Carroll (b. 1966) said in his book The Biggest Ideas in the Universe: Space, Time, and Motion: "The fact that you can say something useful about physical systems while being so ignorant of so many microscopic details is an example of a feature known as emergence." In other words, emergence is not real, but illusory, the product of our ignorance (translated from the Spanish version).

This approach to physics is called reductionism, and it is the opposite of the approach used in Natural Selection, emergence. It is the opposite because reductionism believes that it is possible to exhaust all understanding of a phenomenon through exclusive knowledge of its parts. The best representative of this is the Standard Model of Particles, which reduces all natural phenomena, including sociology, psychology, biology, and chemistry, to a simple interaction between fundamental particles. For example, for the foundations of physics a dog is just a handful of cells, and a cell is just a handful of molecules, and a molecule is just a handful of atoms, and an atom is just a handful of particles. In conclusion, a dog is nothing but a handful of quarks and electrons, bound together by gluons and photons. However, it is evident that atoms have properties that particles lack (physicists can only describe analytically the simplest atom, the hydrogen atom, the rest cannot), and molecules have properties that atoms do not have, and cells have properties that molecules lack, and so on. And, therefore, particle physicists cannot explain how exactly those new properties emerge or are created from those parts. This is precisely what emergence, the existence of synergy, addresses. That the whole is more than the sum of the parts. Therefore, although these phenomena would be the result of a set of elements, their explanation or understanding cannot be exhausted by the exclusive understanding of those constituents, because it would be ignoring or omitting the synergy. This explains why each level requires a new science (particle physics, chemistry, biology, psychology, and sociology) with its own laws; because the levels are not completely reducible, due to the existence of synergy.

A physicist might disagree and say that physics includes emergence because it has thermodynamics. One of the major branches of physics that studies emergent phenomena such as temperature and pressure. Physical quantities or phenomena that are not found in the micro or in the parts, but in the whole. For example, temperature is not at the level of the water molecules in a teacup but emerges at the scale of the teacup when you think of countless molecules colliding with each other. Thus, temperature does meet the definition of emergence; however, it corresponds to a very particular type called weak emergence. This is because the understanding of the emergent phenomenon can be exhausted by the exclusive understanding of the constituents. In the case of temperature, for example, by the kinetic energy of the molecules. If the water molecules in the teacup move faster on average, the temperature will be higher. It is therefore not surprising that this is an emergency that physics does recognize, being analytical or lacking in synergy. This makes it a very basic emergence, being the product of a simple association by aggregation of its components. An association that does not change the nature of the constituents, does not transform them. It is simply the product of bringing together a large number of things, for example, a mound of sand, which does not modify or alter the grains that form it. Thus, a sand mound is not stable or durable because it does not create cohesive bonds between the grains and therefore does not require additional energy to hold these components together. As a result, the grains of sand fly out of the mound as soon as a weak breeze blows. In other words, weak emergence phenomena are very simple. In contrast, the type of emergence we are interested in explaining is strong emergence, which involves the creation of long-lasting connections between parts that build synergy, such as the formation of molecules, cells, multicellular organisms, and civilizations. All are stable and lasting phenomena, because they seek to conserve themselves and, therefore, require energy to do so. Therefore, pure analysis is not enough to fully understand these physical systems, because it is blind to synergy, since the latter is not found at the level of the parts (the micro), but in the whole (the macro).

And although extreme specialization protects physicists from having to recognize strong emergence or synergy, because the vast majority of these phenomena are present in other sciences (molecules, cells, plants, or societies), there would still be one type of emergence that even the most absolutist reductionist physicist would have trouble ignoring. And it turns out to be nothing more or less than the mother of all disruptive emergences, the Big Bang. The moment when everything (energy, space-time, information, and even laws themselves) supposedly came out of nothing. We are talking again about the eternal problem of the origin of the Parmenian paradigm discussed in the previous chapter; for, even if we delude ourselves into believing that nothing genuinely novel has arisen in the universe, there would still remain the undeniable fact of having to explain where the whole universe came from in the first place. What or who produced the Big Bang. Therefore, to exclude the ex nihilo emergence of the entire universe, physicists had to fill in what was before the Big Bang with an eternity, extending the jurisdiction of their laws into the infinite past. Something consistent with Parmenides, since like being, the laws must be uncreated. Therefore, the laws of physics have always been there because they are eternal and immutable, as God commands.

But even if this pseudo-solution could be accepted, physicists would still have to explain the extraordinarily low entropy initial conditions that existed at the time of the Big Bang. In other words, the origin of the universe had to be an incredibly improbable event. The most improbable event that ever happened. Although the reason for this will be discussed in much more detail in Chapter XI, for now all we need to know is that the expanding universe is analogous to an expanding gas. Therefore, if a compressed gas is released under a lot of pressure, as in the Bang of a pump, what is the probability that all the molecules of the gas will by chance re-concentrate at the same point they were before being released? Ridiculously low. This is what physicists mean when they say that the Big Bang had a very low entropy, only that instead of a kilo of explosive exploding, this time it is the whole universe, so the chances are even more ridiculous. So, faced with this absurd probability, theoretical physicists have decided to fill eternity with what is fashionable these days, the new god of physicists, the multiverse. In this way, low entropy is justified, because in the multiverse new universes are constantly being created. As, for example, in the popular theory of eternal cosmic inflation. Something we already know is at the cost of leaving an even bigger (in fact, infinitely bigger) question: where did that multiverse come from? Thus, it (supposedly) solves an already gigantic problem by creating another, much bigger problem. A colossal violation of Ockham's Razor, just to deny the existence of strong emergence.

But why is this kind of emergence so tenebrous and unacceptable to physicists? It is due precisely to the same fear of Parmenides, nothingness, because the fact that there can be real novelty, that something truly new can be created, necessarily implies that it must have come out of nothingness. And nothingness is unacceptable, it is an impossibility in the Western philosophical tradition. Therefore, religion has filled every void of nothingness with God, just as modern physics does with the multiverse. Thus, all chemical, biological, and social novelties did not arise spontaneously, but were planned from eternity according to a divine plan called the laws of physics. Thus, because of this temporal nihilophobia, or the denial of a nothingness before the Big Bang, there must be an eternity; and because of spatial nihilophobia, or the denial of a limit to existence, the universe must be infinite. Two abominable consequences of absolutism, because both eternity and infinity are incredibly paradoxical concepts. For example, consider the implications of an infinite universe due to the omnipresence of the laws of physics. Right from the start, infinity makes it difficult for us to make it consistent with the scientific fact that the universe had a beginning, a Big Bang. A moment where the whole universe was concentrated at a single point. So, what was beyond that point, more points, infinite points, so are there infinite galaxies, did an infinite amount of matter and energy arise at the Big Bang? Then, what was there before the Big Bang? To all this, we must add that the universe is expanding, but to where, if there is nothing beyond it, just infinitely more and more galaxies? And there must be more and more galaxies, because for the universe to be infinite, it makes the most sense for it to be flat, as required by the current standard cosmological model. This is because if it had any (positive) curvature, however minute, it would close in on itself like a sphere. Which is clearly not infinite. Therefore, to avoid the question of what lies beyond, which would imply answering with nothingness, the universe would have to extend infinitely in all directions, with an infinite number of galaxies distributed in a very homogeneous way to prevent the universe from closing in on itself due to the effect of gravity.

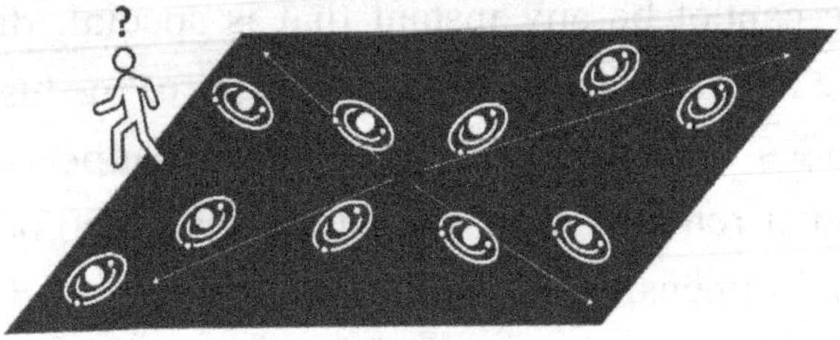

Even assuming this enormous fine-tuning to maintain flatness (homogeneous distribution of matter), which is very difficult to justify for all eternity and infinity, there would still be many, many absurd implications. For example, since there are infinite galaxies, we should eventually find a galaxy identical to ours in every direction, with a star identical to the Sun and a planet identical to the Earth, on which a person identical to you would live, reading an identical book. This is because the combinations of atoms that form us and create all those circumstances, although they are unimaginably enormous, is a finite number; therefore, if there are infinite galaxies, then there are infinite combinations of atoms as well. So, the combination of atoms that forms this moment must repeat itself. And not just once, but infinitely many times. In other words, there would be infinite versions of you doing exactly what you do, but also other versions of you being president, Kennedy assassin, tiktoker, Brad Pitt, and so on. This is the magic of infinity. And if our universe is not infinite (or too), those infinite "selves" will be in those infinite parallel universes then.

But it is not only the absurd consequences of infinity that must be endured, but also those of eternity. Another equally paradoxical concept since it implies the existence of the entire infinite past and future. Infinite moments that must exist simultaneously because, since time must be perfectly symmetrical, there cannot be any instant that is special, distinguishable, as the absolute laws of physics require. In this way, that which divides time into past and future disappears from the present. What remains is an inert time that becomes just another spatial dimension. A view of the world called the block universe, which condemns us all to an immovable, absolute destiny, already determined forever. Therefore, not only the becoming and the present would be a mere illusion, but also free will. Exactly as many prestigious physicists from Albert Einstein, Stephen Hawking and Roger Penrose to Brian Greene, Sean Carroll, Sabine Hossenfelder and Carlos Rovelli believe. Therefore, if this is true, we should not imprison the perpetrators, as this would be immoral, since they would actually be the true victims of the laws of physics, because they were doomed to do what they did, like any other event that has happened and will happen, because it is already written in stone in the block universe. A ridiculous discussion, since we could not change anything no matter how good an argument we have, because destiny is immovable as current Parmenian physics dictates. With the only consolation that the situation you long for already has another "I" in one of those infinite parallel universes, or in one of those infinite galaxies far far away in our infinite universe. A consolation that I do not know if it is of much use, not only because it is not really you, but because the other "I's" live all the misfortunes as well.

In short, great and absurd implications that one would have to tolerate, just because one has a phobia of nothingness. A rather ironic situation, because infinity (and eternity) is supposedly the archenemy of theoretical physicists, because if they find a zero dividing in their calculations, it means that their model is wrong, because it is an indefinite result, because infinity is. The latter is very important, because many people, including physicists, believe that infinity is a clear, solid concept, as if it were any number, but it is completely false. For example, it cannot be a number, because if you add any number to infinity it will still be infinity. So, it does not operate at all as just another number. Furthermore, there are many types of infinities, and the most absurd thing is that they can be of different sizes. And we do not speak of the infinitely small (the infinite decimal numbers between 0 and 1, for example) and the infinitely large (what would be at the end of the sequence: 1, 2, 3, 4... $+\infty$); rather, there are infinities with more quantities or with more elements than other infinities. For example, the great mathematician born in Russia, but naturalized German Georg Cantor (1845 - 1918) proved that the set of positive integers (1, 2, 3, 4, ...) has the same amount of elements (infinities) as these same ones including negative numbers and zero (... -3, -2, -1, 0, 1, 2, 3...). Something that in itself is quite paradoxical. However, the most incredible thing is that he proved that these two infinite sets above would be smaller compared to the set of real numbers (which includes the irrational numbers such as π and e), which would be an uncountable infinity. In other words, the infinity of natural numbers has fewer elements than the set of real numbers, even though both have infinite numbers. A nonsense that reveals that infinity (like eternity) is a purely fictitious, abstract or symbolic concept, not real or physical; that it would not even really exist in the Pythagorean world of mathematics as such, because it cannot even be rigorously defined. And yet, this is exactly what theoretical physicists would be supporting when defending the absoluteness of the laws of physics.

Given all this, it is extremely unfair to exclude the concept of nothingness a priori, because the other option, absolutism, turns out to be even more absurd. An injustice that extends not only to its absurd consequences, but also to how dangerous it can be for science itself. This is because absolutism would attack one of the most essential principles of rational thought, the principle of sufficient reason. A thesis that says that every event or fact must have an explanation, even if we have not yet discovered it, because only in this way can nature be consistent and understandable. Conversely, if nature did not obey the principle of sufficient reason, it would mean that there are brute facts in nature. Things that happen or exist just because, without any cause or justification. The first to explicitly mention this principle was the German physicist, rationalist philosopher, theologian, logician, mathematician, jurist, librarian and politician Gottfried Wilhelm Leibniz (1646-1716). However, the principle has been implicitly present throughout philosophy, since the early pre-Socratics, being the basis of every consistent system of thought. In fact, it is what has allowed the development of mathematics and science in general. Therefore, it is a very difficult assumption to refute epistemologically speaking, but also ontologically, because without this principle the universe would not really make sense, as causality is constantly violated. It would be equivalent to us seeing a dog transforming into a horse, and then disappearing and appearing on another planet. All this without any reason or explanation. So, it would be a very difficult nature to model and understand, as patterns and regularities would be constantly violated. Therefore, it will sound incredible to hear that such a fundamental principle for science is precisely the one that would be violated in current physics. This is not said by the author, but by the famous philosopher, logician and mathematician Bertrand Russell, who defended that there can be science without complying with the principle of sufficient reason, since nature can be described through mathematical laws, which can be explained by other more fundamental mathematical laws, and these by other even more fundamental mathematical laws, until reaching the most fundamental mathematical law of all, the theory of everything. The great dream of physicists. And since that law is the most fundamental, because there cannot be an underlying law that explains it, then, that law would not require explanation. It would be an axiom or a fundamental premise that would be assumed to be true and evident just because, without demonstration. Something that we already know is impossible to obtain in a complete and consistent way, due to Gödel's incompleteness theorem. But omitting that detail, even assuming that it were possible to reduce all existence to a single fundamental mathematical law, such an absolute law would have no cause or reason, having no explanation or justification for its existence. Therefore, this would imply that all existence, the whole universe, would have no sense, because if the other laws and natural phenomena acquire their reason thanks to the fact that there is a more fundamental law that justifies it; then, if the law of laws lacks justification, ergo, this lack of sense is transmitted to all the other laws that it sustains as well. In other words, the whole of nature becomes absurd, because that which sustains all the laws has no reason. All the consistency that is perceived in physics would be simply illusory, because the most fundamental law of all really has none, transferring its meaninglessness to all the other laws that derive from it or depend on it. The whole world would be a brute fact, because it simply exists because it does. This logical conclusion also applies to God or the multiverse, since these are interchangeable with absolute laws, because in essence they are the same. Entities that exist because they do, because they always have been and always will be. Thus, if the lack of explanation for the absolute law transforms the whole universe into a brute fact, the lack of an explanation or justification for the origin of God or the multiverse would do the same. For example, if God is the one who created everything in the Big Bang, what had he been doing a whole eternity before that Big Bang, learning how to make a Big Bang, why did he do it, why didn't he do it before, was he bored by eternity, how did he get there, was it a necessary thing, did he do it out of loneliness, and so on? Infinite questions that we could also ask to the multiverse and to the laws, because all these are dogmatic, and dogma violates the principle of sufficient reason. So, absolutism or nihilophobia can be practical and useful in particular problems; but if it is taken too seriously, as is currently happening in the foundations of physics, it can make absurd not only physics, but also the whole science, being the most fundamental science. A very real danger, as it is currently being demonstrated by the multiverse fanaticism, which is putting cutting-edge the foundations of physics on a par with Marvel movies.

Perhaps you consider the danger of absolutism a bit exaggerated, that we are going a bit too far. But its danger has been proven in the past. A good example of this is in biology, where absolutism almost killed one of the most important theories in all of science, evolution by Natural Selection. It was here that the English naturalist Charles Darwin (1809-1882) made a giant leap forward in that science by overturning the absolutist dogma that species could not change or modify over time because they were created by an absolute God. Of course, when

Darwin published his book On the Origin of Species in 1859, it was met with much hostility, which is not news to many. What may surprise some, however, is that the sharpest and most dangerous criticism of Darwin's theory came not from the church, nor from other naturalists, nor from society in general, but from physics. At that time the absolute worship of physicists was almost indistinguishable from the religious, not only in meaning, but also in form, since the laws of physics were literally the word of God for physicists. A word that had to be absolute, being the will of an absolute being. Thus, it is not surprising now that the most important and influential physicist at that time and, therefore, the greatest authority in his field, the English physicist Lord Kelvin (already cited in chapter III) was a great fan of creationism and a staunch enemy of Darwin. So, for Lord Kelvin, species are immutable, having been created by an immutable god, as well as his laws. Thus, using all his talent, skills and, above all, his reputation, Lord Kelvin set out to destroy Darwin's theory at all costs. Thus, as one of the fathers of thermodynamics, he used his knowledge to show that the Earth was not as old as Darwin believed, so that there would not have been enough time for evolution to take place and produce all this biological diversity. All this despite the fact that there was plenty of geological evidence that the Earth should be at least a few billion years old. For example, already in 1738, the French geologist Benoit de Maillet had estimated the age of the Earth at 2,400 million years, due to already known geological processes and evidence such as sedimentation, fossilized shells on the top of mountains, erosion, and so on. Giving a reasonable time for Darwin's evolution to come into action. Ignoring this evidence, Lord Kelvin decided to prove that geologists were also wrong, so in 1862 he developed a calculation where he estimated the age of the earth at only about 100 million years. Although it was a much longer time than the 6 thousand years estimated from the bible, it was also a completely insufficient time for evolution by Natural Selection to generate all the biodiversity we observe. Lord Kelvin arrived at this result by calculating the time it would take for the Earth to cool down and compared it with the time it would take for the Sun to cool down (ignoring nuclear reactions, obviously, since that had not yet been discovered). An estimate that yielded an age of 100 million years for both. A coincidence that reinforced the biblical idea that the Earth was created at the same time as the Sun by the hand of God.

Fortunately, not all physicists were religious absolutists, for ironically, a student of Lord Kelvin, John Perry (1850-1920), decided to check his mentor's calculation. Of course, he knew that the greatest physicist of his time could not have made a mistake in the calculation; therefore, if there was a mistake, it must be in the assumptions he used. And there it was. The mistake Lord Kelvin made was to assume that heat transfer was homogeneous throughout the globe of the Earth, ignoring the fact that the Earth is made up of different layers that have different conductivities because they contain different concentrations of metals (the heavier metals are concentrated in the Earth's core compared to the lighter ones in the mantle, due to simple gravity). Correcting for this, John Perry came up with a radically different estimate, about 3 billion years old for the Earth. More than enough time for Darwin's evolution to work. So, in the end, the calculation that tried to disprove Darwin's theory ended up being one of its major supports. Obviously, Lord Kelvin never acknowledged it. He was too prestigious to recognize such an ideological and unscientific error, so he continued to deny not only the new mathematical evidence, but still the geological and biological evidence. Thus, one of the greatest advances in science was almost sabotaged by physics, because absolutist faith blinded them to the overwhelming evidence in favor of change and evolution.

However, biology was not the only victim of absolutism, as another important scientific discipline was also attacked. This time an indispensable one for physics itself. We are talking about mathematics, because the abhorrence of nothingness generated a strong resistance and rejection of the introduction of zero into mathematics in the West. The first appearance of this important mathematical concept dates back to Babylonian writings around the 3rd century BCE; and to the Mayans in 36 BCE However, this proto-zero does not resemble the zero we use today because it lacked its own meaning. It looked more like a sign, like the comma in decimal numbers. Instead, zero began to appear as another number in India as early as 1000 BCE, where it was given its own symbol, 0, and its own name: shunya (Sanskrit for nothing). A concept that was later corrected and developed by the Indian astronomers and mathematicians Aryabhata, born in 476 CE, and Brahmagupta, born in 598 CE; who formally described the modern decimal place value system, and the rules for the correct use of zero. A mathematical innovation that quickly became popular, due to its obvious benefits in calculating within the positional system, where each digit acquires its value according to its position. For example, the digit 2 in the number 2, does not mean the same as the digit 2 in the number 25. This is so because in the first case it is in the position that represents the units (two), while in the second case it is in the position of the tens (twenty). However, the problem arose when trying to write the number "two hundred and five," for example, because a symbol was needed to represent nothing in the tens position. For this reason, the "0" was invented, to allow writing the number 205, that is, 2 hundreds, no tens and 5 units. But in addition to its great virtues, it should come as no surprise that the concept of zero arose in India, since for the ancient Indians the concept of nothingness was quite natural, in fact, very important in their belief system. Contrary to the situation in the West, for Hinduism nothingness is the purpose of humanity, its end; for nirvana is the place where everything is erased. In fact, shunya (zero) and nirvana (heaven for Hinduism) are used synonymously. So, for Hindus the universe was born from nothing, and to nothing it will return. An endless cycle where the form of zero possibly represented the cycle of life and death, of being and nothingness; through the serpent that eats itself.

Clearly, then, the spiritual idea of nothingness inspired the Hindu to invent zero, which migrated from India to China and the Middle East, and from the latter, again thanks to Arab philosophers and scientists, to the West in 970 CE by Leonardo Fibonacci, who introduced the decimal system to Europe. However, far from being received with great gratitude and amazement for its obvious and powerful advantages in calculus (by greatly simplifying mathematical operations compared to the obsolete Roman numerals), the "0" was strongly rejected and even banned as being against the nature of God. Europeans believed that by accepting zero, they were also accepting nothingness, and thus the non-absoluteness of God. Fortunately, however, despite the terror of nothingness, the "0" ended up being imposed, due to its obvious and undeniable virtues, together with the Indo-Arabic numerical system. An annexation that not only revolutionized mathematics and science, by making it possible to record experimental observations much more accurately, and to build more sophisticated mathematical tools and models; but also revolutionized society more generally, through calculus in engineering and architecture, but also in finance and economics. Enabling the development of banking and commerce, driving globalization and prosperity around the world. Thus, almost missing one of the most powerful innovations in science; all due, again, to nihilophobia. Thus, although the vast majority of physicists openly declare themselves atheists, they are unwittingly carrying on a long Christian tradition by sharing their devotion to the absolute. A belief that has proved to be very practical and useful so far, but which may jeopardize the future of science, as has happened in the past.

Chapter VI

A True Theory of Everything

As we have seen, physics has been biased throughout its existence by a deep faith in the absoluteness of its laws, which has excluded in this science the rigorous study of emergent and synergistic phenomena. To correct this, not only an axiomatic correction in the hard core of the Parmenian paradigm is necessary, but also a methodological change to be able to approach these phenomena, especially disruptive emergence, since it is impossible to treat them with a purely reductionist-analytical approach characteristic of physics; because, as we saw in the previous chapter, such an approach destroys emergent phenomena by being a top-down approach. That is, it reduces all things to their constituents. Thus, for physics, things are what they are made of. An approach that obviously ignores or denies the other realities that exist at higher scales, such as molecular, biological, and behavioral. These are precisely the scales where emergent phenomena are most evident. For example, consciousness or intelligence cannot be adequately studied using analysis alone, since this phenomenon is not found at the level of neurons, but emerges from the interaction of the set of millions of neurons as a whole. Nor can life be approached using only the exhaustive analysis of complex organic molecules, because this phenomenon emerges from the interaction of many molecules as a whole. Therefore, the emergent phenomenon is lost when things are reduced to their parts, because emergence is precisely what is left over from the sum of the parts. But this is of little interest to the foundations of physics. Hence they call the unification of the four fundamental interactions (electromagnetism, gravity, strong and weak nuclear force) "theory of everything," because they really believe that they can reduce all natural phenomena, including the most complex ones, to a simple interaction between fundamental particles, although there is no demonstration that this is possible.

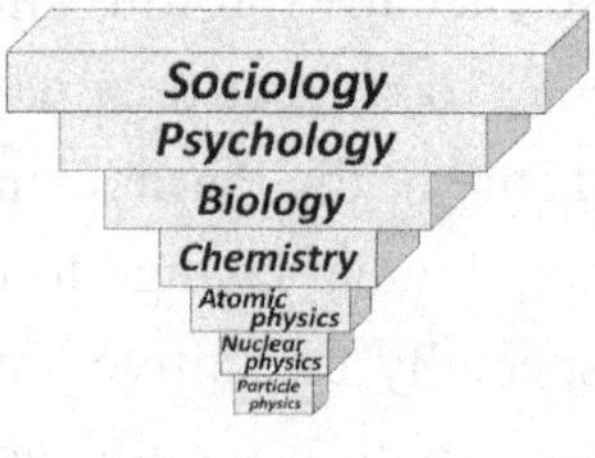

All that the foundations of physics offers is a story that societies are made up of people, and people are made up of cells, and cells are made up of organic molecules, and molecules are made up of atoms, and atoms are made up of particles. However, there is no mechanism or process that explains how these scales are connected. In fact, particle physicists only have exact solutions for a single particle, such as an electron, and for the simplest atom, the hydrogen atom. But from there they cannot calculate anything exact for, say, a more complex atom like carbon. In fact, they do not even have exact calculations for a simple proton, which is a conglomerate of quarks and gluons. Therefore, physicists do not know the exact mass of this particle because they do not know how to calculate it. So, it is not true that a molecule can be fully understood analytically by the Schrödinger equation of the particles that make it up. Much less a cell or a person. You just assume that it is possible in principle, but there is no proof or demonstration that this is true. In fact, there is much evidence to the contrary, for example, the electrons of chemists are completely different from the electron a particle physicist refers to. So, if chemistry is far enough from particle physics, let alone biology, psychology and sociology. But because of the great ignorance of these sciences, again a product of extreme specialization, these higher realities are regarded by the foundations of physics as mere epiphenomena, byproducts, residues. Hence, the reductionist-analytical approach strips complex things of their essence by transforming them into mere aggregates of more elementary things. That is why a pile of sand does not really have an essence, since we can perfectly reduce it to the grains that make it up. That is why we can find hills of all sizes, because they can be modified by removing or extracting grain by grain of sand. And that does not change what is a simple mound of sand, whether there are 85 grains, 113 grains, or a million grains. They are just larger mounds than others. But this is not the case with dogs or trees, because they cannot be reduced to their organs, cells, or molecules. Therefore, they have an essence that makes them unique, identifiable, something that would disappear if we only looked at one part. In this sense, for the foundations of physics, an atom is not in essence an atom, a thing as such, but a set of quarks and electrons; a molecule would not be in essence a molecule, but a set of quarks and electrons; a dolphin would not be in essence a dolphin, but a set of quarks and electrons... I think you get the idea. So, the reductionist-analytical worldview of the foundations of physics sounds as absurd as saying that in biology, multicellular creatures like an elephant or a fly do not exist in essence because only cells exist as such. This is the most fundamental form of life. In other words, only cells have the character, value, or essence of life, while dolphins, dogs, and us would be simple colonies of cells, like a liver, a lung, or an excrement. In this sense, you could not technically kill a dog or a person, because there is nothing living to kill, because there is no essence of life in them. You could only kill the cells that make up these multicellular organisms. Therefore, you would not really be committing murder by shooting someone, but by scratching them. This is the consequence of the exclusive use of analysis, the reduction of all things to their components, the denial of all essence and complexity at higher scales. All because of an approach that preaches that there is only the sum of the parts, nothing more.

Then we already understand the reason why current physics is incapable of explaining the origin of things, because there can be no creation of novelty in a physics that believes that what is truly essential must be absolute. That is, uncreated, eternal and infinite. That is why it cannot answer what caused the origin of life, of consciousness, much less the Big Bang, because everything must have always existed or been planned forever. This radicalization of the reductionist-analytical approach also explains why the current foundations of physics have an exclusively descriptive purpose, confirming what was said in Chapter III; conceiving the universe in a static, passive and inert way, unable to recognize the existence of emergence and change. As if nature were a 4D photograph, by restricting itself almost exclusively to answering how it is, how it behaves, what it is made of; and avoiding as much as possible to address why it is like this and not otherwise. Its reason for being. Hence the intensive and exclusive use of mathematics in this science. A purpose recognized by physicists themselves. For example, the winner of the Nobel Prize in Physics, the Hungarian mathematician John Von Neumann (1903-1957) said: "Science does not try to explain and hardly tries to interpret. It devotes itself above all to making models. And by model we mean a mathematical construction which, with the addition of certain verbal statements, describes the observed phenomena. The justification of this mathematical construction is solely and precisely to be effective." A legalistic definition of science, in the sense that it is limited only to discovering laws through equations, without being interested in explaining such laws and models. This way of doing physics could be anecdotal, because it has worked quite well so far; however, the lack of an explanatory purpose may be the reason for the current stagnation of physics. Let us remember how scientific revolutions occur, through a deep understanding of the paradigm that allows us to modify and correct its hard core. Therefore, it is possible that the current normal science has reached such a point in its development that the function of describing nature is insufficient. Perhaps the time has come to understand, to explain such descriptions. The reason why this is so, because if one does not know what the quantum fields that form the Standard Model of particles are, nor does one know the reason for their twenty or so constants appended by hand, nor does one know what space-time is, nor why the speed of light is invariant, or even what it means to measure in quantum mechanics; then, it becomes more evident why physics has stagnated, because it would be like wanting to find an answer to a question that is unknown.

A devout absolutist will disagree with the above by declaring that the reductionist-analytic approach is sufficient to explain all of reality, including the emergence of complex phenomena. He will cite, for example, the famous Game of Life created by a leading British mathematician named John Horton Conway (born in 1937), who unfortunately died of COVID-19 in 2020. This model is of the cellular automaton type, just like the famous Theory of Everything by the English physicist, mathematician and computer scientist Stephen Wolfram (born in 1959), but which uses graphs, or rather hypergraphs, instead of cells. The difference between these two models is that the game of life has a given flat grid mesh (cells), like a chessboard, but infinitely extended in all directions. Whereas, in the case of graph automata models correspond to a network or mesh of vertices or nodes joined by links called edges or arcs. Something similar to the Pert Mesh. In this way, the mesh is created as the program progresses, which creates more nodes that are joined through the edges. The same applies to hyper-graphs, except that, in this case, the edges can be connected to more than two nodes, allowing to generate closed meshes, and with this, three-dimensional objects are created, and not only two-dimensional figures, as in the case of graphs and cells. But, regardless of these differences, they are all essentially the same, therefore, the Game of Life can be taken as a reference for all of them. So, each cell in the Game of Life has eight neighbors, which can have two possible states: alive (on) or dead (off), which will depend on what is happening around it. If a dead cell has three live cells next to it, it is born. If a live cell has less than two or more than three neighboring cells, it will die from loneliness or overcrowding. With just these simple rules, incredible patterns such as oscillators, spaceships or guns appear.

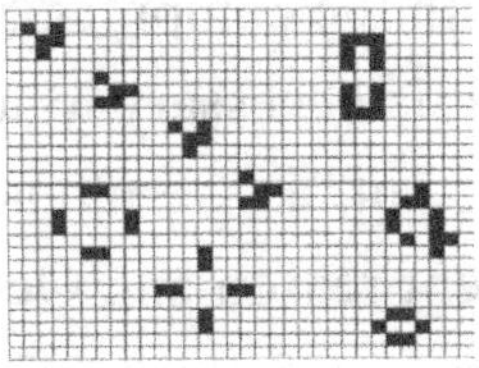

Unfortunately, the image above does not reveal all the extraordinary dynamics of the model, so it is advisable to look for it on YouTube, where there are many examples. To continue the theme, the universe for physicists would be something similar, but in reverse, because we would not start with the fundamental laws or rules, since that is exactly what we are trying to find out, but we would start with the configuration that we already observe in the universe. And from there we reverse engineer to deduce what those simple rules might be that govern the game of the universe. The famous Theory of Everything. However, this proposal has many very important problems, the first is the eternal problem of the origin of the Parmenian paradigm: where did those rules come from, who programmed the universe in that way, why these rules and not others? The second major problem is the very special initial conditions or also known as fine tuning. Who configured the cosmos so finely so that we get everything that exists in the universe, because it is not enough to have the right laws, you also need the right configuration to get everything that we see. This can be clearly seen in the automata models, since, having the same rules, you can get radically different results if you change the starting configurations. In fact, the vast majority of configurations yield nothing interesting. That's why Wolfram says that a new kind of science must be created to investigate and discover which configurations are needed to generate the most striking behaviors. In other words, play with the computer. So, whether by laws or special configuration, it is necessary to invoke god or the multiverse to justify them; but, as we already know, this violates the principle of sufficient reason and Ockham's razor. In other words, they are not very satisfactory explanations.

But even omitting the above, one of the biggest problems with automata models is that their interesting dynamical patterns lack real complexity. They are simply very complicated systems, not complex ones. This is a very important distinction, because an airliner is very complicated because it is made up of millions of parts in a very specific configuration. That is why it is not so easy to build one. But that does not mean it is as complex as the weather, the economy, or a simple double pendulum. So, to understand what complexity means, we need to look at the third revolution in physics of the 20th century (after relativity and quantum mechanics): chaos theory, which, although it is one of the great discoveries of physics, is very unpopular among theoretical physicists. This should not be surprising, since it goes against the absolutist orthodoxy of conserved symmetries and magnitudes. In fact, it is no exaggeration to say that the first physicists who tried to develop this new line of research had a hard time. Not only because it was something new, but also because they were met with great resistance and even hostility from the community. For example, physicists accused them of not doing science. Mathematicians branded them as not doing mathematics. This was despite the fact that they were providing new methods of calculation to solve old, unsolved mathematical problems. So, it is not surprising that these brave pioneers had great difficulty finding work and publishing in scientific journals. A difficulty that was reinforced by the universality of their discovery, since complex and chaotic systems could be found in practically all natural phenomena. This obviously crossed several disciplines and sciences, making it more difficult to find work and publish, since all research departments and journals are highly specialized. A bias that prevented to see the importance and scope of this new research in its beginnings.

A rather awkward beginning, which should be surprising, because chaos was looming from the very beginning of modern physics, in Newton's equations of motion, because Newtonian physics was spectacular in solving two-body problems. The dynamics between the Earth and the Moon, or the Earth and the Sun, and so on. Here, the behavior of the system can be predicted for all eternity, in full accordance with determinism. However, when a third element was added to the system, even a simple asteroid, the equations no longer gave an exact solution, making the prediction simply impossible. This is known as the three-body problem or the n-body problem (with $n > 2$), because as the number of elements in the interaction increases, the problem gets worse and worse. And although in 1890 it was demonstrated that this problem was unsolvable, thanks to the contribution of the incredible French mathematician, physicist, scientist and philosopher Jules Henri Poincaré (1854-1912), the foundations of physics continue to ignore the transcendental consequences of this mathematical fact, clinging to the false hope that everything has been a nightmare, because in essence you can always get a single perfect solution, because the universe has to be clear, orderly, transparent, knowable, certain, beautiful, symmetrical. That is to say, as God commands.

But what exactly did Poincaré discover that terrified and continues to terrify physicists? Nothing less than the difference between the Pythagorean world of mathematics and the real world. In the world in which we live, there is nothing perfect. We are surrounded by irregular figures, far from the idyllic circles and triangles of mathematics. Therefore, just as there is no such thing as a perfect circle, it is not possible to obtain perfect measurements, because our tools or instruments will always contain errors in their observations. This is an unavoidable problem. We knew this long before chaos theory came along. So why is it a problem now? Because it used to be tolerable. As we improved our technology, we also improved the accuracy of our rules. Measurement errors became so small that we could ignore them. For example, if we buy a table that is four meters long, what does it matter if it is actually 3.986 meters long? Or when we navigate with GPS, what does it matter if our device tells us that the destination is 12 kilometers away when it is actually 12,072 kilometers away? Practically speaking, these errors are so small that they do not matter. Thanks to our technology, errors are becoming increasingly irrelevant. This is where the orthodox approach to physics comes in, which is called linear, and it means that small errors in measurements produce equally small errors in predictions. Thus, if the margin of error did not bother us when we measured the initial conditions of the system, it should not bother us when we compute the prediction in the future. Hence the name linear, because the error is proportional or constant.

This was the way physicists believed nature behaved. However, the third revolution in physics in the 20th century revealed that the vast majority of natural phenomena are non-linear. This means that chaotic or complex systems are very sensitive to initial conditions, which means that those negligible errors in measurements that we ignored at the beginning can now grow and amplify as the system evolves, producing errors that are impossible to ignore. Something that also happens in a few cycles or steps in the future. This makes these systems inherently unpredictable, because we should have perfect measurements to get perfect predictions. But as we already know, perfect only exists in the world of mathematics (and not even there); therefore, these unavoidable insignificant errors will be amplified, rendering any prediction we make with them completely useless. And the culprit is iteration. To understand this, imagine a system of only 3 bodies ($n = 3$), for example: the Earth, the Moon, and the Sun. If the Earth interacts exclusively and only once with the Moon, and the Moon interacts exclusively and only once with the Sun, then we would have a linear system. The behavior theoretical physicists love. But this is not what happens, because in reality the Earth interacts with the Moon, and the Moon interacts with the Sun, and the Sun interacts with the Earth again. The same cycle is repeated over and over again, which means that the small error that was made in the measurement or calculation at the beginning of the iteration is added and accumulated in each cycle, creating a positive feedback that amplifies such an error, making it more and more significant.

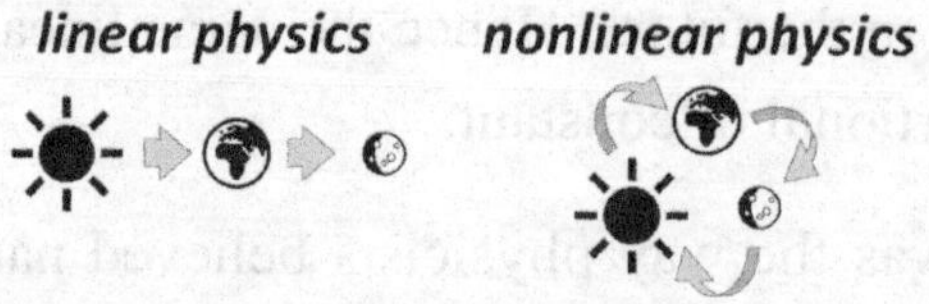

In fact, it is much worse in practice, because each body interacts with the other two at the same time, making the behavior of the system even more complex or less linear. In this way, the prediction we will be able to compute is only good for a limited number of steps, because iteration quickly renders the prediction useless, as the actual result becomes farther and farther away from the prediction. This was observed by the American meteorologist and mathematician Edward Norton Lorenz (1917 - 2008), who immortalized this phenomenon with the famous name of the Butterfly Effect. He discovered it by trying to predict the weather on a computer, repeating a previous simulation, but this time approximating the values to only three decimal places instead of six. Just out of laziness. To his surprise, such an insignificant difference, only one thousandth, produced a large discrepancy with the previous result. For example, if the prediction with six decimal places showed that next week would be sunny, with the approximation to three decimal places the prediction changed to rain. Hence his famous metaphorical phrase, that the mere flapping of a butterfly's wings can generate a hurricane on the other side of the world. A poetic phrase that seeks to represent how a small error in the measurement, by not considering the wind generated by a simple butterfly, can generate great consequences later, by amplifying and transforming into a storm that we have not been able to predict, having omitted that insignificant piece of information. This is why weather forecasts are only reliable for a few days. Not because meteorologists are not good at mathematics, nor because the instruments are not accurate, but because prediction in the medium term is technically impossible, since the weather is a chaotic system. In fact, with the best precision technology, we can currently measure with a degree of accuracy of about six to eight decimal numbers (one millionth of a centimeter, for example). Impressive precision by our everyday standards. However, even with this impressive certainty, this allows us to predict, in most chaotic systems, only about 10 to 20 steps into the future. After that, the error in the measurement accumulates and becomes so amplified that the calculation becomes useless. An unpredictability that pertains not only to large and complicated systems such as the weather, but also to very simple systems. Recall that only three elements are needed to make a system chaotic. Even a system as simple as the double pendulum can manifest an impressive chaos.

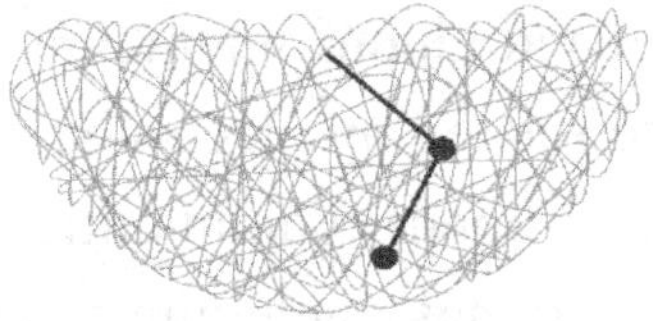

Therefore, even if the behavior of the system is described by a very simple and perfectly known equation or law, as in the case of the harmonic oscillator of the pendulum, insignificant errors in the measurements of the initial situation, which elude any measuring instrument, very soon render the calculation or prediction useless, since this insignificant error, due to iteration, expands to such an extent that it makes the system fundamentally unpredictable. Thus, the fact that a system is complicated does not necessarily mean that it is complex, since, as we have seen, even very simple systems can be complex; because complexity is not essentially related to the number of components, but to whether or not it has linear behavior (among other things, as we will see later). And although Parmenian theoretical physicists hate nonlinear systems, precisely because it prevents them from calculating; unfortunately for them, these systems are practically omnipresent in nature, since the vast majority of natural systems are formed by three or more components that interact intensely with each other, generating loops. In fact, the entire universe is an n-body problem, because everything is interacting with everything. That is why iteration is inevitable, since nothing is isolated in the universe. Everything is related in some way. Compound particles like protons and neutrons are made up of particles bound together intensely by the strong nuclear interaction. And as already stated, the mass of the proton cannot yet be known theoretically, precisely because it is the product of an interaction of "n" bodies, due to the multiple types of quarks and gluons interacting strongly within it. While molecules are related by electromagnetic forces, living things by biochemical bonds, communities by traditions and social norms. And all things, including fundamental particles, are related by gravity, by the fabric of space-time that allows them to be. That is why the reach of gravity is infinite. In other words, chaos is the permanent, the inevitable; however, most theoretical physicists continue to believe that nature is perfectly linear, stable and predictable, due to their absolutist fanaticism. They stubbornly try to study the universe as if it were a two-body problem, adding them countless times to obtain a linear result. In fact, the few linear systems that physicists can cite are not really linear, because they do not last forever. They become nonlinear eventually. The solar system, for example, may seem very stable to us, very solid, because it has maintained the order and number of planets for hundreds of millions of years, yet in the future, some planets may fall toward the sun, or be ejected from the solar system, lost in the cold void of interstellar space. Possibilities that are impossible to predict (don't worry, it won't happen too soon). Therefore, linear behavior would be temporary, contingent. In contrast, chaos is ubiquitous, for we can find it in the turbulence of fluids and gases, in the mundane drops falling from a faucet, in the beating of the heart, in the stock market, in brain activity, in the fluttering of a flag, and so on. And yet, the foundations of physics focus only on the parts, the particular, and forgets the whole, the connections between the elements that make up the whole. This is where chaos arises.

In this sense, Chaos Theory would be the first true theory of everything, since its models are applicable to a vast number of systems and natural phenomena that transcend all sciences. From particle phenomena to celestial, biological, meteorological, economic or sociological dynamics. Phenomena that are very different in the analysis, because they are composed of very different elements; but that, in the global, their behavior is very similar. Something that can be discovered and rigorously observed through a completely new approach to fundamental theoretical physics, the synthesis. A method that looks at the system as a whole. A good example of this is the strange attractor of chaos theory, a method of studying the behavior of complex systems by plotting them in a finite region, even though they manifest infinite complexity. This seems paradoxical because although the curves describing the behavior of the system are infinite, they do not leave any particular region, nor do they cross or intersect. In other words, the same pattern never repeats itself. That is why analytical methods are completely useless, because locally the chaotic or complex system is unstable and unpredictable, but globally it is very stable and predictable. An ordered chaos.

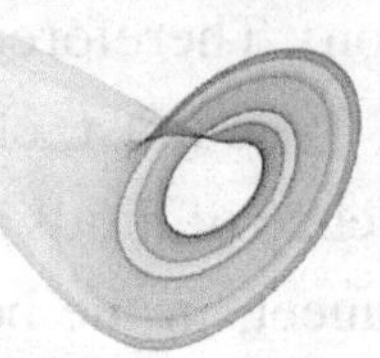

Then it is clear that the complicated and the complex are two very different things, because the latter can be present in both simple and complicated systems. Therefore, many of the natural phenomena that are genuinely complex cannot be naively reduced by traditional physics or by cellular automata or graph models, since the latter, as we already know, are essentially linear. However, although complexity theorists considered themselves great rebels and revolutionaries, they ended up surrendering to the orthodoxy of the Parmenian paradigm, concluding that nonlinear systems are as deterministic as any other traditional physical system. This equating between chaotic and linear systems was done perhaps because of being immersed in the Parmenian paradigm, or as a desperate measure to be accepted by the physics community, or both. Regardless of the reason, this took away all the impact that this third revolution in physics should have had, leaving it in the shadow of relativity and quantum in notoriety and importance, since they would essentially be the same. Only now it is necessary to differentiate between two types of determinism: the strong and the weak. The first is characteristic of linear systems, where the errors remain proportional and, therefore, it is possible to make a prediction as accurate as your measuring instruments allow. In contrast, weak determinism is characteristic of nonlinear systems and, therefore, it is not possible to make very accurate predictions in the future. However, this unpredictability would not be ontological, but exclusively epistemic, because fundamentally there remains a single history, trajectory, or outcome. However, we cannot know it due to our technological incapacity, which prevents us from obtaining all the information necessary to make a perfect prediction. Therefore, that uncertainty of chaotic phenomena is totally illusory, because in principle, if one could make perfect measurements of the initial conditions of the nonlinear system, with a precision of infinite decimal digits, one could know with perfect accuracy the entire infinite future of the system. In this way, one could argue that the universe is a gigantic game of life, as Wolfram postulates with his hypergraph hypothesis. One that appears complex and unpredictable only because it is incredibly complicated, with so many components so small (those hyper-graphs), that one cannot obtain sufficiently precise measurements, generating that unpredictability that we experience. Then all this complexity that we see in nature would not be real, in the sense that there would be no strong emergence, no synergy; but only emergence and weak determinism, by virtue of the fact that all things would be perfectly reducible to simple rules (laws) and to very particular and special initial conditions. Therefore, in principle, it would be possible to go back or predict any moment, the whole past and the future of the universe, if we had these absolute fundamental laws (theory of everything). Something that would be impossible, but only technically, because we would have to measure and calculate very rigorously a very complicated configuration of countless elements. In principle, however, everything would be perfectly known and determined. So, in the end, the physicists of complexity and chaos gave in, leaving a great revolution always unfinished, drowned under the absolutism of the foundations of physics.

Unfortunately for modern parmenideans, this absolutist hope that unpredictability is only epistemic (weak determinism) vanished completely with the advent of the second revolution in physics, quantum mechanics. We now know that errors are not only made by our measuring instruments, but are intrinsic to nature itself, due to Heisenberg's uncertainty principle, which, as we saw in Chapter IV, forbids us to have perfectly precise values between two simultaneously conjugated or related parameters. For example, if we increase the precision of position, we make the value of momentum or velocity more and more indeterminate, to the point that if we have perfect precision on the first, we will have an infinite error on the second. The same thing happens when we seek precision in duration, in time versus energy, and vice versa. So, the premise that even though systems are chaotic, if we had perfectly precise measurements we could still make perfectly reliable predictions is simply false, because perfect precision does not exist, not even in principle; because uncertainty at the most fundamental scales is not epistemic but ontological, as quantum mechanics has shown and proved. In other words, the problem is not that we do not know, but that nature itself does not know.

As if that were not enough, quantum mechanics showed that it is impossible to derive the principles and phenomena of quantum mechanics from a classical model or theory, including cellular automata or graph models, because of Bell's inequalities theorem, formulated in 1964 by the Northern Irish physicist

John S. Bell (1928-1990). A proof that has been tested experimentally since the 1980s, confirming time and again that quantum reality cannot be classical. It is precisely for this extraordinary result that the authors of these first experiments were awarded the Nobel Prize in Physics in 2022. So, it is simply wrong to believe that at the most fundamental level there are only certainties, a simple network of well-defined graphs with simple rules, because the most fundamental theory we have, quantum mechanics, is inherently non-deterministic, whether it is strong or weak determinism. Therefore, even if there is another even more fundamental theory on even smaller scales, such a theory or reality cannot be absolute or deterministic, because it is impossible for quantum mechanics to emerge from it, since it is essentially a non-classical theory. In conclusion, all the complexity we observe is real, not apparent; because those insignificant errors that amplify and make the universe truly unpredictable are not only due to our fault, to our ignorance, because we do not have the sufficiently precise technology to make rigorous measurements; but those errors are ontological, because nature at its most fundamental level is truly indeterminate and random.

Thus, the methodological tools of Chaos and Complexity Theory are not mere mathematical gadgets or effective models, but truly necessary tools, since they are the only way to study nonlinear phenomena. Thus, the approach of chaos and complexity theory is essential for a complete understanding of nature. An approach that we will call emergentist-synthetic or bottom-up, which would complement the traditional reductionist-analytic or top-down approach of linear physics. This means that physics should study nature not only from what it is made of, from its components, but also from its connections and relationships, because natural phenomena are not only linear, but also complex and chaotic. In fact, every thing manifests this double face, because from its scale each one would be a system, so it has to be studied under the reductionist-analytical approach; but from a larger scale the same thing becomes components of another system, so it is necessary to use the emergentist-synthetic approach. Therefore, all things are essentially systems at one scale and components at another. An epistemic unification that would not only allow us to understand and predict complex systems, but would also have the potential to solve the conundrum with which we began this chapter: to explain strong emergence, or how complexity arises, through an underlying assembly mechanism that explains the emergence of new relationships and thus the formation of new structures. A possible explanation that requires quantum mechanics. This will be explained in more detail in the next chapters, but the central point here is that the hope that strong emergence was just an illusion disappears when determinism dies; for only determinism could guarantee that all of nature can be perfectly predicted or reduced, at least in principle. But if this is not the case, as quantum mechanics proves, strong emergence can be a perfectly real phenomenon, thanks to "nonlinearity" and "ontological randomness." The essential ingredients that would allow the creation of a real novelty.

This epistemic unification would not only help us to explain strong emergence, but could also bring as an additional benefit: the unification of all sciences, because it would mean being able to consistently connect and relate all phenomena of different scales and complexity, such as particles, atoms, molecules, living beings, consciousness, and culture. It would therefore be able to explain how these more complex realities emerge from simpler realities. How chemistry emerges from particle physics, how biology emerges from organic chemistry, how psychology emerges from neurology, and how sociology emerges from psychology. This simple story is finally explained by current physics by providing us with a bottom-up mechanism that describes the emergence of the higher scales from the lower ones. But since the latter is missing, science is completely disintegrated or fragmented. The result is a world view with the same characteristics. A society (sociology) disconnected from people (psychology), which are disconnected from other living beings (biology), which are disconnected from complex molecules (chemistry), which are disconnected from the particles that make them up (physics). And all of these are separate from the space-time that contains them (general relativity). Therefore, current science does not give us a description of an integrated and complex world, but of many simple worlds separated from each other. Each world is an isolated science with its own laws. This should not come as a surprise, since this reductionistic-analytical approach fits in well with the extreme specialization of the current academic system, promoting the emphasis on the parts, on the particular, rather than on the whole. That is why each science and discipline is only responsible for describing its small world, because it is not interested in obtaining a deep understanding of the whole of nature, in knowing its reason or origin, nor how the different natural phenomena are integrated in the macrosystem that we call the universe. Each science only wants to describe its small part of the reality. It does not matter if they contradict each other, because they do not have to deal with it because they are separate.

If you think that all this about unifying two approaches as different as emergence and reductionism is utopian, because it turns out that they are contradictory in themselves (bottom-up and top-down), it should be mentioned that there is already a successful case in science where these two approaches coexist harmoniously and complement each other. This science is again biology, since in it we find the reductionist-analytical approach, which provides the understanding of the parts that make up biological systems, but also the emergent-synthetic approach, which provides the understanding of the origin and evolution of such biological systems. Therefore, biology demonstrates and proves that, in order to really understand nature, both approaches are needed, since each one provides indispensable knowledge that the other cannot provide; because in the universe there are no isolated elements, hence the importance of synthesis, but neither are there relationships without elements, hence the importance of analysis. Therefore, it is not enough to know only the constituents, what it is made of, but it is also necessary to know its organization, its structure, the relationships between those parts. That is why biology, unlike physics, can explain the origin of its systems, because the emergent approach allows it to do that. A mechanism called Natural Selection. Therefore, biology can indeed explain the relative and even disruptive emergence of biological systems, perfectly complementing reductionist analysis; because Darwinian evolution cannot accurately describe or explain the particular phenomena of biology, such as how blood, bladder, lungs, brain, pollination, fish fins or bird wings, spider venom, and so on, work. To explain these particular phenomena, exact biology is needed, which works with the reductionist-analytical approach to accurately describe such phenomena by breaking them down into their parts. Examples are molecular biology, cellular biology, marine biology, genetics, botany, neurology, physiology, zoology, epigenetics, epidemiology, and so on. However, all these branches of biology would have no meaning or coherence if they were not contained within the general framework of Natural Selection, which integrates and explains them. In other words, exact biology provides a very precise description of particular biological phenomena, the way they are; but the origin and meaning of these phenomena is taken care of by Natural Selection, through its emergentist-synthetic approach.

In this way, biology shows us that epistemic unification is not only feasible, but indispensable to a deep understanding of how nature works. Therefore, biology would teach physics how to do science, because current physics would be exact physics, analogous to exact biology, which is essentially reductionist-analytic. However, it would be necessary to create a mechanism equivalent to Natural Selection that would unify the descriptions or models of exact physics; but above all, that would explain the origin of such phenomena. The origin of physical laws. This sounds so radical to today's physicists, but it only shows how absolutist today's physics is, because what this book proposes makes perfect sense. If everything evolves, why not the laws? Wouldn't that be an excellent explanation for why we have such rules? A reasoning that makes so much sense that apparently, I am not the first to have it. For example, Harvard University professor of astronomy and mathematics, but also American philosopher and logician Charles Sanders Peirce (1839-1914) said, "To suppose universal laws of nature, capable of being learned by the mind, but remaining inexplicable and irrational, is not a justifiable position. Uniformities are precisely the kind of facts that need to be explained. Law is, par excellence, that which requires reason. Now, the only possible way to evaluate the laws of nature and uniformity in general is to suppose that they have been the result of evolution." An argument from authority that perhaps does not have so much force, because he is not a physicist as such. So, let's try one of the most respected physicists of the 20th century, the British physicist and Nobel laureate Paul Dirac, who said, "At the beginning of time, the laws of nature were probably very different from what they are now. We should therefore assume that the laws of nature are constantly changing with time, rather than remaining uniform throughout space-time." Or consider another legendary physicist, the American and fellow Nobel laureate in physics Richard P. Feynman, who said in his Lessons in Physics I (Mechanics, Radiation, and Heat), "There is another kind of problem in the sister sciences that does not exist in physics; we may call it the historical problem. How did this come about? A problem that applies to biology, geology, or astronomy [...] There is no such thing as the historical problem in physics. We do not ask how these laws of physics came to be, nor do we imagine that they change with time. That they were different in the past than they are in the present. Of course, it may be possible, and when we know how, the historical question of physics will be linked to the rest of the history of the universe. And then physicists will be talking about the same problems as astronomers, geologists, and biologists."

As you can see, the crisis in physics today may be deeper than physicists imagine or want to admit. A paradigm that has served very well, because it has been practical to assume that the laws have always been there, and have always been that way, because it was necessary first to have a good description of how the universe works. However, perhaps the time has come to address the historical question of why it is so, what is the origin of such laws; if one really wants to understand the evolution of the whole universe and, above all, what happened at the Big Bang. Therefore, it would be necessary to change the strategy and adopt a less mechanical view of the universe in favor of a more organic one, using a more Darwinian approach. One that is compatible not only with emergence, complexity, and chaos theory, but also with quantum mechanics, thanks to the fact that Darwinian mechanics is also ontologically non-deterministic. Thus, if Darwinian evolution uses the natural randomness of genetic mutations to evolve living species, one could use the natural randomness of existence itself at its most fundamental level to evolve the laws of nature. This would allow for the emergence of true novelty, neither determined nor foreseen, but spontaneous. In the same way that new living species emerge in the biosphere, explaining how complexity at the macroscopic scale (classical physics) arises from simplicity and randomness at the smallest scale (quantum physics). An epistemic unification that would allow the ontological unification not only of physics, but of all sciences, by explaining the emergence and evolution of all natural phenomena. Something that can only be achieved by abandoning the Parmenian paradigm that abhors change, emergence, and chance.

CHAPTER VII

A MORE RIGOROUS SCIENCE

We are beginning to have all the necessary ingredients to make the next leap in the foundations of physics. We have already seen that philosophy would allow us to find out what it is to do science, what a paradigm is, and how we can replace it, through a qualitative, critical and rigorous analysis of the fundamental assumptions at its hard core. We also identified one of the dogmas that would be most necessary to replace: the premise of the absoluteness of the laws of physics. We also added another approach, one that is urgently needed to incorporate chaotic and complex phenomena, and the possibility of explaining strong emergence or true novelty. However, we still need an additional epistemic ingredient, because if the goal is to develop a new paradigm, then we need to develop a scientific criterion for its evaluation; because if it is possible to develop one alternative, then three or a thousand can be developed. So, it is important to ask: how do we make sure that the new paradigm is better? Because the idea is to make physics better, not worse. And we know from Chapter II that Popper's current falsification-based definition of science is inadequate for this purpose, since it was not designed to apply to a complete paradigm or theoretical framework, assuming that only normal science exists. Therefore, if we want to make science more complete, not only functioning at the stage of normal science, but also including the stage of crisis and, above all, the stage of revolution, then another criterion of scientificity must be developed that is capable of being applied to theoretical frameworks or SRPs.

But before proposing and explaining what this new criterion of scientificity would be, we must first review what it means to do science today in order to better understand what is needed. The first criterion of scientificity, which is not often mentioned because it is taken for granted, is internal consistency. This criterion requires that the theory or system of ideas be logical in itself, with no contradictions among its own postulates. This was the criterion of the ancient Greek philosophers, who generated explanations of nature through rationality. The problem, as we already know, is that a mathematical theory or construction, no matter how consistent and logical it may seem, is no guarantee that it has any relation to reality, because it may well be pure fiction. In fact, every day an enormous number of scientific theories are refuted and rejected, theories that are internally very consistent, that make a lot of sense in themselves, but that contradict observations and experimental results. In other words, they are not reliable representations of nature. To ensure that the logical and coherent symbolic construction has some connection with the real world, there is the second criterion of scientificity, which we will call empirical consistency. A criterion that has its highest representation in Popper's falsification principle, which dictates its implementation. This is science today. The problem, as we saw in Chapter III, is that we have already discovered that neither mathematics nor experiments can guarantee that a particular model is a truly valid representation of nature. Something that was demonstrated in that chapter by the principle of underdetermination and the interpolation and incompleteness theorems. An inadequacy that becomes even more apparent when one tries to evaluate not only particular models, but entire theoretical frameworks or paradigms. For example, many physicists say that string theory is not scientific because, although it is mathematically consistent (in quotes), it is not falsifiable for the reasons we have already explained. A criticism that string physicists defend by saying that the theory is a theoretical framework, like general relativity and the Standard Model of Particles. So how can we know whether string theory is really a scientific theory, if the current definition of science does not serve to evaluate theoretical frameworks or paradigms?

Again, it is useful to quote Chapter III, because there it was mentioned that Thomas Kuhn believed that there are three main criteria that scientists use to prefer one theory over another: simplicity, precision, and broad scope. These are principles that different scientists have to interpret and evaluate models, but it is not known which one is the most important. Therefore, it is valid to choose any of the three, depending on the preferences and tastes of each person. Something we also saw in this chapter, applied to the examples of the Ptolemy and Copernicus models, and the strategies against the COVID-19 pandemic. In these cases, we realized that simplicity and accuracy are really equivalent criteria, as the interpolation theorem shows. Therefore, one cannot say that one is more valid than the other, which creates these conflicts in science, since some prefer simpler theories, while others prefer them more rigorous, adding more parameters. In fact, we will now see that both criteria are really dangerous, since they have the tendency to fall into a fallacy that is unfortunately widespread in today's academic and intellectual world, the fallacy of incomplete evidence, also popularly known as cherry picking. This is one of the most dangerous fallacies that can exist, because there is no worse lie than a half-truth, since it can very well disguise itself as truth. And this is exactly what happens when the criteria of simplicity and precision are abused, because in the former it is very easy to pick the data that one considers (conveniently) to be the most important, and say that here is a model that fits and explains them very elegantly. An argument that theoretical physicists love, who preach that their great guide in their research is mathematical beauty. That is why they love to add more and more symmetries to their theories, to make them even simpler. Even though it is known that such symmetries cannot exist, they omit or downplay other data and evidence that does not fit their elegant theory.

On the other hand, the accuracy criterion can fall into the cherry-picking fallacy, where a very accurate model is presented that fits a very specific set of data exceptionally well, but beyond that the model does not fit anything else. In other words, the model is overfitted by adding a large number of parameters that have no empirical support, or even what they would actually be, what they would represent. But it doesn't matter, because the model works beautifully in that narrow energy range or for that particular phenomenon. The rest, of course, is left out. So, the emphasis on simplicity and accuracy encourages this mistake, the arbitrary selection of data or range of applicability, and supports the confirmation bias in such models. A very common mistake in science, unfortunately, because it is encouraged by a phenomenon, we have already criticized several times in this book, extreme specialization. We should not be surprised that scientists have a strong incentive to fit their models to the data in their small area of expertise, even at the cost of being inconsistent with adjacent areas.

Fortunately, the third criterion postulated by Thomas Kuhn avoids precisely the previous problem, because the principle of broad scope is by definition an anti-cherry-picking criterion, by requiring precisely the opposite. If currently, under extreme specialization, there is a perverse incentive to select only the data set that favors one's model, under the new paradigm this would be forbidden, since it would be evaluated precisely by the least amount of evidence that would be left out of the scope of the model. This would prevent us from being self-deceived, biased by our assumptions that may work relatively well in our area of expertise, but may in fact be completely wrong, contradicting theories and evidence in other sciences, or even other disciplines within the same science. In addition, this new scientific requirement would solve several problems we have discussed in previous chapters. For example, the broad scope supports the new approach discussed in the previous chapter, emergence and synthesis, because now we must also consider the whole and not just the parts. An approach that would help us better understand synergistic, complex, and chaotic phenomena. It also gives meaning to what is meant by the condition of physicality or realism that a mathematical model should satisfy.

Recall that this refers precisely to the fact that some mathematical solutions of a model are not necessarily real, since they may contradict other models and physical evidence. Therefore, even though it is a mathematically allowed solution, it would be suppressed by not respecting the restrictions imposed by the rest of scientific knowledge. Something that would again be natural and obvious under this new criterion of scientificity.

Another advantage of the principle of broad scope would be to give justification and meaning to the integration of philosophy into the scientific process. As was also said in Chapter III, philosophy is essentially explanatory, it seeks to understand. In contrast to mathematics, which seeks to describe, to model. To continue this point, there is great confusion in fundamental theoretical physics about what it means to understand, precisely because of the lack of philosophical skills. In fact, for many physicists, understanding means calculating. That is, if you can do the calculations, then you have understood the phenomenon. For example, one understands Einstein's theory of gravity very well, because one can solve its equations and obtain various solutions from it, even though one still does not know what space-time is, why the universe expands, what singularities are, why light always travels at the same speed in a vacuum, or what time is. This strange definition of understanding is easy to refute, as we have all been through school and had the subject of mathematics, and there we have been or known someone who always got good grades, but that, however, did not necessarily mean that they really understood what they were doing. He just mechanized the whole process and was very good at applying it. Precisely what is promoted at school and even at university, because if you stop too long to think about what an equation is, what a function is, why there can't be a square root of a negative number, or why you can't divide by zero, then you fail the exams because you didn't learn in time to solve the mathematical problems. The best example of this is that in many schools it is still taught to solve an equation by saying that something that is subtracting or multiplying on one side must be passed to the other side by adding or dividing, and vice versa. Something that makes no sense in mathematics. And if you still doubt that understanding is not calculating, then, you will have to recognize that, the ones who understand mathematics the most would have to be digital calculators and computers, because they are the best at calculating. Faster than anyone else by far. However, we know that believing that is absurd.

But if there is one branch of physics where the word "understand" causes real confusion, it is in quantum mechanics, where many physicists have resigned themselves to the fact that they do not really understand this theory, because, according to them, they know how to do the calculations. In fact, in the same context of the multiple interpretations of quantum mechanics, one hears another definition of what it would mean to understand, which would be to relate such a phenomenon to something everyday or known. This is a famous excuse that physicists have used to say that there is no problem with quantum mechanics, that we are simply not familiar with it. In other words, our primate brains have not evolved to quantum scales, so we will never have a familiar experience with superpositions or entanglements, and therefore we will always believe that we do not understand quantum mechanics. So, physicists really do understand quantum mechanics because they can solve the equations, but we think they don't because understanding is a feeling, a matter of getting used to it, of having an intuition about things. That is why many physicists today deny that there is any problem with measurement, because they have gotten used to the peculiarities of quantum mechanics, compared to the geniuses who developed the theory at the beginning of the 20th century. Something quite sad, and even pathetic, because the truth is quite the opposite. Many physicists today understand quantum physics even less than the pioneers, precisely because they do not believe that there is a problem of interpretation of the theory. A fact that will become quite clear later, especially in Chapter XII of the third part of the book.

So what is the epistemic mistake that physicists make about what it means to understand? They confuse it with the concept of knowing. The latter simply refers to being aware of a datum, fact, or event. Here we can use the concepts of memorization, mechanization, or habituation, as we have just seen. Understanding, however, has to do with taking the known data and relating it coherently to the rest of the available knowledge. And not just with the information that is directly or closely related, not even with all the information within a discipline, but with the other disciplines as well. This means that understanding has a holistic character (synthesis), because the more degrees of connection (relationships) one has, the better or deeper the understanding will be. This can be seen very well in network theory, since the number of related data and information grows exponentially with the number of links in the network. From this feature comes the popular idea of the six degrees of separation, which says that you can relate two people anywhere in the world, using at most only six contacts. For example, the cousin of the friend of the neighbor's acquaintance. This explains why it is so difficult to obtain a deep understanding of any phenomenon, because the greater the degree of depth sought, the greater the number of facts and data that must be coherently related. Therefore, the greater the understanding, the greater the number of consistent relationships between all the information. A single logic that relates a large amount of data that we perceive as a great explanation or cause of things. Returning to the previous example, with a depth of only six degrees one could already relate all the people on the planet. Therefore, a deep understanding requires systematically connecting a lot of information into a single logic, which is obviously difficult. Something that again would naturally be promoted under the criterion of wide-ranging scientism, because it means just that, to synthesize or coherently integrate a phenomenon or information with the rest within a worldview. So, if we really want to understand nature, we cannot limit ourselves only to each specialty, but we must bring together all the sciences. This requires us to have to coherently relate phenomena of different degrees of complexity at different scales. A deep understanding that we long for, but that we have not been able to achieve, due to the atomization of science.

In fact, the above would also explain why it has been impossible to provide a concrete definition of science that applies to all scientific disciplines. For example, we are taught in school that science is any discipline that uses the scientific method; that is, anyone who makes observations, formulates hypotheses, performs an experiment to test them, and draws a conclusion from the results. The problem is that the method is used not only in science, but also in other non-scientific activities, such as some philosophical and artistic inquiry. Even the method can be found in some everyday inferences, such as a baby learning that things fall down when you let go of them. Here the baby observes, hypothesizes, experiments, and concludes that things fall every time he drops them on the floor. So, is the baby doing science? No one would say so, especially since many animals do exactly the same thing when they learn. To this we must add that there are sciences that are recognized as such, but do not apply the method, for example, astronomy, anthropology and paleontology; which none of them do experiments, they only make rigorous observations of nature, without creating controlled environments. Therefore, what is it that science has that other non-scientific disciplines do not have? This dilemma is known as the demarcation problem. A single criterion that determines what is science and what is not. It was believed for a while that Popper's falsificationism was such a criterion, but as we already know from chapter II, this also has its problems. Therefore, the search for such a criterion continued, but all attempts have failed. Currently, no one is trying anymore, and the most accepted are multi-criteria lists that evaluate, for example, what is studied, how it is worked, what type of training, philosophical assumptions, and so on. For example, Mario Bunge's proposal consists of 10 points, which in turn are subdivided into a series of much more complex conditions. In other words, he has given up on providing a definition of science, and now only the sciences are recognized separately. This is not surprising in today's age of extreme specialization.

However, under the new criterion of scientificity, this epistemic dream could be fulfilled, because the mistake that would be made would be to conceive of science as something limited, delimited, separated, since in reality science means understanding nature. All of it. Therefore, the key to discovering what science is was in synthesis, not in analysis; because science is holistic by nature, because there are not several worlds, but only one world, whole and coherent. Of course, each science has its peculiarities and differences, because the world has different manifestations with different degrees of complexity. However, regardless of their differences, a science remains a science if, or only if, it remains consistently integrated with all the others. Simply because there are not several worlds, but only one. So, science must integrate all existing phenomena in a coherent way, regardless of their scale or degree of complexity. This is the great difference between science and other non-scientific disciplines, such as religion or politics, because the latter do not seek to integrate with the others, quite the contrary. If they tried to do so, they would be forced to change, to update, to replace themselves, to be consistent with new discoveries and research. This is exactly what non-scientific disciplines do not want to do, because they want to protect their beliefs, their dogma, their faith, their ideology. On the other hand, the only knowledge that abandons its beliefs and strives to falsify its theories with empirical evidence is science. And it does so (or should do so) because it cannot fall into the fallacy of cherry picking, since it is obliged to recognize all the available empirical evidence, and not only the one that suits it (confirmation bias). This is exactly what this new scientific criterion seeks to emphasize and promote. In this way, the scientific method would not be the definition of science, but rather an inevitable consequence of its true definition, which is the correct understanding of nature, all of it.

This new criterion of scientificity is a very significant change in the way science would work today, but it does not create any problem of incompatibility with the criteria already used by science today, because it simply applies the criteria of internal and empirical consistency in a general or holistic way, and not only locally or limited to each specialty. Therefore, a truly scientific theory would not only work in its niche, but should also not contradict the rest of the confirmed theories and evidence. Or at least as little as possible. Let's remember that the universe is whole and consistent as a whole. That is why the strategy of unification worked in physics, such as the integration of all electric and magnetic phenomena under Maxwell's theory of electromagnetism. Newton did the same with celestial and terrestrial mechanics. Einstein unified space and time with his special theory of relativity, and even unified the concepts of particle and wave with his explanation of the photoelectric effect using Planck's constant. Meanwhile, Paul Dirac took special relativity and unified it with quantum mechanics. A quantum theory of electromagnetism was later unified with the weak nuclear force in a model called electroweak in the 1960s by Sheldon Lee Glashow, Abdus Salam, and Steven Weinberg, who won the Nobel Prize in Physics in 1979. And physicists are continuing this program, trying to add the strong nuclear force to the electroweak to get a new theory called grand unification or GUT, and then unify it with Einstein's gravity to get the theory of everything. A program that makes perfect sense for Darwinian Physics, except that it differs in how, but above all in what is understood by this "everything," since current physics arbitrarily excludes the rest of the complex phenomena at higher scales, as we saw in the previous chapter. A distinction that lacks justification, because its explanation is simply sociological, since the sciences are disconnected due to extreme specialization. However, we continue to live in a single universe where all phenomena occur simultaneously and harmoniously, regardless of which science they belong to. Contradictions and paradoxes exist only in our minds, in our theories, not in nature. So, what the new criterion does is to take this scientific fact seriously, to break down this artificial epistemic barrier that we have built, and to include the other sciences in order to respect the unity of the universe.

In this spirit, let us call this new criterion of scientificity holistic consistency, which essentially means applying internal and empirical consistency globally, because we must not forget that a half-truth can also be a lie. Therefore, the only way to ensure that the paradigm is as valid as possible is for the theory to be as consistent as possible, encompassing all other theories and evidence from other sciences. For this very reason, holistic consistency would not be a dichotomous criterion like the falsification principle (the theory is either false or not), but a gradual or ordinal one, in the sense that it does not have an absolute value but a relative one, used to compare framework theories or candidates for paradigms, such as string theory, since empirical consistency is useless for their evaluation. And if there is only one framework theory, it would perhaps indicate the need to search for and create alternatives due to its low degree of holistic consistency. In other words, it would be a warning that we may be in a crisis. Therefore, the new criterion makes science even more rigorous, because it does not replace or devalue the previous criteria of scientificity (internal and empirical consistency), but it is an additional criterion that must be fulfilled. Thus, for a theory to be truly scientific, according to this new definition, not only must it not contradict itself or the experiments, but it must also not contradict the rest of scientific knowledge that has gone through the same filter.

This makes holistic consistency work perfectly well with Popper's falsification principle, because it does not detract from its role, since holistic consistency cannot disprove specific theories, at least not directly. However, holistic consistency can do what the falsification principle cannot, which is to compare rival general theories in order to judge which might be better. Framework theories that the available evidence cannot yet falsify, because they are compatible with two or more alternatives (interpolation theorem). A complementarity between falsification and holistic consistency is confirmed by the fact that they specialize in different stages. Popper's principle specializes in the stage of normal science, where it is necessary to perfect an already established paradigm, while the principle of holistic consistency is most useful in the stage of crisis and revolution, when candidates for the new paradigm are tested. Thus, the new criterion of scientificity cannot directly refute string theory, for example, but it can evaluate it in relation to other alternatives that seek the same goal. To be the next SRP of physics. A criterion that bases its validity on the premise, very difficult to refute, that the universe is one, consistent, logical and comprehensible. In other words, that the universe as a whole is coherent, without contradictions. A condition that seems quite obvious, but that in practice science forgets.

However, if we are going to broaden the definition of science, then it will be necessary to broaden what it means to be a scientist; for what we have at present are essentially specialist scientists who use a fundamentally reductionist-analytic approach and are judged by Popper's falsification principle. However, if we apply the new criterion of scientificity, we will also need generalist scientists, who would mainly use an emergent-synthetic approach and would mainly be evaluated according to holistic consistency. A second type of scientist does not currently exist because of the extreme specialization of the current academic system, which, as noted in Chapter II, was designed to do normal science. To develop and refine a paradigm, not to change it. However, far from being antagonistic because of their different approaches, these two types of scientists would complement each other perfectly, since each would satisfy a need of the other. To see this, imagine scientific activity as the process of solving a puzzle. In this sense, the specialized scientists would have the indispensable role of finding the pieces of the puzzle of nature. The facts and evidence that confirm certain essentially descriptive models. As is currently the case. While the generalist scientists would have the role of taking the pieces produced by the specialists in each discipline and science, to try to reconstruct the complete puzzle, a world view of the universe as a whole, from an increasingly fragmented but at the same time more accurate knowledge, from the hands of the specialists. A collaboration from which both sides would benefit enormously. The generalists, of course, benefit from the work of the specialists because they get their pieces of the puzzle from them. But the specialists would also benefit enormously from the work of the generalists, since they would provide very important clues to the discovery of new pieces, especially those at the frontiers of each discipline or science, thanks to a general map of all scientific knowledge. Pieces that are usually the most valuable because they connect different lines of research; but, for the same reason, they are the most difficult to obtain because they require the knowledge of several disciplines to obtain them. In this way, a virtuous circle would be created between these two types of scientists in order to provide us with an increasingly reliable and complete scientific view of nature as a whole, without losing its rigor and accuracy. And as science advances, so does technology, driving the development of new technological applications that inevitably follow every scientific advance.

Many will think that the integration of a generalist academic background is very radical and utopian; however, the truth is that this was the normal situation for much of the history of science. For example, Nicolaus Copernicus was a mathematician, astronomer, lawyer, physicist, Catholic priest, governor, diplomat, and economist; Leonardo da Vinci was a painter, anatomist, architect, paleontologist, artist, botanist, scientist, writer, sculptor, philosopher, engineer, inventor, musician, poet, and urban planner; Louis Pasteur was a chemist, physicist, mathematician, and bacteriologist; Isaac Newton was a physicist, theologian, inventor, alchemist, and mathematician; Benjamin Flanklin was a politician, polymath, scientist, and inventor; Archimedes was a physicist, engineer, inventor, astronomer, and mathematician; and so on with a long list of scientists, many of whom are quoted throughout the book. They are many because being a generalist was the norm from the beginning of science. From the time of ancient Greece to the 19th century. The really new and rare phenomenon is the current extreme specialization, which is only a brief academic experiment that began at the beginning of the 19th century, a product of the need to support the industrial revolutions. In this way, activities in factories were separated in order to increase productivity, which naturally led to specialization in the training of workers to adapt them to these specific activities. Therefore, far from being a very progressive proposal, the training of generalists would be very conservative, since it would mean a return to pre-industrial education. Something that is not only very necessary and valuable for the progress of science, but also a right that should exist; because just as there are many people who are fascinated and love to specialize, there are also people (including me, of course) who love to know everything. Therefore, a lot of talent is wasted, and a lot of frustration is caused by creating only one type of scientist. A wasted talent not only in the normal stage of science, by helping to find the valuable pieces that specialists are looking for, but especially in the crisis stage, since specialization reinforces the paradigm, making it even more difficult to change it. On the other hand, generalists would help to find out what is wrong with the current paradigm, what does not fit, because they have the whole map of science and can show where the problem might be. Therefore, generalists would be an essential help to overcome the crisis phase. This is proven by the fact that new paradigms are usually the result of the union of two lines of research that were initially thought to be unrelated.

In addition to the practical advantages of a generalist academic education, it is also necessary on principle, since the division into sciences and disciplines is completely artificial. Nowadays it is normal to see scientists from different sciences, and even from different disciplines within the same science, who cannot understand each other because of the great differences in their methods, approaches, and jargon. Something that should be of great concern, because in nature everything is united in a perfectly harmonious way. For example, you have physics because you are made of particles and atoms, but you also have chemistry because of your molecules, biology because you are alive, psychology because you have consciousness and personality, and sociology because you are an individual integrated into different groups, organizations and institutions. And all these aspects or dimensions operate simultaneously, without separation or boundary, without contradiction. It is surprising, then, that the unification of science and all natural phenomena seems to many scientists to be so insane, even pseudoscientific, because the nature that science itself tries to reveal is not fragmented. On the contrary, it is highly integrated. However, extreme specialization has convinced us that physics is a very different science from biology, and biology is a very different science from psychology. We have come to believe that there are separate sciences, that it is impossible for them to communicate, let alone understand each other. As if each science studied a different universe with its own laws. This is truly ridiculous. There is no denying that this assembly line strategy in the academic system has been incredibly successful in terms of the progress and development of each individual science over the last two centuries. However, it has been at the expense of their divergence, making them increasingly unrecognizable from each other. It has given us an increasingly mutilated and disintegrated view of nature. A process of fragmentation that is increasing with time, because it is not enough that each science continues to diverge, but that the different disciplines within each science also diverge. Again, this is not to deny the need for specialization, since it is undeniable that it has brought amazing progress to each science. However, it is ideal for generating an incredible amount of data, but not so much for generating knowledge and understanding, since the latter requires connecting, integrating, and relating these data in a coherent and systematic way. Exactly the task that generalist scientists would have.

This need to reform the academic system is urgent because, like any radical position, extreme specialization becomes dangerous and creates problems for the very activity of science, undermining its credibility and reputation. This is because the increasing distance between sciences and disciplines creates serious inequalities between them, because they do not supervise or monitor each other. So, they end up regulating each other. And we know that this is not a good idea, neither in business, nor in politics, nor in science, because it is inevitable that at some point they become corruptible. For example, we all know that there is a big difference between the hard sciences (physics, chemistry and biology) and the soft sciences (psychology and sociology). In fact, there is a crisis of replicability (an essential condition for science) in studies in sociology, medicine, psychology, and psychiatry; due to bad practices and lax statistical methods that allow researchers to get the results they want (confirmation bias) and not the results they should be getting. As a result, virtually half of the publications in the soft sciences cannot be replicated by other researchers. But even the hard sciences are not without problems. In fact, we can take the foundations of physics itself, supposedly the most reliable and rigorous science, as an example, since the same confirmation bias may affect, for example, the string theorist community. This is because, as the only experts in their field, there is little incentive or opportunity for non-string physicists to evaluate these investigations, because it takes a lot of preparation to deal with all the mathematical scaffolding that string theorists use. Time that does not exist in academia, where you have to publish or die. So, they can't get out of their specialty. So, this makes the string theorists their own judges, creating a gigantic bias in favor of their theory, where instead of having a scientific spirit, that is, a critical one, and always trying to falsify their hypotheses, they try just the opposite, to verify them, creating conjectures and dualities, where often the only evaluation criterion is whether it is consistent with string theory or not. This is a very human behavior, since no one wants his life's work to be proven wrong. But it is a very unscientific attitude. Hence the importance of correction, evaluation and supervision by other scientists. Something that applies only within the discipline, but should also apply outside of it. Let us remember that each discipline and science is immersed in its own paradigm, in its own bubble; therefore, the supervision of researchers outside the specialty would be essential to evaluate and correct deviations in proper scientific action. Something that is far from happening under the current extreme specialization.

However, the importance of a generalist scientific education goes beyond science itself, for it can be essential in meeting the great challenges of our civilization, which can only be addressed in an interdisciplinary manner. For example, global warming is a very complex problem that needs to be addressed simultaneously and consistently in its climatic and ecological dimensions, but also in its political, economic, technological, commercial, cultural and many other dimensions. If we leave out technological solutions, changes in society's habits, laws, policies, the transition to a green economy, or any other necessary measure, we will not really solve the problem, especially not in the timely and effective manner that is needed. And so it is with virtually every problem we face as a civilization: pollution, the ethical use of biotechnologies such as cloning and gene editing, obesity, migration, social networks, our relationship with animals, poverty, sustainable economic growth, artificial intelligence, new pandemics, population aging, and so on.

These are problems that inevitably need to be approached from multiple sciences and disciplines in order to solve them properly. But we are far from that, precisely because of this fragmentation of knowledge. Therefore, to compensate for this major disadvantage of specialization, it is essential to compensate by training another type of expert, the former generalist, who will help the former to remember that we live in one world and not in many parallel worlds.

In summary, the new criterion of scientificity of holistic consistency makes the definition of science more complete and rigorous. Complete because it would address the three stages of development of a science (revolution, normal science, and crisis) and not just one (normal science) as in the current definition of science. And it also makes it more rigorous, because it does not replace or diminish the existing criteria, but adds an additional one, one that does not even pretend to be the most important of the three, since a scientific theory must first have, as a basic condition, internal consistency. In other words, it must be a systematic and structured set of coherent ideas that do not contradict themselves. Next, it must have empirical consistency or experimental support to ensure that this system of ideas or mathematical model refers to real objects or facts. And finally, it must have the highest degree of holistic consistency, to ensure that the theory or model does not describe a small part of reality in a forced or illegitimate way, at the cost of creating inconsistencies in all the rest of scientific knowledge.

Only by respecting these three criteria of scientificity will we have the maximum possible certainty that science is really approaching the truth, and that it is not just an effective model rather than a reliable representation of nature. All this thanks to the fact that the paradigm has been synthetically validated (holistic consistency), after having been analytically falsified by means of experiments (internal and empirical consistency). A new definition of science, which will mean that in order to advance in one field or specialty, it will be necessary to push or drag along all the others. A criterion that again seems too revolutionary or utopian, but of which there is already a good example of its successful application. Again, we are talking about one of the best scientific theories, Darwin's theory of Natural Selection, because the theory of evolution is not only internally consistent with the rest of exact biology, but also externally consistent with other sciences, especially geology, chemistry (genetics), psychology (evolutionary psychology), medicine (Darwinian medicine), sociology, and chaos and complexity theory. Achieved a high score in holistic consistency. The highest so far. However, there is one science with which it is still particularly incompatible, and that is physics. Therefore, the time has come to combine these two theoretical frameworks. This opens up a great opportunity to find the solution to the great problems that the foundations of physic are currently facing, and at the same time to re-establish a unique and integral scientific worldview that will benefit all of science. And, logically, society.

Second Part

A Truly New Physics

Chapter VIII

Ex Nihilo Omnia Fit

The time has come to apply all that we have learned in the first part of the book and to dare to propose an alternative to the current paradigm. To accomplish this ambitious goal, we will be guided by a model that was used several times in the first part of the book as an alternative to the Parmenian paradigm. We are talking about Darwinian mechanics. A paradigm that has many differences from the Parmenian mechanics that has been used in physics until now, one of the most important being that Darwin's theory is essentially teleological; not in a religious way, of course, but by following a purpose shaped by Natural Selection. Specifically, all living things seek to survive and reproduce. Darwin's theory is therefore holistic, because Natural Selection does not act on the level of genes, but on the final product. Nature selects directly on the phenotypes, or specific characteristics of individuals, and not on the genotype, or the particular configuration of genes that produces those characteristics. In other words, it acts on the "effect," not the "cause." For this reason, Charles Darwin did not need to know the genes to formulate his theory. That was the job of precise biology and its specialists, such as molecular biologists. Therefore, the emergent-synthetic approach is said to be explanatory because Darwinism, being teleological, can give a reason, a meaning to evolution and the diversity of life, thanks to the existence of a purpose, a struggle for survival and reproduction that drives it.

Another feature that distinguishes Darwinism from modern physics is that it is monistic. As we saw in Chapter V, this is a characteristic of the ancient pre-Socratic philosophies, which attempted to explain nature by postulating the idea that the multiplicity or variety of things came from a single principle called the arch. All these theories were based on the idea that things have the capacity to change, to transform into other things. Now, modern monism is called the theory of evolution by Natural Selection, because again we start with a single principle. A rather simple one. In fact, the simplest possible form of life, from which all the impressive variety of complex life that has existed in these 4,000,000,000,000 years of the planet's history has evolved. A richness that is hard to believe came from a single organism that biologists named LUCA, as we saw in the same chapter.

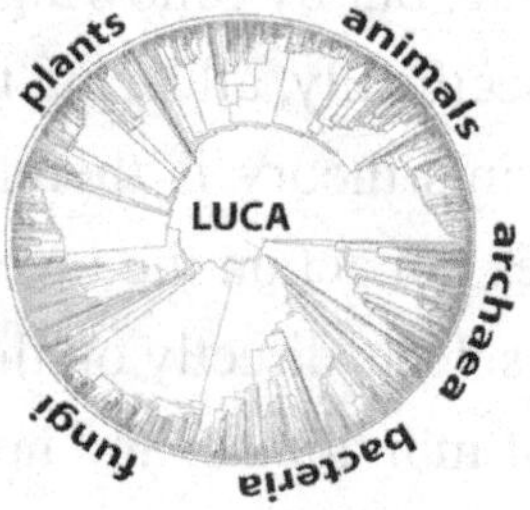

From dinosaurs, plants and whales to bacteria and fungi, we all come from the same organism. We are all children of LUCA, the biological arch. However, we already know that this solution of explaining the "many" from the "one" did not please Parmenides, who flatly denied the logical possibility of change and emergence. Therefore, after him, all monistic theories were discarded and gave way to pluralism. The belief that the "many" can only come from the "many." Thus we went from a single arché to the postulation of multiple arches. For example, we have Empedocles of Agrigento (about 494-424 BCE), who postulated that the arché is the four elements: water, earth, air, and fire. Or Anaxagoras of Clazomene (500-428 BCE, approximately), who postulated that the arché is infinite, since all things inhabit all things, albeit in different proportions. Thus, a stone is not a cat, because it has a greater percentage of stone than of cat. While the most famous pluralists, Leucippus and Democritus of Abdera (5th century BCE) also postulated infinite arches, but this time based on the idea of the atom. The unit or building block of reality. And although they are infinite arches, because there are infinite types of atoms, their metaphysics could be classified into two groups: the void and these atoms. Current physics is a more sophisticated version of the latter pluralism, since we have Einstein's general relativity describing this vacuum through space-time, while the atoms would be the particles or excitations of the various quantum fields of the Standard Model of particles inhabiting this vacuum. So today we have a pluralistic physics that shares its devotion to the absolute and its aversion to change, emergence, and nothingness with the ancient Greek absolutist philosophers.

Since this Parmenian strategy has stagnated in recent decades, we will try something different. We will return to the original path, monism, to develop a truly new physics, while remaining consistent with what is officially known about physics (empirical and holistic consistency). To this end, we will introduce Darwinian mechanics into physics, with the goal of finally solving the problems of time and strong emergence. This means that we have to postulate an arché like the ancient pre-Socratic philosophers. A titanic task, because this arché will have to be very special, because we will have to solve the problem of the origin also according to the principle of sufficient reason, in order not to repeat the same error of current physics: the regression ad infinitum. Then we have to postulate a First Cause of really everything, which, moreover, does not allow us to wonder about its reason, its origin or its cause, so as not to ask endlessly why? Like in the case of God, the multiverse or absolute laws. So, the challenge is enormous, since not even the original Darwinism (evolution by Natural Selection) has been able to achieve this, because Darwin's theory can perfectly explain the emergence of all forms of life from LUCA, but Natural Selection is not able to explain the origin of this first form of life. That is why LUCA is the arché of Biology, because although we are absolutely sure that it had to exist because of all the scientific evidence available, especially through genetics, it is still a hypothetical entity that is taken for granted because its origin cannot be explained. That is why this problem is kicked to other sciences, especially chemistry. A problem that they have obviously not solved either, but which relieves biologists of responsibility. A strategy that will not work, however, if we apply Darwinism to physics, because there is no more fundamental science to kick the problem to. That is why some physicists, who are more faithful or aware of their religious origins, throw this problem to theology, saying that the origin of the laws is God, as Newton or Leibniz thought. But if we want to solve this problem in a serious and rigorous way, the only valid and legitimate way to do so is to address it within physics itself, offering a natural mechanism capable of explaining not only how the manifold emerges from a single arché, but also how that arché emerged in the first place.

The time has come to deal with the worst nightmare of physicists and theologians, nothingness, because it is the only way to avoid regression ad infinitum, and also to comply with the principle of sufficient reason. The great challenge of Darwinian Physics, then, is to postulate a mechanism ex nihilo (strong) emergence that explains how everything could really have come out of nothing. In other words, it seeks to finally explain and justify the simple account offered by physics: that a society is made up of people, which are made up of cells, which are made up of molecules, which are made up of atoms, which are made up of particles. In this way, each reality or scale is connected in a coherent way. A goal that sounds totally ridiculous only because we are immersed in a Parmenian paradigm that unjustly abhors nothingness, even change, emergence and randomness. However, the very etymological origin of the word "nothing" gives us hope that perhaps the idea is not as absurd as it is thought, for the word "nada" (nothing in spanish) comes from the Latin nata, the feminine form of the participle natus, meaning "born" or "originated," from the verb nasci, meaning "to be born." So, the idea that everything was born from "nada" does not seem so far-fetched, at least etymologically.

But before we see how this could be physically possible, it would be useful to know why the appearance of nothing is considered something impossible or ridiculous within the Parmenian paradigm. Hard as it may be to believe, perhaps the reason is simply a matter of definition, since nothingness has been defined on several occasions as "not being something" and not simply as "not being" as Parmenides defines it in the strict sense. A definition that has perhaps recently been reinforced by the dominant language in the West today, English; because "nothing" means literally "no thing." And by defining it in this way, the concept of "nothing" becomes obviously illogical and meaningless, since by defining it as "not being something" it automatically becomes "something" by giving it a name that contradicts its own definition. This is very similar to the self-referential paradoxes we saw in Chapter III. This is why many philosophers and physicists declare that "nothing" cannot exist because it is a logical or paradoxical impossibility. To this lack of clarity or precision in the definition of nothingness we must add that it is constantly confused with another concept called "emptiness." If we go back to Parmenides' definition, nothingness is defined as "non-being," while "emptiness" would essentially be "not having," being defined in terms of quantity, the lack of things. For example, a room is empty because there is no furniture in it. However, this does not mean that the furniture does not exist, because it may be somewhere else. Rather, to say that there is no furniture means that it is nowhere because perhaps it has not yet been invented or has become extinct. Therefore, nothingness is a concept more related to time, to one's own existence, to the now, while "emptiness" has more to do with space, with locality. The "not being" there, empty of it. A confusion that is aggravated by the fact that we do not understand very well what emptiness means either, forgetting that it is a relative concept, not an absolute one. For example, a room is empty because it has no furniture, but it is still full of air. So, vacuum depends on what you mean. Something that particle physicists know very well, because in quantum mechanics it was discovered that vacuum can be something very frenetic and complex, representing quantum fields without stable particles. So, it is not surprising that there is so much mysticism about nothing, because it is not even very clear what "vacuum" means.

In any case, whatever the reason, this mistake was not made by the ancient Indian philosophers. As we saw in Chapter V, Hindus have a completely different view of nothingness than Westerners, because for them nothingness refers to the "oneness" from which the multiplicity of phenomena would emerge. According to Hinduism, all things, even the gods, emerge as different manifestations of an ultimate and underlying reality called Brahma, who is not just another god, but the background of being, pure and undifferentiated. Thus, unity would be nothingness, while the whole would be the manifold. By this definition, things would be mere determinations or particular configurations. To use an analogy, Brahma would be the sea, while things would be the various waves that appear and disappear on its surface. A nothingness that inspired the great German philosopher Arthur Schopenhauer (1788-1860) to study the noumenon, postulated by another great German philosopher, Immanuel Kant (1724-1804), who defined it as the unknowable, but which the brilliant Schopenhauer made knowable. He even gave it a name: the will. A fundamental force that moves all things, which is not rational or ordered, as Hegel, Newton or Einstein would say, but rather the opposite. A chaotic, random and wild force; similar to the arché Apeiron (without limit in Greek) of the pre-Socratic philosopher Anaximander of Miletus. An amorphous and undefined element. Then, like Brahma, the Apeiron would be a valid "nothing" by rigorously fulfilling the condition of "not-being," since these philosophers defined "being" as the multiple, the separate, the determinate; therefore, nothingness must be the opposite, the unity, the totality, the indeterminate. In this way, at least in the East, they were saved from the tyranny of absolutism by accepting and recognizing nothingness, by defining it correctly.

Unfortunately, this philosophy never developed into a science and remained exclusively in the philosophical-religious realm. But if we learn the lesson left to us by the ancient non-absolute philosophers and strictly define nothingness as "not being" rather than "not being something," then nothingness can logically be possible even if it were something, because what makes nothingness to be nothingness is the opposite of being, and not being the opposite of "something." Then the key to continuing would be to find out what being really is, since nothingness is its negation, its opposite. Therefore, we cannot find out what nothingness is unless we know exactly what being is. To do this, we have to find out what is the purpose of existence, its meaning, because let us remember that Darwinian Physics has to be teleological, just like Natural Selection. So, what would be the purpose of being itself, which includes inert physical things like atoms, molecules, planets and galaxies, what would be the equivalent of survival and reproduction for existence itself? A simple answer would be to continue being or existing, to preserve itself, to maintain itself, to stabilize itself, in the same way that a living being is something that seeks to continue living in itself (survival) and/or in its descendants (reproduction to pass on its genes). So, starting from this clue, we can see that the beings we observe are characterized by having a certain size or occupation in space and a certain duration in time. And it seems that the more space and time beings have, the more existential they are. And vice versa, the less space and time they have, the less they are, or the more they are nothing. It is a relationship that makes sense. For example, a simple fly is very small compared to us and lives only a few weeks, so it seems to us to be less existential or more nothing. That is why we kill it at the first provocation. On the other hand, the fly seems much more existential than a bacterium, which lives for a few hours, and this seems much more existential than a neutron in a vacuum (which exists only for a few minutes) or a muon (which disappears in 10^{-6} seconds), and this seems more existential than the Rho meson, which disappears in 10^{-24} seconds.... Until we come to the least existential thing, a virtual particle, which remains or "is" (exists) for such a short time that it is impossible to observe directly, so that its existence is practically virtual. Thus, by redefining existence, no longer in a dichotomous category (exists or does not exist), but as something gradual, we can realize that beings have different degrees of existence or being. Therefore, we can also rigorously define nothingness in the same terms. Thus, the less space-time a being occupies, the more nothing it is. And if we take this logic to its extreme, we have that "pure nothingness" would have to be the smallest and shortest possible length and duration. This space-time definition of nothingness is obviously much better than its Western version, which is constantly confused with "emptiness," which is a more Newtonian nothingness, because it would be a nothingness that excludes space and time by considering only content. A nothingness within space-time. On the other hand, the new nothingness integrates space-time because it is a nothingness of space and time. Therefore, it would be a true nothingness because it is truly all-encompassing.

So, we have a rigorous definition of nothingness. But what exactly does the space-time graduation of beings mean? To find out, we will use mathematics, specifically geometry, because for this branch of mathematics, having space or length means having a spatial dimension (dr), while having time or duration means having a temporal dimension (dt). Thus, we will define existence or being as anything with $dr > 0$ and/or $dt > 0$. In other words, having some space-time dimension. Therefore, the greater the extent of dr and/or dt, the greater the degree of being. And conversely, the smaller the extent of dr and/or dt, the greater the degree of nothingness. In this sense, we have never observed anything in the universe that is absolute, i.e. with a $dt = \infty$ (eternal), nor with a $dr = \infty$ (infinite); because everything we know has a finite and transitory extension, for example: the earth, stars, galaxies, and even black holes, which were thought to be the only eternal things, will eventually disappear due to Hawking radiation. Even the universe would not be saved from its finiteness by having a beginning, a big bang; and by having a size due to its shape (if it is closed) and its expansion. Therefore, all things that exist would be limited in space and time. But they are also not total nothingness ($dr = dt = 0$), because all things we know have some extent or length and duration. If we add to this the fact that things evolve into ever more durable or stable states (due to the second law of thermodynamics); and, at the same time, become ever larger or more massive (due to gravity), we could conjecture that things naturally tend to become ever more existential, in terms of stability (duration) and massiveness (size or weight). A tendency that may even apply to the universe itself as it ages ($+\Delta$ temporal degree) and expands ($+\Delta$ spatial degree). Therefore, everything seems to indicate that the universe began at one moment ($dt = 0$) and one point ($dr = 0$), the big bang. And from there it has evolved towards eternity ($dt = \infty$) and infinity ($dr = \infty$). So, we seem to be in full transition. In a process that began in nothingness and is moving towards absolute existence. A singular origin, very similar to what is postulated in cosmology, where at the beginning of the universe, the Big Bang, there was a singularity, according to general relativity. Something very similar to the nothingness of Darwinian Physics, a space-time point ($dr = dt = 0$). Something without time dimension (duration) or space dimension (length). A point, but not in the sense that it is a particular point, as if it were a point in space or time, but that all of space and time was that point at the time of the Big Bang, including all the trillions of galaxies with their hundreds of billions of stars. But that since then it has expanded to the size of the universe we have today. A size that not only continues to grow, but is growing faster and faster due to the existence of dark energy. So, incredibly, at the moment we are not too far from the orthodoxy of current physics.

However, before we go on to see how Darwinian Physics can fit perfectly with observations and experiments (which we will see in the third part of the book), we must first understand what exactly nothingness means, especially if the space-time point makes sense. This is because, until now, the point has been a purely abstract concept. No one has ever observed a point in the real world. For example, atoms, which would be points so tiny that they cannot be seen with the naked eye, have a complex structure and shape because they are made up of several layers of electrons around them. Even its nucleus, which is about 100,000 times smaller than the atom itself, is made up of protons and neutrons, which in turn are made up of quarks and gluons. Far from a point, then. And yet Darwinian Physics postulates the physical reality of the dot by saying that nothingness is real. So, what is a point? Let's start with the easiest aspect to understand, in quotes, its spatial punctuality. Actually, it would be a big mistake to try to define the spatial point in terms of size, since one could always postulate an even smaller size. It would be like postulating the greatest number. Something completely useless, because you can always add something else to that number to get an even bigger number. And vice versa, if you postulate a size as the smallest possible, you can always subtract or divide it by a number to get an even smaller size. So, it makes no sense to think of size as absolute, because it is an inherently relative concept. For example, what is tiny to us, a grain of sand, is a huge mountain to a bacterium. Therefore, postulating a size to define a point is not the solution. The same applies to the other aspect of nothing, its temporal punctuality. Following the logic above, it would mean that nothingness has to last only one moment, because if it lasted two moments, it would mean that $dt > 0$. So, nothingness has to last the most ephemeral moment possible only once. But here we repeat the same dilemma as in the case of the spatial point: how long does this single instant last? And again we would make the same mistake, because again it makes no sense to define an absolute time, because time is also relative. For example, one could say that for geological phenomena, such as the formation of a mountain range or the Grand Canyon, 100 years would be an instant for them, but more than a lifetime for us. Or a second is an instant for us, but for a shrew it might be a minute, and for a bacterium it would be a whole day. But for certain subatomic particles, a second would be perceived as an eternity.

But what makes sense is to stick to the mathematical or geometric definition, since the point is defined in terms of lack of dimensions. This means that when applied to the point in space, it means that in the nothingness there are no directions, therefore, it is not possible to advance or move. In other words, you can't go anywhere because there is not up or down, no left or right, no forward or backward. When this happens, there is a spatial nothingness. It does not matter how big this point is, or that this nothingness was "everything" before existence, because the concept of size is meaningless, because there is nowhere to go and nothing to compare it with. The same is true of the temporal point, which means that it remains only an instant, since it cannot extend into the past or the future, since that would mean having $dt > 0$ and thus having a temporal dimension. Recall that the opposite of temporal nothingness ($dt = 0$) is eternity ($dt = \infty$). Therefore, it must be when or if ephemeral, transient. A moment with a clear beginning and end, from which it can no longer be segmented since it must lack any extension into the past or the future.

It is this lack of dimension that is the real reason why it makes no sense to speak of size and duration in the nothingness, because these concepts necessarily imply having to count. In the case of duration, how many instants occurred; and in the case of size or length, how many kilometers, meters, nanometers, or some other segment of space is there? In this way, one can compare and say that something lasted longer or is smaller. However, this is precisely what loses its meaning in nothingness, not only because there is nothing to compare with, but the very concepts of duration and size lose their meaning when there is only a single unit of time and space, which is by definition indivisible. So, there is no dimension, because everything is and is at one point, because you cannot go in any direction, neither in space nor in time, because there are no more points, only one, nothingness. In short, if there is nothing to count, then there is nothing to measure, no distances, no times.

Then nothingness is a space-time point, because it makes no sense to speak of size and duration in absolute terms. However, for the same reason, it is possible to assign a size and duration to nothingness, but in relation to the whole. This is because nothingness will be the unit of space and time by which our entire universe will be constructed. In other words, nothingness will always be the smallest and the shortest, not because it has an intrinsic size and duration, but because it will always be the block by which all space and time in the cosmos will be created. The space-time unit of the universe. So, no matter how long a fundamental moment lasts, if it makes sense to say that such a moment will remain the shortest moment for us, because it is the moment with which all other times of all other things will be built. The unit of time that will create all the dynamics of the universe. Even if it made sense to consider that this temporal atom would change its duration, we would never know about it. For example, if time in the entire universe stopped for a day, a month, a decade, or a million years, no one would know about it. It would be as if nothing had happened, because no one would notice it, because everything would be at a standstill. It is only possible to measure or observe relative changes in relation to other changes, not absolute changes. Thus, if fundamental time could fluctuate, being millions of times larger or smaller at each instant, it would be completely irrelevant because no one would notice. The same is true for the space point, because if everything in the universe were to double or triple in size: stars, planets, people, atoms, and so on, no one would notice, because it is impossible to detect, since everything remains proportionally the same size, just like our rules of measurement. That is why we can say that the universe is expanding, because it only happens on a large scale, between galaxies, but not within galaxies. However, if expansion were occurring at all scales, it would be impossible to measure such expansion. The same thing happens with time. Therefore, although the point in time could be of any duration and even fluctuate, for us it would still be the same atom of time, the smallest possible instant.

So, although it makes no sense to talk about size and duration when we are talking exclusively of nothingness, since there is nothing to compare, no place and no time to go to, we can assign a size and duration to it in relation to us, since it is the building block of all nature, the true atom of existence. In this sense, we can define nothingness as a Planck space-time point (Planck length = $1{,}616... \times 10^{-35}$ meters, and Planck time = $5{,}391... \times 10^{-44}$ seconds). The smallest space and time imaginable in physics. A hypothesis shared by the vast majority of physicists, especially those working on the quantization of Einstein's general relativity, such as loop quantum gravity theorists. So, it is a pretty convincing definition. Now, having defined what nothingness is, several important questions arise, for example, does this mean that nothingness lasted only a moment? This is very important, because if nothingness lasts one Planck time, then a single second fits into more Planck times than there have been seconds in the entire history of the universe so far. So, does this mean that the nothingness that existed before the Big Bang lasted only one Planck time, and what was there before that was nothing plus nothing? Absurd. But before I throw the book away, Darwinian Physics answers these questions by making this space point oscillate between two states, similar to a coin tossed in the air (heads and tails). Thus, if the spatial point can be exchanged between one state and another, it can be a temporary point by spending an instant in each of these two states. This solution may sound very ad hoc, but it is just the opposite. It is the inevitable logical consequence of the very definition of nothingness, because the first consequence of the temporal point is that there cannot be a past or a future, because if they existed, there would be an extension in time, a temporal dimension ($dt > 0$). Therefore, for it to be a temporal nothingness, there can only be an instant, the now or the present ($dt = 0$). This has big implications, because if only the present exists, then the future cannot exist, but neither can the past. Remember, if in spatial nothingness you can go neither up nor down, neither left nor right, neither forward nor backward, then in temporal nothingness there can be neither past nor future because there is no temporal dimension in an instant ($dt = 0$). This is very important, because if we have an absence of past, then there can be no causality, because there is no previous moment to cause. And without causality there can be no determinism, and without determinism there is only randomness. In other words, in Darwinian Physics, the ontological or real randomness measured in quantum physics laboratories around the world is simply a product of the absence of a past. The fact that there is no before conditions or determines the present moment. This means that nothingness is equivalent to randomness, which in turn is the product of the absence of causality; much as darkness is the product of the absence of light, or cold is the product of the absence of heat. A logic that would be very consistent with the definition of nothingness, since it would be the absence of existence; the latter being equivalent to causality or regularity.

Thus, in one fell swoop, we have the explanation of quantum uncertainty from the definition of nothingness, by explaining the reason why ontological randomness appears at the smallest scales of space and time, at the quantum level, simply because it would be closer to the spacetime point or nothingness. Exactly what has been demonstrated and proven in quantum mechanics by Heisenberg's Uncertainty Principle. Something that cannot be understood in Parmenidian physics, since its absolutism abhors not only change and emergence, but above all randomness and uncertainty. Recall that in the Aristotelian principle or in absolutist logic one can only be (100%) or not be (0%), nothing in between. Therefore, ontological randomness is simply unacceptable with the determinism of absolute physical laws, because the latter allows only values: 1 (is) or 0 (is not), but not fractions. The problem is that the position of an electron, for example, can be 17% here, but 20% there.

Theoretical representation of the electron position

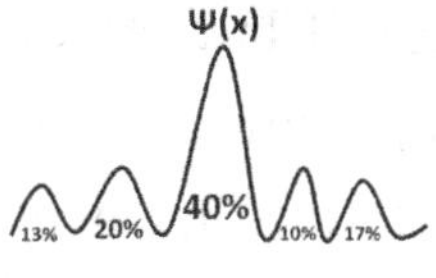

In other words, the position of the electron is not absolute. It is not 100% here and 0% there, at least not before the measurement (this will be discussed in more detail in Chapter XII), but it is here and there at the same time. This is what physicists call superposition. Now it makes sense what the French mathematician Joseph Bertrand (1822-1900) said: "How dare we speak of the laws of chance, isn't chance the antithesis of all law?"; because ontological chance has no place in absolutist logic. That is why probability has been considered only as a mathematical contrivance, something fictitious; necessary only because of our ignorance of very complicated systems, with many elements, where it is technically impossible to know all their parts with sufficient precision. A classic example would be the millions of molecules that make up a gas, whose speed, position and direction must be known. Since it is obviously impossible to obtain such information, the only thing left is statistics, i.e. thermodynamics. Or even in something much more mundane, such as predicting the outcome of each toss of a coin or a die, it would be necessary to know the velocity, the spin, the inclination, the air resistance, the temperature, the pressure, the altitude, and so on. Therefore, in the current paradigm, uncertainty and probability would be an exclusively epistemic phenomenon, the product of our ignorance, because what is truly real or fundamental is absolute law. Pure certainty. The problem for physicists is that, unlike absolute laws, ontological uncertainty and chance have irrefutable empirical or experimental support. On the other hand, the absoluteness of laws has never been observed or proven, because it is simply impossible to prove, since it would have to be tested for all eternity and at every point in the infinite universe, as we saw in Chapter IV. Therefore, the absoluteness of laws has always been an assumption. On the other hand, every time an atom or subatomic particle is measured, it proves that randomness is ontologically real. This is an empirical fact that physics has not yet been able to assimilate because of the Parmenian or classical paradigm in which it is immersed.

Returning to the justification of the two states of nothingness, if in nothingness is a temporal point; then, it has to be random, due to the absence of a past to cause. And for just that to occur, the spatial point must have more than one state. Let us remember that the opposite of a temporal nothingness is eternity, a temporal absolute, and the latter is equivalent to absolute certainty (determinism), for if something is eternal it is because it has happened and will happen forever. In other words, there is no doubt, literally, of such an event; because we will always know what will happen, having an infinite frequency. So, "absolute certainty" is equivalent to "eternity," and these two are equivalent to there being only one possibility. That is why determinism always consists of a single story, an event that will happen with 100% probability, because any other possibility would have 0% probability of occurring. In other words, those other possibilities do not exist. This means that an absolute existence is equivalent to a single option or state, since, if infinity means that one is everywhere, then eternity means that one is everywhere at all times. From all the above it follows that, if nothingness is the opposite of absolute existence, it must be the opposite of absolute certainty as well, so nothingness must be random. And for the latter to be true, it requires, yes or yes, more than one possibility. In fact, it's not even good enough to have zero possibilities, because that would also be an absolute certainty, a 0%. An event that would never occur. Therefore, in terms of certainty, zero and one possibility are equivalent ($0\% \equiv 100\%$), because a possibility "A" with 100% of being is the same as any other possibility "not A" having 0% of being. The former implies the latter, and vice versa. In conclusion, for randomness to exist, and with it a temporal nothingness, it is inevitable that there are at least two possibilities. Hence the explanation that nothingness must have two states or faces. However, we could continue with this logical deduction, because if eternity is equivalent to absolute certainty, then it is equivalent to absolute permanence as well, to the impossibility of change, as we saw in the previous chapters. So, again, if nothingness is the opposite of absolute existence, then, nothingness must be the opposite of absolute permanence. And this means that there must be mutability, that it must constantly vary at every instant. A change that must also be executed randomly, in order to fulfill the condition of being the opposite of absolute certainty as well. Something that is only possible if nothingness has more than one state.

nothingness	*being*
Ephemeral	Perennial
Stochastic	Accurate
Mutable	Static

So far from being arbitrary, the two-state condition is completely necessary to fulfill the definition of nothingness. However, why only two states, why not three, four, or a million, if all of them satisfy the condition of brevity, randomness, and variability? The simplest explanation is that "two" is the minimum number to satisfy such a requirement; however, there would be a more fundamental reason why nothingness must have two states, and this again comes from quantum mechanics. This reason is the necessity or requirement to interfere with itself. As we already know, in this branch of physics objects can be in a superposition of states, many of which are not observed because they interfere with each other. This is one of the properties of waves in the wave-particle duality of quantum systems. This behavior explains the interference pattern observed in the famous double-slit experiment (which will be discussed in more detail in Chapter XII). So, it is a fundamental property, vital to quantum mechanics, but which makes sense in Darwinian Physics, because interference with itself to cancel each other out is a much more obvious necessity in the case of nothingness, since it is essential to be just a nothing. Therefore, in order to fulfill its nature, the number of states of nothingness must be even, so that it can completely interfere with or cancel out itself. So, whether it is 2, 4, 8, 32 or 134568435246 states, you can always group them into two groups. Therefore, one can simplify and say that nothingness must essentially have two states, because only in this way can it fluctuate, oscillate, and annihilate with itself to be nothingness. This would then be a fundamental property of quantum mechanics, which in turn would be explained by Darwinian Physics, thus proving the intimate connection between this pillar of physics and the new paradigm. However, we could go further with this justification, since this property of nothingness would explain not only the peculiarities of quantum mechanics, but also the reason why our universe is strongly dual: matter-antimatter, cold-heat, man-woman, light-darkness, live-dead, on-off, true-false, chaos-order, 1-0, and so on. A binary nature strongly integrated into physics through waves, like the swing of a pendulum or any other oscillator. A bipolarity that would have its origin in the random fluctuation between these two states of nothingness.

Despite the above, it may still seem very strange to us that a spatial point can randomly oscillate between two states or have two faces; however, quantum mechanics once again confirms these characteristics of nothingness, because this branch of physics not only proves that randomness is a real phenomenon, but also that quantum points exist. In the Standard Model of Particles, the most current theory of quantum mechanics, the constituents of matter are defined as point objects, without structure, dimension, or extension, since they would not be formed by anything else. For example, the electron is formally described as a point because it is a fundamental particle. A pointiness that has been experimentally confirmed to an accuracy of 0.0000000000000000000000000001 cm (or 10^{-28} cm). But despite being a point, it has an intrinsic property, the spin or angular momentum; which, curiously, can have only two possible values: up or ½, and down or -½. That is why the spin of the fundamental particles does not represent a real spin, since, being a point, it cannot have an axis of rotation, since it has neither length nor extent. In fact, physicists do not understand exactly how spin is created. They have simply assumed that it is an intrinsic property of quantum particles. This makes the fundamental particles a perfect natural analogy of what nothingness would be like, which becomes even more ideal when we add that the determination of spin happens in a truly random manner, since each spin has exactly a 50% probability of appearing. A behavior that remains a complete mystery to physicists, who do not understand how it is generated. Something that, as we already know, should not surprise us, since ontological randomness goes against their absolutist paradigm. However, the behavior of the electron and other quantum particles would be entirely expected for Darwinian Physics. In fact, the comparison with the electron is quite apt, since it helps us to conceive of nothingness as it really is, because it would be a great mistake to imagine it as a point with two states in a classical or Laplacean way, as if it were really a coin with two sides. We must not forget that nothingness is the quantum object par excellence, not because quantum properties must be applied to it, but because nothingness would be the origin or cause of all quantum phenomena. Then, like the electron, nothingness would be a punctiform and indivisible object that nevertheless possesses two states because they are in a permanent quantum superposition, fluctuating and annihilating each other. Therefore, however strange the description of nothingness may be, it is the result of the logical consistency of its own definition. A description that not only has its phenomenological analogues in quantum mechanics, but also has the potential to explain all its peculiarities, which have remained incomprehensible under Parmenian logic.

So we have a rigorous definition of nothingness that is not only internally consistent, but also empirically consistent thanks to the support of quantum mechanics. However, there is still a very important question to be answered: how could everything have come out of this nothingness? A question that seems to make no sense, but, although it sounds incredible, it is the same properties that define nothingness that make it inevitable for existence to emerge from it. We know that each of the two states of nothingness must have a 50% probability of appearing, precisely because it is completely random. Then, by chance, it can and will happen that a state repeats itself successively, generating a past, a permanence, and thus an existence. In this case of the temporal type ($dt > 0$). So, in this context, if the goal is to obtain a total nothingness, why not alternate the states permanently, since this would guarantee a perfect nothingness by ensuring that each state lasts only a instant? Although it may seem like a good idea at first, this alternative does not make sense because it would be inconsistent with the very definition of nothingness. If there were this perfect alternation between the two states, it would mean that there is a pattern that is being fulfilled or obeyed. A plan, a rule, or a law (as physicists would say) that causes or determines such an alternation. But this is the opposite of nothingness, not only because permanence is against its nature, but also because such a law, plan or rule would have to always exist in a "before" in order to constantly cause or determine such alternation. But how can there be an eternal plan that causes it before if there is no past in the nothingness? It makes no sense. But that is not the only problem, for it would also violate the condition of spatial nothingness. If there were such a law of permanent change, it would have to be encoded somewhere, so it could not be true nothingness either, this time spatial, because that law would have to be written down somewhere. In conclusion, the only way the spatial point can also be a temporal point is if the alternation between the two states of nothingness is perfectly random, with exactly 50% probability for each state. There can be no statistical deviation, no matter how small, not even 50.00000001% or 49.99999999%, for example, because it would favor the occurrence and thus the existence of one of the two states. This would imply that there is a law that produces such a preference, which in turn produces the temporal and spatial inconsistencies already mentioned. This again justifies the thought of the French mathematician Joseph Bertrand: "How dare we speak of the laws of chance, isn't chance the antithesis of all law?

Then the condition of pure randomness is indispensable for nothingness to be a true nothingness, because randomness is nothingness. Therefore, perfect randomness is necessary to obtain perfect nothingness. However, this itself seems to create a problem, because randomness allows the emergence of its existence, given the possibility of the same state repeating itself consecutively and remaining more than an instant. This could lead us to question the validity of nothingness, because if it is so easy for its existence to appear, is it fair to still call it nothing? Despite its defect of nature, we can still consider it a true nothingness, because if successive repetition means existence, then interruption or change by the other state means its death, its extinction. An excellent analogy to understand this is again the tossing of a coin, where there is (in principle) an equal probability of heads or tails. The probability of the same side of the coin appearing two times in a row is 25% (50% × 50%), three times in a row is 12.5% (50% × 50% × 50% × 50%), four times in a row is 6.25% (50% × 50% × 50% × 50% × 50%), five times in a row is 3.125% (50% × 50% × 50% × 50% × 50%), and so on. Here we see that each new consecutive repetition halves the probability of that happening, which can be calculated as follows:

$$\mathbf{P(n\ tails)\ or\ P(n\ heads) = (\tfrac{1}{2})^{n}\ or\ 2^{-n}}$$

(where "n" is the number of times it is repeated)
consecutively one of the two sides of the coin)

This is the trick of nothingness to be a true nothing, because although existence is likely to emerge from it, it will not remain or exist for long, for if the sequence of successive repetitions is interrupted, it will mean the end of that existence, its death, and the return again to nothingness. And we have seen how the probability of continuing to exist decreases exponentially the more it has existed. This means that the existence that emerges from nothingness is incredibly ephemeral, so much so that it is completely negligible, or "nothing," worth the redundancy. Remember that time is quantified by the changes of state or oscillations of nothingness, so each "successive repetition" is a unit of time gained by that existence. And we have said that it is probably a Planck time (5.391 x 10^{-44} seconds). So, it is an unimaginably small moment. Then the existence that emerges from nothing is incredibly fleeting, a few Planck times of life. For example, the probability of something emerging that lasts as long as a muon, a subatomic particle that decays in just 10^{-6} seconds, would be incredibly small. This is because it would be the equivalent of repeating the same state 10^{38} times in a row, a 1 followed by 38 zeros (the Planck times that fit into 10^{-6} seconds). Then the probability of something existing for the time that a fleeting muon lasts (using the coin formula above) would be 2 raised to the above number (2^-10^{28}). A percentage that all scientific calculators give as 0, because it is incredibly small, with many zeros as decimals. So, if an existence with such an insignificant and ephemeral lifetime as that of a muon is so unlikely to emerge from nothingness, then it is an extremely reliable nothingness, since the enormous amount of times it will remain nothing. This is because, if we calculate the frequency towards infinity, the average duration for each state tends to be only one instant, i.e. 50% (according to the large number theorem). Therefore, thanks to randomness, nothingness manages to be an incredibly stable nothingness.

However, now the question is reversed, for if the nothingness of Darwinian Physics is so solid, how could the whole universe have come into being from it? It seems absurd, doesn't it? Especially a universe that, according to physicists, has existed for 13.8 billion years and will continue to exist for at least several trillion years. A universe that is not only very old, but also unimaginably large, at least 90 billion light years in diameter, just in our observable universe. One that continues to expand and grow faster and faster. If we put all the Planck times (n) that the universe has had and will have, an unimaginably large number, into the above formula (2^{-n}), the probability that the same state of nothingness has repeated itself so many times is simply ridiculous. Since it makes no sense for our universe to be so lucky, there must be a natural mechanism that has helped the universe to exist for so long. A process capable of transforming randomness into causality, nothingness into existence, and that existence into something larger and more complex. It is time to talk about the Darwinian mechanism of physics.

CHAPTER IX

FIRST MODEL OF DARWINIAN PHYSICS

At the end of the 17th century, the German physicist, philosopher, mathematician, jurist, historian, and politician Gottfried Leibniz asked himself the question: "Why is there something rather than nothing? More than three centuries later, we will again ask this question; but this time, we will not resort to God, nor will we use its modern equivalent, laws or the multiverse. On the contrary, we will offer a creatio ex nihilo explanation not only of energy and matter, but of space and time itself. For if the only way to explain the origin of life is from non-life, that is, from chemistry, then it also makes sense to suppose that the only way to explain the origin of all existence is from non-existence, that is, from nothingness. A hypothesis that we will try to justify below by showing that nothingness is not only possible, but also necessary and intrinsically creative.

We saw in the previous chapter that nothingness is unimaginably stable; however, by definition, it is not absolutely stable. Therefore, it is more accurate to say that the nothingness of Darwinian Physics is metastable. A state that means that it is incredibly safe or reliable, but that it can eventually decay. And the state to which it can decay is, ironically, what we call the universe. To understand how this incredible phase transition is possible, we must first look at the mechanism that would allow it, the same mechanism that explains how Natural Selection works. This mechanism, which Darwin discovered, essentially works in three steps. The first is reproduction, which is why biology is discrete, because reproduction consists of the creation of individuals (asexual or sexual). In this sense, biology would also be quantum, because just as nature does not produce ¾ of a human, ½ of a dog, or ⅛ of a bacterium (at least not in a healthy way), neither can you get ¾ of an electron, ½ of a photon, or ⅛ of the Higgs boson in quantum mechanics. They are always whole numbers, quanta. The next ingredient of Natural Selection is mutation, the genetic changes that living things undergo that allow biological variability. A process that occurs in a truly random manner. Curiously, the same way quantum systems behave. So, at least in these first two steps, Darwinian mechanics is quite similar to quantum mechanics; therefore, the great challenge will be to find the equivalent of the selection mechanism, the third and last step of darwinian theory. In this sense, what is the criterion of selection in physical reality, what would be the analog of survival in physics? According to Darwinian Physics, the analog would be existence itself.

To understand how this mechanism of physical selection would work, we must first overcome the absolutist or Parmenian worldview represented by the Aristotelian or classical logic we saw in Chapter IV. A logic that was believed to be the only possible one, and that is based on a unit of information called a bit, which can only take the value "1" (is or true) or "0" (is not or false). This means that in this logic there is no ontological or real chance, since things are (100%) or are not (0%). No intermediate values (principle of the excluded third). However, what Aristotle did not know (which we cannot blame him for, since he did not have the technology to discover it) is that quantum mechanics proved that there is not only a third possibility, but also a fourth, and a fifth, and a sixth, and even infinitely more possibilities; since not only the possibilities 1 (100% or true) and 0 (0% or false) are real, but also the infinite decimal numbers in between, such as 0.9, 0.003, or 0.9989. For example, the spin of an electron can be up (1) and not up or down (0), but these are not the only possibilities that exist, since the spin can be simultaneously in a superposition between all combinations of these two states, for example: 0.5 up and 0.5 down, but also 0.02 up and 0.98 down, or 0.673 up and 0.327 down, and so on. Thus, Aristotelian or classical logic (bit) is not useful to represent nature at its most fundamental level, because it excludes randomness or nothingness, since it is based on an absolutist assumption or prejudice in the sense that, ontologically speaking, there are only certainties.

The previous point is very important because many philosophers defend their great Parmenian principle by arguing that quantum mechanics cannot disprove the principle of non-contradiction because it is simply impossible to disprove. As explained in Chapter IV, in order to disprove anything, you need this principle, so you cannot disprove something that you need to disprove. For example, to say that quantum systems are indeterminate or indefinite requires the principle of non-contradiction, because it means that the quantum system cannot be both determinate and indeterminate. So, to say that quantum systems are not deterministic requires the principle of non-contradiction. This is a half-truth, which, as has been said, is a dangerous lie; the fruit of a very recurrent confusion in science, but surprisingly also in philosophy. The error consists in constantly confusing two very different aspects: the ontological and the epistemological. When the book says that Aristotelian or classical logic, based on the principle of non-contradiction, is false, it is said in ontological terms. That is, that nature itself, at its fundamental level, is not absolute or certain. Before quantum mechanics, there was no such problem, because in the Parmenian paradigm it was assumed that epistemic principles were also ontological principles. And that was the mistake. Confusing the map with the territory. On the other hand, now we have to make this distinction, separate the ontological from the epistemic, because it is true that in order to affirm that reality is fundamentally not classical or Parmenian, we need the principle of non-contradiction, because we are saying that something is and is not something else. This is correct, because the principle is being used in an exclusively epistemic way; because ontologically it is being affirmed that something can be and not be at the same time. In contrast, before, a purely epistemic principle was used to affirm something about the world: that nature is absolute. An idealistic position, because it tries to force reality to behave in a way that is compatible with its Aristotelian beliefs. Whereas the position of the book is realistic, because it tries to grasp nature as it is and, from there, to adapt our way of thinking about it. Therefore, what is criticized is not the use of the principle of non-contradiction, but its abuse.

Having clarified the above, we can proceed with trying to represent how the world really is, which, according to the latest research and scientific evidence, cannot be represented by an Aristotelian or classical logic (bit), since, according to Quantum Mechanics, nature is inherently probabilistic and uncertain. Therefore, we must use another kind of reasoning, one that is truly fundamental. And this would be quantum logic, which is already effective and operational in the first quantum computers that work with qbits. Qbits, unlike bits, can have not only 1 or 0 as their value, but all values in between, unleashing an enormous computational power. Then, in this logic, "x" can be "A" and "non-A," since it can be in a superposition of states between "A" and "B," for example: 0.5·A and 0.5·B; or 0.12457·A and 0.87543·B, and so on. Just as Schrödinger's cat could be alive and dead at the same time. All this is due to the fact that quantum systems, such as the fundamental particles and atoms that make up material things, exhibit just such behavior. Everything therefore suggests that the fundamental code of nature is not written in binary form, but in qbits.

Then, from this new logic, more fundamental or valid than classical or Aristotelian logic, we will build a very simple model. A pilot model that will represent nothingness through the simplest possible system. One that many physicists would say is the harmonic oscillator. But in reality, the simplest system is really the "bit," which would be anything that has two states: up-down, open-closed, on-off, live-dead, or 1-0. But as we just discovered, the bit is the simplest of the classical systems. In other words, it is the simplest existing thing, whereas what we are looking for is to represent nothingness, its opposite. Fortunately, physicists have already discovered the way to represent it, the qbit mentioned above. We will return to this unit of quantum information in more detail later, because it is the key to understanding how nature really works; but for now, we will work with the classical version of the qbit, which is much simpler, just for the sake of simplicity. Remember that we are going to build a pilot model, since the purpose of the book is simply to demonstrate the feasibility of Darwinian Physics. So, with the above in mind, the classical version of the qbit is called the "pbit" or probabilistic bit. Here the system can take not only the values 1 and 0, but also the infinite decimal numbers in between. The big difference to the qbit is that the qbit has not only all the values of the pbit, but also negative values and complex numbers (numbers with a real and an imaginary part). To be even more technical, the qbit works with probability amplitudes and not only with probabilities like the pbit. This difference is crucial because it allows you to interfere with each other as a wave. But this will be explained in more detail in the third part of the book. So, since the purpose of the book is again to build a simple model that will help us better understand how this new mechanism of Natural Selection applied to physics would work, it will suffice to use the classical version of qbit, the pbit, or also known as Bayesian logic, fuzzy logic, which is an extension of classical logic that allows working with degrees of truth instead of an absolute or binary truth (true or false). The big difference is that these logics have usually been used to deal with epistemic uncertainty and imprecision, in the sense that it is the product of our lack of information or our ignorance. In contrast, we are now going to use them ontologically, because nature itself is fundamentally ignorant and ambiguous. In terms of Darwinian Physics, we would say that it is nihilistic. Having made the previous point clear, we will begin to construct this pilot model by returning to the example of the coin from the previous chapter.

$$P(H) = 2^{-t}, \text{ where } t \in \mathbb{N}$$

Here, P(H) represents the degree of being or existence, measured by the probability that a state of nothingness will repeat or "exist" up to "t." Therefore, the parameter "t" or the number of instants will represent the temporal extent of existence, measured by the number of "successive repetitions" that have randomly occurred. Thus, "t" corresponds to a discrete number positive or greater than zero (belonging to the set of natural numbers or $\mathbb{N}$). Thus, according to the given definition of nothingness, the "minimal existence" or the most insignificant "thing" that can exist is when $t = 2$. In other words, when one of the states of nothingness has been repeated once. The probability of this happening is 25% (if $t = 2 \rightarrow P(H) = 0.25$). If we want something more existential or with a longer age, e.g. at $t = 3$, we would see that the probability drops drastically to 12.5%, and at $t = 4 \rightarrow 6.25\%$; and at $t = 5 \rightarrow 3.12\%$, and $t = 6 \rightarrow 1.56\%$, and so on, as happened in the coin example.

However, all of the above concerns only nothingness. Therefore, the turn has come for us to analyze what is the nature of being itself, of existence itself. To do this, we must recall what was said in the previous chapter, that the existence that emerges directly from nothingness is of the temporal type, since the successive repetitions of a state of nothingness, resulting from chance, generate the growth of the temporal dimension ($+\Delta\, dt$), in terms of its permanence. Thus, the greater the number of successive repetitions of the same state, the more such a state will have existed. Therefore, "t-1" will correspond to the past, to the previous repetitions. And as was said in the previous chapter, this has consequences, because if the presence of randomness is a product of the absence of the past, then when the past is created, causality is also created. The opposite of randomness. Therefore, if nothingness seeks to make both states appear equally by giving it the same probability (preference) to remain nothing, existence will do the opposite, it will seek to favor the state that repeats itself by increasing its probability of reappearing. And it achieves this thanks to the repetitions themselves, to the emergence of the past that generates causality. In other words, existence seeks the opposite of nothingness, "not to change," to maintain the same state, while nothingness seeks to "change" its state constantly, to remain nothingness through randomness. Therefore, the more successive repetition is the product of chance, the more extensive will be existence or the temporal dimension, and therefore the more past or causality. This will determine or influence the next random fluctuation of nothingness, causing it to move further and further away from total randomness, from nothingness, by increasing the probability that the same state will repeat itself again. This, in turn, will create more past, which will further increase the probability, and so on. In other words, a positive feedback loop in favor of existence ends up being produced by the appearance of more and more past, the product of the randomness of repeating the same state of nothingness several times in a row.

The mechanism that will represent this causal iteration in the pilot model will be the equation of a circle of radius one. The justification for this choice is again the use of a classical version of a quantum phenomenon, just for simplicity. In this case, the equation of a circle is the simplification of the graphical representation used in quantum computing to visualize only the qbit, which is called the Bloch sphere. A sphere which, as mentioned above, is made up of complex numbers. Since we are only trying to construct a simple pilot model in this book, the circumference equation is a good classical simplification of this Bloch sphere. But the justification is not only that it is a classical representation of nothingness itself, but also that it is the best imaginable representation of existence itself. This is because the circle is not only the symmetrical object par excellence - not for nothing did the ancient Greek philosophers believe that God or the Absolute was round, hence the orbits of the planets in Ptolemy's geocentric model had to be perfect circles; but also because the equation of circumference is the elementary representative of all classical wave motion, which in turn is the greatest representative of a pattern, being an oscillation, a back and forth that is always repeated with the same regularity. This is especially true in physics under the harmonic oscillator, since all physical phenomena can essentially be reduced to a wave, from waves in the sea, sound waves, seismic waves to gravitational waves in general relativity and in the rest of the fundamental forces that are transmitted by waves in quantum fields, as in the case of light, which is a perturbation of the electromagnetic field.

A physical truth that applies not only to forces or interactions, but also to matter itself, since all things are ultimately made up of quarks (protons and neutrons) and electrons, which are also described in a wave-like manner by the wave function (Ψ) in the Schrödinger equation (we will see this in more detail in Chapter XII as well). Quantum waves, which are closely related to nothingness, since the equation of the circumference of radius one is also used to perform the calculations of the probabilities, thanks to the fact that it naturally represents the percentages thanks to this radius one. In this way, it gives values between 0 (0%) and 1 (100%), since it is not possible to go beyond the radius one of the circumference. A property that is particularly exploited in quantum mechanics, which describes all particles, both matter and interaction, as essentially waves of probabilities. Only in this case, the circumference is a bit more sophisticated, of course, by adding complex numbers (C). Real numbers that have an imaginary part (i) due to Euler's identity. Exactly the basis for the construction of the Bloch sphere.

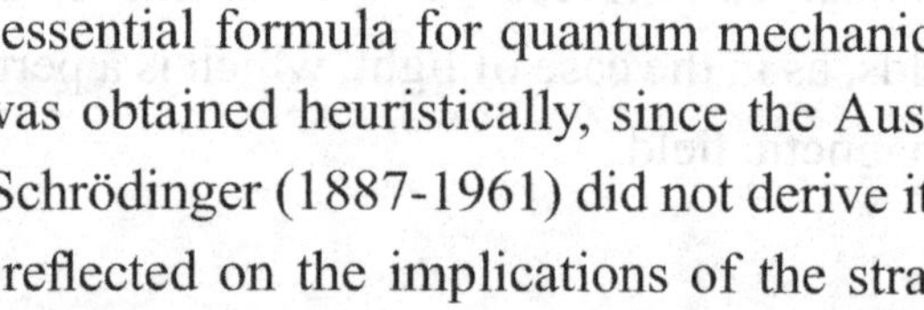

An essential formula for quantum mechanics, it should be noted, was obtained heuristically, since the Austrian physicist Erwin Schrödinger (1887-1961) did not derive it rigorously. He simply reflected on the implications of the strange duality proposed by the French prince and physicist Louis-Victor de Broglie (1892-1987), who stated in his doctoral thesis that every quantum particle is also associated with wave behavior. This was in response to another radical idea proposed by Einstein a few years earlier, that the wave of light also behaves like a particle, the famous photon (the photoelectric effect). Both ideas won Nobel Prizes in Physics. So, if waves seem to have a corpuscular side, then perhaps particles behave like waves as well.

An idea that was soon proven by electron diffraction. In 1925, in a moment of inspiration during a romantic Christmas vacation in a hotel in the Swiss Alps with his lover, Erwin Schrödinger invented a wave equation to describe the wave behavior of quantum particles. An equation that, to everyone's surprise, worked like a charm, although no one knows exactly why. Something that could change in the new paradigm, since the wave behavior would have a nihilistic origin, due to the random oscillations of nothingness. An explanation that also has the potential to explain the reason why existence in its essence is built by regularities and symmetries. Exactly the subject that will be discussed later. So, for all of the above, it is believed that the circumference equation is an excellent representative of the essence of existence, besides being a perfect simplification of Bloch's sphere.

Equation of the circumference: $(x - a)^2 + (y - b)^2 = r^2$.

If the center of the circle is $(0,0) \rightarrow a = 0, b = 0$.

and if $r = 1 \rightarrow 1^2 = 1$

then we have that: $x^2 + y^2 = 1$

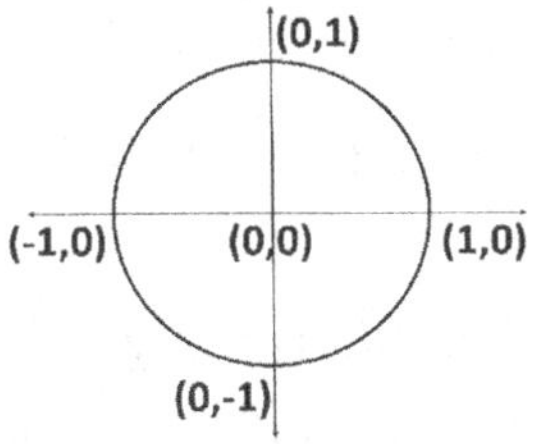

Above we have the formula for a circle with radius one, centered at point 0,0. However, as the formula now stands, it is of no use to us for the model, because if "x" is to represent time ($x \rightarrow t$) and "y" is to represent the degree of causality - which from now on we will call causal inertia or Ǝ ($y \rightarrow$ Ǝ) - then the path of "x" must always be positive, because time cannot be negative, just as it makes no physical sense for spatial distances, kinetic energy (motion), or temperature to be negative. Therefore, the formula of the circle cannot be centered at the point (0,0), because the path of time (t) would be between -1 and 1, with a negative half. To correct this, we must move the circle one unit to the right, toward the point (1,0).

If x = t, y = Ǝ, and the center moves from the point (0,0) to the point (1,0), then:

$$t^2 + \exists^2 = 1 \rightarrow (t - 1)^2 + \exists^2 = 1$$

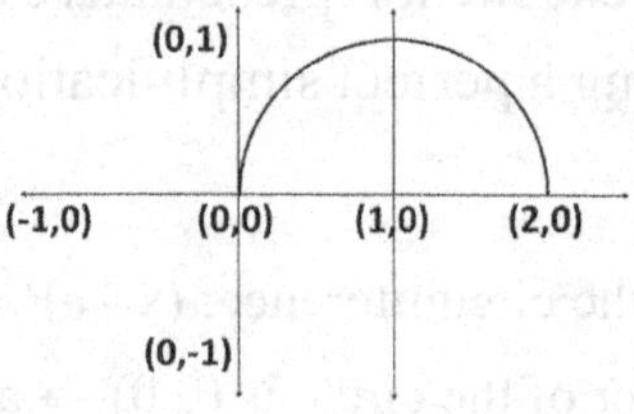

Clearing the degree of chance or causal inertia (Ǝ) to let it be a function of time (t), we obtain the following expression:

$$\exists = \sqrt{1 - (t - 1)^2}$$

However, the formula is not yet finished, because let us remember that "t," time, extends over the set of natural numbers (1, 2, 3, 4, 5... ∞). The number of Planck times there have been and will be in the universe, which, remember, are very many. Just one second has more Planck times than there are seconds in 14 billion years. However, the "x" representing "t" has as its path up to 2, which corresponds to the diameter of the circumference. This would mean that the maximum duration of existence would be 2 instants, exactly the minimum existence (a successive repetition). Since the universe has obviously lasted much longer than 2 Planck times, to represent this we have to divide "t" by another expression, a very, very large constant, which we will symbolize by "g."

$$\exists = \sqrt{1 - \left(\frac{t}{g} - 1\right)^2}$$

Thus, thanks to the denominator "g," "t" can represent many natural numbers. Not all of them (i.e. infinity), because "g" is not infinite. But since "g" is an incommensurable number, it would give time enough leeway to allow for an existence as old as our universe. But what exactly is "g"? Keeping in mind that this is only a pilot model, we could say that "g" allows nothingness to exist. This is because nothingness has been defined as non-being or the negation of being, so for there to be nothingness there must be its opposite, absolute being, which must be denied. Precisely what allows "g," because if $t = g \rightarrow \exists = 1$, an absolute certainty. Thus, for total nothingness to exist, there must also be total being, so there must be a maximum time, a limit to the successive repetitions that our universe can have to reach a maximum degree of being ($\exists = 1$). Exactly what the constant "g" allows. This can be clearly seen in the above formula, because the purpose of causal inertia ($\exists$) is to increase the probability that a state will be repeated consecutively, or continue to exist. If there is very little past, "t" will be very small.

And since "g" is very large, the causal inertia will be very weak, because the ratio "t/g" will be very close to zero, so "Ǝ" will also have a value very close to zero. In other words, the probability of continuing to exist or repeating again would be very close to 50% or pure nothingness. But if "t" is significantly large, say a third of "g" (t = g/3), there would be much more causality, specifically a "Ǝ = 0.74," moving significantly away from nothingness or pure randomness. Therefore, "g" seems to be the maximum distance that existence could be away from nothing, because when "t" gets the number of repetitions of "g" (t = g), the causal inertia is total (Ǝ = 1). That is, it obtains 100% certainty of continuing to be. Causality or perfect determinacy. This is similar to maximum entropy or thermodynamic equilibrium. A state of the universe where it is completely dead because nothing changes, because everything is frozen. Here the past is equal to the future because there would be no way to distinguish between yesterday and tomorrow. It would be eternity. Curiously, the most probable end of the cosmos predicted by physicists, the thermal death of the universe, due to the second law of thermodynamics. Something that will happen in about one googol of years (10^{100} years or a 1 followed by 100 zeros of years). So, in this sense, "g" would be equal to the number of "t" or Planck times that would fit into that googol of years. A monstrously large number.

Then, with the above, we would already have the two mechanisms that make Darwinian Physics possible: the force of change (nothingness) and the force of conservation (causal inertia). Now we have to make them interact, make them influence each other. To do that, we have to combine them in the same pilot model:

$$Nada: P(H) = 2^{-t}; \quad \text{Inercia Causal: } \exists = \sqrt{1-\left(\frac{t}{g}-1\right)^2}$$

Juntando ambas expresiones:

$$P(H) = \left(2 - \exists(t)\right)^{-t}$$

$$P(H) = \left(2 - \sqrt{1-\left(\frac{t}{g}-1\right)^2}\right)^{-t}$$

Here we have the first pilot model of Darwinian Physics. A very simplified one, of course, since the real model would have to be based on the Bloch sphere (qbit) with complex numbers and other fundamental elements of quantum mechanics. However, for the modest purpose of this book, which is to demonstrate the feasibility of Darwinian Physics, the above model based on the pbit and the circumference formula is more than sufficient. Then we have that the causal inertia tries to subtract the base 2 from the power to reduce it to 1, since the path of Ǝ(t) goes between 0 and 1 when "t" approaches "g." This means that the task of causal inertia is to reduce the number of possibilities of nothingness from 2 states to just 1. Remember that absolute existence means absolute certainty. So, it makes a lot of sense that "Ǝ" wants to reduce the possibilities to just one. A single possibility where there would obviously be 100% probability, independent of time or "t," since 1 raised to any natural number results in 1.

If $t = g \rightarrow \exists(t) = 1$, then:

P(H): $(2 - 1)^{-g} = 1$ or 100%.

Therefore, as "t" increases, not only does the force of nothingness (the probability that the other state will occur) increase, but also the force of being, which influences the repetition of the same state. So, there is a constant tension between these two forces. A struggle between randomness and variability versus determinism and permanence. Similar to what happens in Natural Selection, where we have a constant struggle between the stability of the environment (geological time) and the intergenerational variability of organisms resulting from genetic mutations. So, it is a good emulation of Natural Selection applied to physics. The problem for us, however, is that this struggle begins very unevenly, because the force of nothingness is far more powerful in the beginning. This is because causal inertia grows very slowly, so its effect is practically negligible at first. The culprit is the unimaginably large "g." Therefore, when existence begins to evolve Darwinian, the ratio "t/g" is incredibly small, so the effect of ∃(t) is also small. This makes it so that, despite the mechanism on which existence relies, nothingness is still incredibly nothingness in the beginning, by keeping the probability of the state repeating very close to 50%. So, the intrinsic property of existence, causal iteration, does not make much difference at the beginning. Which is logical, since we are starting from nothing, so it makes a lot of sense that the strength of mutability and randomness is far superior. Remember that "g" represents the distance from absolute existence to nothingness, so if "t" is very small relative to "g," it is because you are still very close to nothingness. Then the mechanism of existence or causal inertia becomes relevant only when existence is relevant, when "t" manages to be a significant percentage of "g" and you are far from nothingness. The problem is that to achieve that distance or seniority, one relies almost entirely on the randomness of nothingness. For example, consider an extremely small "g" compared to what the original should be, a g

= 1,000. For example, for $\exists(t) = 0.62$, the same state of nothingness should be repeated 215 times in a row (t = 215 or 21.5% of "g"). Considering that the probability remained at 50%, i.e. did not gradually increase to 62% (just to simplify the calculation), the probability of this happening by chance is 0, followed by 63 zero percent ($2^{-215} = 1.8991x10^{-65}$). Obviously a ridiculously small probability (even if you increase the probability gradually up to 62%, the result does not improve much). And this considering an incredibly small "g" as an example. Therefore, considering the true "g," nothingness remains an extremely stable nothingness, despite the fact that there is a mechanism that promotes the emergence of existence.

Fortunately, incredibly unlikely does not mean impossible. For if nothingness is the default, the unimaginably small probability that "t" will become relatively large relative to "g" becomes a certainty if we wait long enough. In other words, nothingness is incredibly stable, but not infinitely stable. This is what we meant by metastability above, because after an unimaginable number of chances, nothingness can decay into a stable existence. Something that will happen eventually, because if chances are simply the random oscillations that nothingness performs to remain nothing, then in principle there will be infinite chances for existence to have sufficient antiquity. So even the most ridiculously small probability must occur in the face of infinite chances. Obviously, we are talking about an unimaginable number of chances, beyond what we can even count in our universe, since any number would be insignificant compared to the number of chances that had to happen for us to exist. Big numbers, but in practice irrelevant, because if we are talking about the random fluctuations of nothingness, there can be no one waiting for the miracle to happen, because we are talking about before existence, before any possible place and time.

Then the miracle happened. What in physics is called the Big Bang. However, the genesis of Darwinian Physics is much more modest than the spectacular explosion or expansion described in current physics; because far from a mechanical beginning, where in a single instant all matter and energy that has existed and will exist was spontaneously created (a practically divine event), the Darwinian genesis is rather an organic Big Bang, much simpler (in fact, the simplest possible). An extremely progressive and gradual growth of existence, like all Darwinian evolution. In fact, a pertinent comparison would be to equate the current big bang theory with the creation hypothesis, where God created the entire universe in just 6 days, only 6,000 years ago; whereas the big bang of Darwinian Physics would be equivalent to Darwin's theory of evolution, where all species emerged progressively from LUCA during more than 4 billion years. In fact, the beginning of the universe was so gradual and simple that the lactation of the cosmos was purely temporary. That is, there was still a spatial point or nothing, except that instead of randomly oscillating between two states, only a single state appeared in succession due to a very lucky statistical deviation. Like flipping a coin and getting the same side several times (and consecutively). This continuity is something, it is time, it is causality. Therefore, there is a temporal extension that grows with each successive repetition. A beginning of the universe consisting of a point that has remained in the same state many times.

Obviously, our universe is not like that, it is not only temporal, because our universe is not only old, it is also immense. So how can Darwinian Physics create space? To find out, we have to go back to the pilot model, where P(H) decreases with each new "t," while exactly the opposite happens to Ǝ(t) with each new successive iteration. As we said, Ǝ(t) does not have much effect at the beginning of existence, since there is very little past or causality. However, if we are lucky, existence can generate a positive feedback (causal iteration) intense enough to reverse the trend and counteract the "force of nothingness," causing P(H) to stop decreasing and start increasing for the first time. The moment when the "force of existence" begins to overcome the "force of nothingness" What is this point at which fortune begins to smile upon the cosmos? To find out, we need to derive the above formula and see when it becomes zero. This is when the change in trend occurs, since the derivative represents the reason for the change at a point on the curve, its slope. Then, clearing the variable "t" in the derivative, we find that this change in trend occurs at a little more than 32% of "g."

$$\mathbf{d/dt\ P(H) = 0 \rightarrow t = 0{,}32223...g}$$

This means that at this point the force of permanence and the force of change become equal, because if $t < 0.322...g$, the force of nothingness is greater, i.e. P(H) has only gone down. However, if $t > 0.322...g$, the force of existence is greater, so P(H) begins to increase. Then, just at that moment ($t = 0.322...g$), P(H) reaches its minimum value, when both forces are neutralized, when they balance each other, creating what we will call in Darwinian Physics a holistic equilibrium.

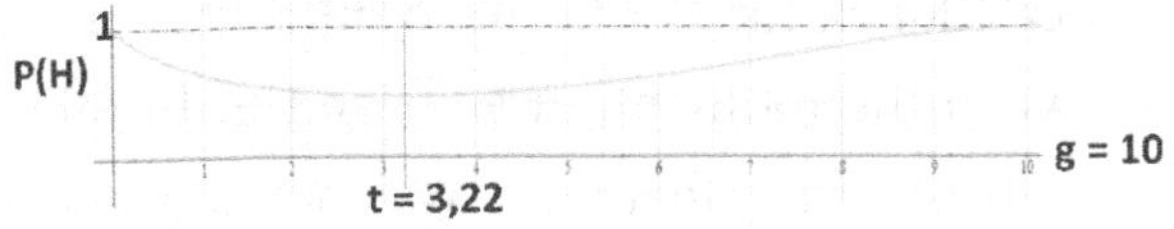

In the figure above you can better see this change in the trend of P(H) plotted with a g = 10 (just as an example). Here you can better see how the curve of P(H) reaches its lowest point (approximately) at t = 3.22 and then rises. However, this inflection point represents not only the strengthening of existence with respect to nothingness, but also something much more transcendental: the birth of a new type of existence. To understand the reason for this, we need to go deeper into the concept of existence, because so far, we have focused almost exclusively on the definition of nothingness. And here again, etymology would help us, because a very appropriate concept to refer to "everything" is cosmos, a word that comes from the Greek kósmos, meaning order, regularity. A word invented as a counterpart or antithesis to chaos, from the Greek kháos; which corresponds exactly to nothingness in Darwinian Physics, representing randomness, uncertainty and irregularity. The definition given to existence, as the successive repetition of a state of nothingness, then represents the essence of what the cosmos means, since from it can be derived all the characteristics that define being, such as: continuity, certainty, stability, permanence, conservation, invariance, symmetry. Concepts that generally mean not changing, maintaining or continuing the same state or continuing to "be" the same pattern. Therefore, existence is essentially a regularity, something we can identify in the chaotic noise of nothingness. Mathematically, it would be having zero differences between two successive instants (t_n - t_n-1 = 0). For example, if we have 111 (t_0 = 1, t_1 = 1 and t_2 = 1), we know that it is an existence, even if it is a very short conservation (of three Planck times), because there was no change or variation between the momentum t_0, t_1 and t_2. In other words, $t_0 - t_1 = t_1 - t_2 = 0$, showing that there was permanence, continuity, or invariance. This itself is mathematically equivalent, only more effective, if we use the derivative. If the derivative is the ratio of the change at a point on the curve, then if it is zero, it means that there was no change at that point. The curve neither goes down nor up, because at that point both forces are equal (holistic equilibrium). The same thing happens when we throw a ball into the air, because before it goes down, after it goes up, it remains motionless for a moment, floating in the air. At that moment, the force of throwing the ball up is neutralized by the force of gravity pulling it back down.

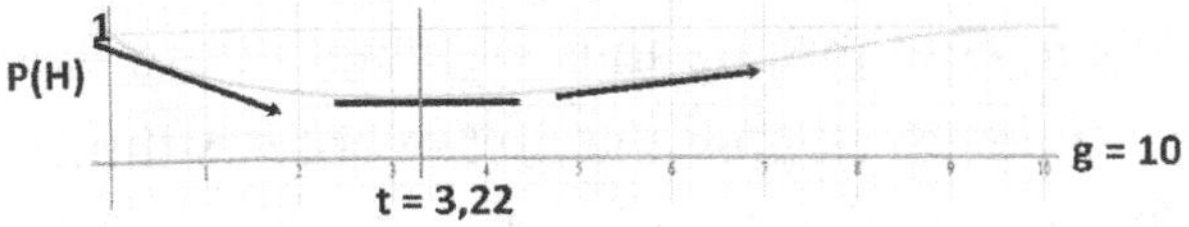

Therefore, this point of holistic equilibrium would represent a new type of existence, since it corresponds to an invariance, a permanence, a conservation. One that did not come directly from nothing, because it is not the fruit of a successive repetition of one of its states, but the product of a repetition of the intensity of chaos and of the cosmos. A new existence that seems to be incompatible with the first one, since the latter wants to continue to exist in order to increase even more its causal inertia. Especially when its probabilities finally begin to improve. In other words, the first existence does not want to exist up to the holistic equilibrium (t = 0.32...g), but far beyond. It wants to reach eternity (t = g). Obviously, this will break such an equilibrium and destroy the new form of existence (similar to how gravity rebounds a ball that has been thrown upward). Then how can the new existence be preserved or "continue to exist" if the first existence destroys it by continuing to exist? These seem to be two incompatible kinds of existence. The solution to this dilemma is to create your own parallel existence. Don't worry, I'm not talking about a parallel universe. What is meant is that if the first existence was represented by the space-time point of nothingness (all there was), then the new existence creates its own point, but a larger one, like a small bubble that surrounds the space-time point. In this way, the new existence allows the existence of the first one to continue by locating itself in its own point or bubble, where it will behave similarly to the space-time point, fluctuating or oscillating randomly between two states. And if one is lucky again, this new existence may also reach its own holistic equilibrium, thus creating a third existence or bubble, this time enveloping the first bubble. A third existence that also vibrates so that, if it is lucky again, it can reach a new holistic equilibrium, creating another new existence, and so on. As if they were different resonances, amplifying the length of the vibration of nothingness.

Again, we must remember that we are talking about a pilot model, which is also quite simplified. Therefore, the real justification for holistic equilibrium and the expansion of space is again to be found in quantum mechanics and its Bloch sphere. In fact, under this justification, Darwinian Physics might even have the potential to explain the dimensionality of the universe. Something that no physical theory has been able to do. Thus, the number of spatial dimensions is a free parameter in physics, especially for string theorists who love to play with adding more dimensions to their equations. In contrast, Darwinian Physics would be the first theory to predict the number of dimensions we have. This is possible through the most faithful representation of nothingness, the Bloch sphere. As mentioned above, this is one way to visualize a qbit, but it is also one of the best representations for understanding quantum mechanics, especially its fundamental pillar, Heisenberg's uncertainty principle. All of this will become clearer in the chapter on quantum mechanics in the third part of the book, but in the meantime, to understand how expansion and spatial dimensionality would work, we should know that the indeterminacy principle precludes having absolute certainty about two conjugate or simultaneously related variables. The classic example is the variables of particle position and momentum, or energy and time, but it also applies to spin, an intrinsic property of quantum particles. Like nothing, spin has two states or values, traditionally referred to as "spin up" and "spin down." Thus, prior to measurement, spin is in a superposition between these two values, with a 50% probability of being "up" or "down."

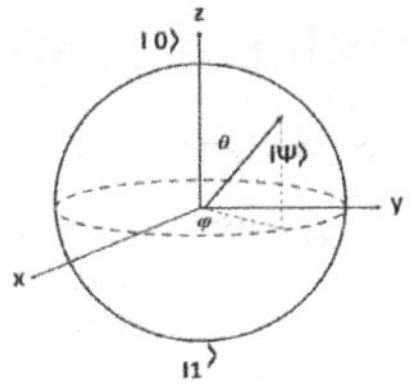

As can be seen in the figure above, the spin would correspond to an arrow that could point from the center to any point on the surface of the sphere. A direction that can only be determined by first selecting an axis, whatever it may be. For example, if we arbitrarily choose the "z-axis" to measure spin, we decide to measure spin by pointing up or down. Then where the arrow was before the measurement would indicate the probability of which result could be obtained. That is, if it was pointing more up than down, it means that it is more likely to measure the spin "up." Or the other way around, the more it points down, the more likely it is that we will observe the spin "down" state. However, when you measure collapsed Ψ and get 100% certainty of "up" or "down" because we see it, that's it. There is no more up-down superposition, only a concrete result.

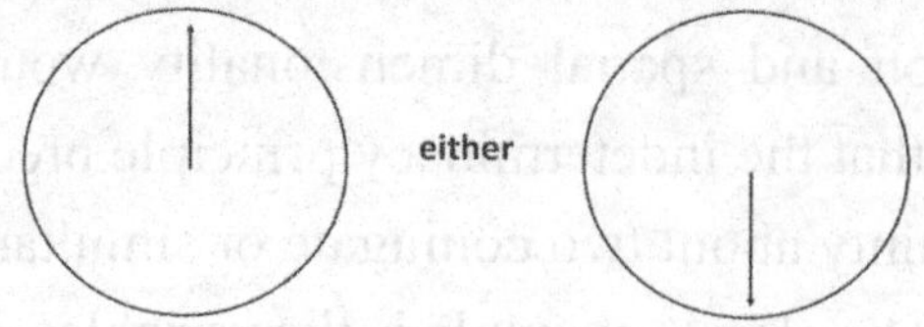

Therefore, the Bloch sphere is never observed, only the spin 100% "up" or 100% "down" is measured. We will never see a quantum superposition of e.g. 80% "up" and 20% "down." However, a dilemma arises here: how is it possible for us to have perfect spin determination when measuring, for example, 100% spin "up," when this clearly violates Heisenberg's uncertainty principle, which prevents us from being perfectly certain of anything? This is solved by quantum mechanics by adding another component to the spin. Thus, the 100% certainty of the spin "up" or "down" is at the expense of the total indeterminacy of this other component of the spin. In other words, Heisenberg's principle of indeterminacy (or nothingness) forces spin to have another axis or dimension, so that determinacy in one of its two possibilities comes at the cost of uncertainty in two other possibilities in that other dimension. And what would this other dimension be? Exactly, an axis perpendicular to the previous axis. Therefore, if the z-axis is traditionally referred to as "up" and "down," then the second dimension will be "left" and "right," traditionally referred to as the "y-axis." This is why spin in quantum mechanics is defined as a vector formed by two dimensions: z-axis (up-down) and y-axis (left-right).

Something that can be seen perfectly in the Bloch sphere, since the spin as a vector or an arrow is constructed by the distance between these two axes. In other words, it is the Pythagorean theorem, since the arrow would be the hypotenuse, for which the two legs are needed to determine it.

Returning to the subject of dimensionality, it is this second axis that allows us to continue to respect Heisenberg's principle of indeterminacy. Again, this can be easily seen in the Bloch sphere, because when the arrow points completely "up" or "down," it is completely aligned with the z or vertical axis, so it is right in the middle of the "y" or horizontal axis; therefore, the arrow is not tilted one way or the other, which is equivalent to 50% "left" and 50% "right." In other words, complete neutrality with respect to the y-axis. And vice versa, if the spin is measured on the "y-axis," it will be 100% "right" or "left," which means that the arrow will be completely aligned on the horizontal axis, so it will not tilt up or down, which again corresponds to complete neutrality with respect to the "z-axis": 50% "up" and 50% "down." Thus, the certainty of the spin on one of the two axes is at the expense of the complete uncertainty of the other axis. This is exactly what Heisenberg's Uncertainty Principle means. Then, since the condition of nothingness or temporal point obliges one of the two states of nothingness to be determined so that it lasts an instant and thus allows the existence of the present (punctuality or temporal nothingness), the same obliges nothingness to have two perpendicular dimensions (opposites), so that such determination is at the expense of the total indeterminacy of the other axis. Therefore, once again, this dual property of nothingness is confirmed, which must have two states in order to be simply nothingness and to fulfill its condition of uncertainty and randomness. Only this time, this property is derived from quantum mechanics itself.

To recapitulate, up to this point we have time (present or point in time) and two spatial dimensions ("z" and "y" axis). However, as you may have noticed, a third spatial dimension is missing; because our universe is not two-dimensional or Flatland. Therefore, in order to explain this extra spatial dimension, we must resort to the other key element of the Darwinian physical model, the holistic equilibrium. A principle that makes this spatial dimension appear in a vertical or fractal manner. To understand what the latter means, we must continue with the main feature of quantum mechanics, the Heisenberg uncertainty principle. A principle that was not actually invented with quantum mechanics, but discovered many years before, because it is an inherent property of waves. And what is a wave? An oscillation. Exactly what nothingness does. And as physicists have known for centuries, waves can be broken down into their harmonics, or spectra. This is the principle behind telecommunications technology, not only how information is processed in sounds, but also in images. Thanks to this principle, we know that a perfectly regular wave, or a wave made up of a single frequency, would be heard as a beep. A very simple or basic sound. The problem is that if this were really true, such a beep would have to last forever. Since this is not realistic, waves are actually packets of waves because they have a beginning and an end. This means that in the window of time that the wave lasts, the different frequencies constructively interfere with each other. But in all the other times, the different frequencies interfere destructively. That is why there is no wave. So, if the waves are superimposed in the right way, they will almost completely cancel each other out, leaving only a peak, a concentration in a small region. This means that a wave packet is actually a superposition or summation of different wavelengths (λ) or frequencies (v) with a correct offset. The transition from one description to the other is called the Fourier transform. So, if you have only a blip or flash of light that happens for only a brief moment in time (i.e., it is or was there), this means that it contains many different frequencies or wavelengths (a very extended spectrum) because they destructively interfere to leave only that blip or flash. And conversely, if there is a well-defined or unique λ or v, then the wave must be spread out over all of space by lasting forever. This is rather like the Heisenberg principle, because it is, since it is a property of all waves, including the quantum wave function or Ψ. Then, by gaining certainty in the location or duration of the wave (a more pronounced peak), one loses certainty in the frequencies (v), since it would be conformed by a greater number of them. And vice versa, the more certainty we have about the frequency (v), having a concrete one, the more extended or persistent the wave will be, losing certainty about its size or duration. It is the same principle that applies to the typical variables of quantum mechanics, such as position (size) and momentum (frequency), which are used as a classic example to explain Heisenberg's uncertainty principle.

So if the above principle applies to waves and quantum, then it must also apply to nothingness. And it does, because since nothingness is a space-time point, something completely localized in space and time, then this must be at the cost of having maximum indeterminacy (nothingness) with respect to frequency (v) or wavelength (λ). Just as in the case of a quick flash of light or a short blip of sound. In other words, a perfectly narrow or point wave packet. This means that in the nothingness, all frequencies annihilate or destructively interfere with each other. Something very consistent with the nature of nothingness. So, what the holistic equilibrium allows is for only one of those frequencies to be conserved and materialized, with higher frequencies or shorter wavelengths more likely to materialize first, requiring fewer oscillations or number of successive repeating states of nothingness to form. In other words, a short wavelength of only 3 Planck times (50% x 50% x 50% x 50% = 12.5%) is more likely to be generated than a wavelength of 5 Planck times (50% x 50% x 50% x 50% x 50% = 3.125%). Therefore, holistic equilibrium conserves waves or oscillations of increasingly larger wavelengths, which we will call holons. These, as they emerge, envelop the space-time point of nothingness, which oscillates at the maximum frequency or fastest vibration.

Now we have seen that Darwinian Physics can explain the spatial dimensionality of our universe, because if the very properties of nothingness create two spatial dimensions or axes, the holistic equilibrium initiates the expansion of the cosmos through a third dimension created by the emergence of larger and larger holons that accumulate layer by layer, forming a structure similar to matryoshka nesting dolls. This is where the name holistic (holon + system) comes from, as these layers are technically defined as holons, representing a part of a whole and a whole at the same time. Then, through the emergence of holons, space grows in a way very similar to the growth of a tree, which can be observed by the concentric rings that form across the trunk as it grows. A structure of concentric spheres (holons) having a holoarchy whose center would be nothingness itself. A space-time point that we will call Holon Zero (H_0), from which temporal existence emerged. While from the first holon (H_1) the emergence of space begins, thanks to the second holon (H_2), the third holon (H_3) and so on. The generation of greater and greater harmonics from the quantum fluctuations of nothingness.

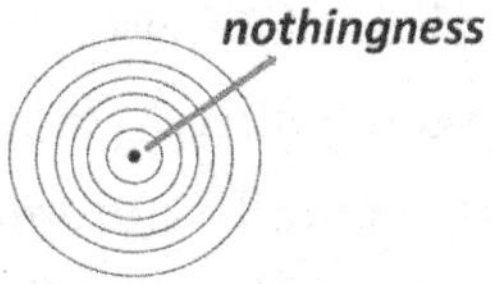

This means that naming the spin components as up or down would not be accurate. Remember that the axis is chosen arbitrarily. You choose the direction of the axis, so traditionally experimental physicists have placed the magnets and the apparatus so that the electron spin measurement is either up or down with respect to the apparatus; but it could be left and right or any other direction and tilt. Therefore, to be rigorous, the spin would have to be left-right and front-back. Whereas the third spatial dimension would actually be up and down. And as we have just seen, the origin of this third spatial dimension is completely different from the other two. Which is curiously consistent with the fact that up-down turns out to be a completely different direction than left-right and front-back. In other words, up-down is not symmetrical because of gravity. In fact, one clock moves slower compared to another clock that is one centimeter higher (something that has been experimentally proven), because space-time is different with respect to height (this will be explained in much more detail in the third part of the book, especially in the chapter on gravity). However, this difference, this curvature of space-time, is so minuscule on a small scale that its effect is quantum negligible. This is why physics currently says that the three spatial dimensions are equivalent, interchangeable, because they make no distinction between up and down by not including gravity. In fact, they take this equivalence even further by saying that time is also another spatial dimension, the fourth dimension. However, in Darwinian Physics, not only is there a clear distinction between space and time (spatial and temporal punctuality of nothingness), but the up-down direction is fundamentally different from the two right-left and front-back directions, having different origins. For this reason, we will classify them differently, calling the right-left and front-back directions horizontal dimensions, whereas we will call the up-down direction the vertical dimension, which is fractal in nature due to this emergence of larger and larger holons. We will see later that this vertical dimension will be the key to explaining the curvature of space-time on a large scale. In other words, it will be the origin of gravity.

But to return to the subject of this chapter, holistic equilibrium causes not only the expansion of space, but also the expansion of time. But beware, this is not said in the way one would traditionally expect. To properly understand what Darwinian Physics means by the above, we must first clarify what time is, something physicists have always tried to avoid doing. To do this, we must deal with an even more basic and elementary concept, change; for there can be no time without change. As we have said, nothingness is, by definition, pure change; but more importantly, it is chaotic change, in the sense that it is a completely random alternation between its two states. The existence that emerges from nothingness, on the other hand, would be ordered change, in the sense that it is invariant. The product of a repeating pattern, regularity, continuity, or symmetry. This gives time this predictive property, since the past can be used to know what will happen in the future, thanks to the fact that there is a natural tendency to repeat what was, to preserve such a pattern. Therefore, it can be assumed with some confidence that what has happened before will happen again. But the most extraordinary thing physicists will notice about time in Darwinian Physics is that it is presentist. That is, there is only now, because, remember, nothingness is a point in space-time. So, although there has been a lot of talk about the expansion of the time dimension and the growth of the past, the truth is that only the last instant is real. The last change or repetition of the state of nothingness. This is because this past is not used in the sense that physicists use it, as a dimensional equivalent to the spatial dimensions. The fourth dimension. On the contrary, in Darwinian Physics, the temporal extension is in the very present, existing as an intrinsic property that we have called causal inertia (Ǝ), which gives the "now" a certain stability. In other words, there is only ultimate repetition; because the future is not yet, it has not happened; while the past is included in the present in the form of "memory" (causal inertia). For example, our childhood is nowhere, but is in the form of the changes our bodies have undergone, and in the memories encoded in the synapses of our neurons in the "now." Therefore, if these synapses are damaged, for example, this past, these memories, are also damaged, as unfortunately happens in the case of people with Alzheimer's disease.

A temporal position radically different from that of current physics, because in the Parmenian or Laplacean paradigm the simultaneous existence of all the past and future is postulated, due to the fact that there is no privileged instant, different from the others, which can act as "now." Let us recall that the classical principle par excellence is the conservation of information or unitarity. In other words, every instant has exactly the same meaning, neither more nor less. However, the very instant that denies the eternalism of current physics is the only one that has been measured, because if one of our instruments had observed an instant that is not the "now," then it would officially be a time machine, since it would be measuring the past or the future. Something that apparently has not been published in any science magazine, only in science fiction and comic books. In this context, Darwinian Physics easily explains the reason why we are not overpopulated with travelers from 1,000, 10,000 or 100 million years in the future visiting us all the time, because there is simply no future, nor a past to go to, because those earlier moments no longer exist, since that past was destroyed when it was transformed into a new present. Therefore, there is neither a past nor a future to go to, because time is not a place or a special fourth dimension, but simply changes in the "now." Changes that are ordered or with some degree of "order" (as opposed to pure noise, chaos, or nothingness) that can manifest in a translational (movement) and/or transformational (evolution) manner. Therefore, the concept of past used in this book can be defined as "properties" that things have in the present that follow a pattern, an order. In other words, the past is information contained in the present, such as photographs, archaeological remains or fossils, and even laws of physics. Therefore, if such information is destroyed in the present, the past is also destroyed. In fact, this definition of the past is so important that it is the accumulation of information that occurs in each new "now" that is essentially responsible for the growth of the cosmos. For as the past accumulates in the present, it causes the now to grow, to evolve. Therefore, it is time, in a presentist way, that is responsible for the evolution of the universe from the singularity of the Big Bang (the nothingness or the space-time point) and its expansion until today. A much more reasonable and economical explanation, saving all those infinite instants of past and future that classical or Parmenian physicists need. Therefore, at least in terms of time, Darwinian Physics, although very exotic in the eyes of modern physics, is already much more empirical and conservative.

Having clarified what we mean by time in Darwinian Physics, what exactly does holistic presentism mean by the expansion of time? It refers to the fact that it is the present itself that grows, specifically that larger and larger presents emerge. This must be so, because if there were only the last "t" or successive repetition of a state of nothingness, then it would mean that there would be only one Planck time at a time, one change of nothingness. The shortest possible moment. This is obviously absurd, since our perception of "now" is much, much larger than that insignificant instant. Remember that there are more Planck times in a second than there have been seconds since the Big Bang. Therefore, if we were to perceive the world in Planck times, a second would technically become an eternity. This is where holarchy comes in, because the emergence of new and larger holons also means the emergence of larger and larger "now." Remember that new holons vibrate or oscillate, just like nothing, so it makes sense that as the holon gets larger, the wavelength of the vibration gets larger. Just as the sound gets deeper (longer wavelength) as the guitar string or church organ pipe gets longer. Therefore, this frequency, this tick-tock of each holon, determines the now or the present for that reality. Thus, the larger the holon, the greater its wavelength and the greater the extension or duration of the tick-tock and therefore the greater the "now. Therefore, holistic evolution is responsible not only for the expansion of space, but also for the growth of the present; where the smallest present, the fundamental present, is the oscillation of H_0, which would represent the fluctuation of nothingness itself, which probably occurs in Planck's time. The following moments would be its harmonics. The present of H_1 would be a larger oscillation belonging to a larger holon. And again, the present of H_2 would be an even larger oscillation, and the present of H_3 would be even larger, the present of H_4 would be even larger, and so on.

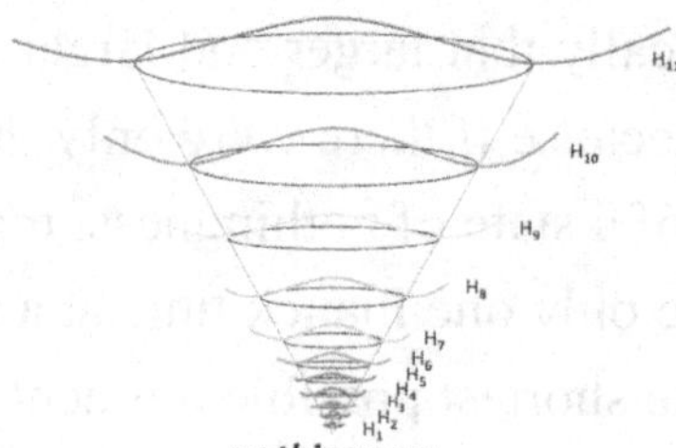

Then, since time or change can only come from a single place, nothingness (because existence seeks just the opposite, not to change), it means that to form a present of H_1, a certain amount of fundamental presents of H_0 must be accumulated. Enough to form H_1. The same happens with the presence of H_2, which must accumulate a certain amount of presents of H_1, and so on. In this way, nothingness would be pumping time, ordering changes to the rest of the holons, allowing ever larger presents to form. Therefore, the expansion of time would be vertical, originating in the nothingness and propagating to the higher holons. It would not be a horizontal phenomenon, as traditionally thought in physics, in the sense that there is a continuous line of "nows" or presents, side by side, extending into eternity in both directions (all of the past and all of the future). In contrast, the presents would be arranged vertically, one above the other. So, there would always be presents, only some larger or more permanent than others.

To better see how holistic presentism brings physics back to sanity with respect to time, let's take an example. A person's present corresponds to approximately 200 milliseconds. This would be our "now." The fastest time or the shortest change we can react to or be aware of. But, for our present or instant to occur, many cellular processes must take place, especially in our neurons. From the moment an external stimulus is detected by our senses, until that signal reaches the brain through the nervous system, the neurons analyze it, generate a response, and that response is sent to the muscles to be executed, for example, to dodge a ball that is headed right towards our face. Therefore, for our present to exist, many cellular presents must occur first. However, for the present of each of those cells to occur, many chemical reactions must also occur, especially in the production of proteins. In other words, the present of a cell is made up of many molecular presents. And for chemical reactions to occur, many atomic presents are needed, and these subatomic presents, and so on. Until, according to Darwinian Physics, we would arrive at nothingness itself. Then, we, as complex beings, are the result of a superposition of multiple holons: a human layer, a cellular layer, a molecular layer, an atomic layer and a subatomic layer, until we arrive at nothingness itself. And each of these layers has its own present. In this sense, we are interholonic beings, analogous, again, to matryoshkas or Russian dolls.

A temporal holoarchy that does not apply only downwards, because we can do the same exercise above upwards as well. The present of the planet would correspond to each year, because it breathes along with the changes of seasons in each hemisphere (it inhales CO^2 in summer through the production of plant material, and exhales CO^2 in winter with the decomposition of that plant material). The solar present would be eleven years, according to its activity cycle, and a solar year would be 250 million years, when it completes one revolution around the galaxy. Meanwhile, galaxies revolve on galaxy clusters, and these on galaxy superclusters, and so on until reaching "g." The formation of the present of the whole cosmos, when the universe itself completes its own present. One that would still be in process, fortunately. In this way, Darwinian Physics reveals a universe with a structure similar to a clockwork mechanism, where the present of H_0 (nothingness) would be equivalent to the engine that moves all the major gears. Continuing with the analogy of the clock, nothingness would be, for example, the gear of the deci-seconds; while the present of H_1 would correspond to the gear of the seconds, the present of H_2 would be the minutes, the present of H_3 would be the hours, and so on. In this way, larger and larger presents and higher holons are being constructed from the lower presents and holons, along with a universe that ages and grows, and that also becomes more and more dynamic, varied and interesting.

This new cosmovision of time would be supported not only by our common sense, but also by the foundations of physics itself, which provides much evidence in its favor. All of this will be explored in depth in the third part of the book, but as a preview, we have said that the present is the vibration of the holons themselves. But what is it that vibrates, what do you mean by this wavelength, this wave? Well, this wave would correspond to light itself. Although it sounds very incredible, it would actually explain in a natural way all the peculiarities of light, for example, the reason why it comes in discrete packets called photons, which would correspond to those present in each holon. A characteristic that fits perfectly with the fact that the most energetic photons are those with a shorter or smaller wavelength, since they would belong to lower or smaller holons, and therefore closer to nothingness. And since time is created in the nothingness, it makes sense that the fastest, shortest, or most energetic changes would occur closer to the nothingness. The less energetic photons, on the other hand, are those with longer or larger wavelengths. This is consistent with the fact that they belong to the larger holons, the higher realities that are farther away from the nothingness. Remember that there are photons of meters, kilometers, and even the size of a galaxy, such as very low energy radio signals; along with photons the size of a molecule, an atom, and many smaller ones, such as high energy X-rays and gamma rays. Then time would be waves pumped out of nothingness, which we perceive as light of different wavelengths. This explains the most surprising property of light, its invariant speed. Something that no physical theory has been able to explain. It is only assumed as a principle in Einstein's theory of relativity. On the other hand, Darwinian Physics provides a very natural reason, and that is that light moves at the speed of the fundamental present, that is, the fluctuations of nothingness. Therefore, nothing can go faster than light, because nothing can go faster than time itself.

But the presentism of Darwinian Physics would not only fit with special relativity. As we already know, Darwinian Physics gets along quite well with quantum mechanics, and in the case of presentist holoarchaism it is no exception, since it is fully consistent with Heisenberg's uncertainty principle. Specifically, with the fact that uncertainty increases as we move down to smaller or more precise scales of space and time. As always, this will be elaborated in Part III, but for now we will say that this principle of quantum mechanics (or theorem) is perfectly compatible with holistic presentism, since it would be equivalent to descending into smaller or lower holons that have shorter presents (more precise times). This compatibility is even more obvious with the relationship between energy and time, since the shorter or more precise the time scale, i.e. the shorter or more precise the present, the more indeterminate and therefore the greater the energy can be. Something that makes a lot of sense as we approach the space-time point of nothingness. Furthermore, holarchy could explain why these probability clouds that we call electrons are arranged in layers (orbitals) around the nucleus of atoms, since they correspond exactly to these holons. In other words, we would have an explanation for the famous quantum jumps that electrons make when they go from one orbital to another, since there is literally nothing between the orbitals, so they cannot go between them. The same holoarchical explanation for the fundamental element of the Standard Model of Particles, the quantum fields that fill all space. Thus, electrons are identical to each other (as is every quantum particle) because they are particular manifestations of the same object. For the Standard Model, a quantum field; for Darwinian Physics, a holon. In fact, quantum mechanics is already holistic, since it is not valid to analyze only an isolated part of a quantum system, but one must always consider the system as a whole. This is why quantum mechanics is non-local and produces such impressive phenomena as entanglement, because the present manifests itself from below and simultaneously in the whole holon. All this will be explained in more detail in the next part of the book.

And of course, if Darwinian Physics is valid, it cannot only be consistent with special relativity and quantum mechanics, it must also be consistent with general relativity. And as we have seen with the inclusion of the vertical dimension, there is great hope that it will be, because holoarchy can explain the fundamental element of Einstein's gravity, space-time. This is due to the fact that each holon has its own present, its own time. Therefore, this intimate relationship between space and time that Einstein discovered is due to the holon-present relationship in Darwinian Physics. Then everything indicates that this third vertical spatial dimension, product of the accumulation of holons, would have all the characteristics to explain Einstein's gravity; because the emerging holons would be essentially two-dimensional, inheriting this duality of nothingness, product of Heisenberg's uncertainty principle. On the other hand, this third spatial dimension would be special because, unlike the horizontal spatial dimensions, it would not be symmetrical. In other words, it is the same to go left or right, forward or backward; but it is not the same to go up or down, due to the difference in size between holons and their presents, which is larger the further they are from nothingness. A difference that we curiously find in reality and call gravity. A gradient in the holarchy that would explain the Lorentz contraction (difference in size of the holons), the time dilation (difference of the presents between the holons) and the curvature of space-time.

On the other hand, at a smaller scale, where there is very little gravity, physics becomes two-dimensional or flat, as rightly described in quantum field mechanics, due to the low accumulation of holons resulting from the smaller scale. All of this will be explained in much more detail in the third part of the book. But for now, suffice it to say that this simple pilot model has the potential to explain the emergence of space-time and the properties of quantum mechanics in a unified way. In other words, it would make it possible to achieve the holy grail of physics, quantum gravity or the theory of everything.

And if Darwinian Physics really is the candidate to be the theory of everything, then it must solve the great mystery of modern physics, dark energy. An accelerated expansion of the universe that puzzled physicists in the 1990s because the effect of gravity should slow down the initial momentum caused by the Big Bang, just as a ball thrown into the air slows down as it rises. An observation that no one knows why it happens. It was simply modeled by adding a new parameter to Einstein's equations of gravity, the famous cosmological constant or Λ. However, as we already know, all the evidence suggests that dark energy is not constant, but is growing with time, because the acceleration of the universe today is more intense than in the earliest universe. This is the only possible solution. Now, this unexplainable phenomenon for current physics would be something easy to explain under Darwinian Physics, in fact it would have been a prediction, because the emergence of new holons explains this expansion of the universe, because it is literally the creation of more space out of nothing. This is how dark energy is interpreted in current physics, because there are galaxies moving away from us faster than the speed of light. Something that clearly violates Einstein's special theory of relativity. But if the galaxies are not moving, but it is the space between the galaxies that is growing, then it perfectly explains the reason why the most distant galaxies appear to be moving faster than light, because there is simply a lot of space being created due to the fact that there is a lot of distance separating them from us. So even though it is a ridiculously small amount of space that is spontaneously created in one kilometer, about the size of a proton; over hundreds and billions of light years, that amount of space coming out of nowhere accumulates enough to push the more distant galaxies away faster than the speed of light. This creation of space out of nothing makes dark energy the most difficult problem in modern physics, because it is a phenomenon of creatio ex nihilo. Something completely unacceptable in the current paradigm, but which would be naturally explained in Darwinian Physics, due to the conservation of the resonances of the fluctuations of nothingness, which would literally inflate the universe with new holons. A phenomenon that is very far from being explained in the current paradigm, where it is normal to have an absolute, eternal or static universe, as Einstein wanted. On the other hand, in Darwinian Physics it is normal to describe a finite universe, with a beginning and in expansion. Exactly what recent cosmological observations indicate.

Again, all of the above will be explained in more detail in Part III, but there remains a very important open question for the pilot model of Darwinian Physics, since our universe is not only ancient and gigantic, but also very varied and diverse. How, then, could multiplicity have arisen from unity, complexity from simplicity? Again, the Darwinian mechanism comes into play, because holistic equilibrium serves not only to expand the cosmos, but also to increase its complexity. To understand how this is possible, we must first understand what we mean by complexity. To find out, let's use ourselves as an example. We are somewhat complex because we need a delicate balance in our temperature of 36°C to stay alive. If we raise or lower that temperature by even 3°C, we upset that balance and put our lives in serious danger, either from hypothermia or fever. Other examples are the stars (like our sun), which are the product of a balance between gravity pulling inward, compressing and concentrating gas, and nuclear energy pulling outward, exploding. Or our planet's atmosphere, which maintains a delicate balance of CO^2 concentration. Too much concentration and our planet would become an inferno like Venus, or a little less and we would freeze in a new ice age. And so on with many other examples in living things, such as glucose balance, blood pressure, acids, water, calcium, salt, and so on. In conclusion, we can summarize complexity as homeostasis. A balance between two extremes, between two opposites.

Far from being a very strange definition for physics, this holistic equilibrium of complex systems is very well known, but obviously by another name: critical point. The moment when a phase transition occurs in a system, for example from liquid to vapor in water as its temperature rises. However, this concept is being used more and more in science in general, as such behavior has been discovered in several very different systems, such as chemistry, biology, geology, economics, or sociology. These are very different phenomena, but they all operate around this point, where small changes in some critical environmental variable lead to drastic and discontinuous changes in the behavior of such systems. A simplified system typically used to represent this phenomenon is known as the Ising model, which visualizes the iron atoms of a magnet as arrows indicating the direction of each atom's spin. In this way, the system can be easily represented as if it were a chessboard, where each square would have an arrow pointing up or down. This allows us to understand very well how a magnet works, because when it is cold, it means that all the arrows are pointing in the same direction, creating this magnetic property. But if the same magnet is heated, the arrows will start to rotate due to the extra energy. This will cause the arrows to become misaligned, pointing in different directions in a chaotic manner, neutralizing each other, causing them to lose that uniformity and therefore that magnetic property. It is as if the arrows had values of "1" and "-1," so if we add them by flipping a coin to choose which of the two values to add, the tendency will be that the same amount of "1" and "-1" will appear very close to each other, so the sum will be very close to zero. This is exactly the value of the magnetic property that a hot magnet would have. Then you get a phase transition, going from a very homogeneous configuration (cold magnet) to a totally disordered one (hot magnet). However, right in the middle of this transition is the critical point where the arrows or spins form similarly oriented clusters along the lattice or system. If the sizes of these clusters were measured as a function of scale, the data would show a so-called power law, where the dynamics at one scale mirror the dynamics at other scales. This phenomenon is also known as scale invariance, which is another way of saying self-similarity or fractality. Precisely what the emergence of holons from nothingness represents, thanks to holistic equilibrium. In fact, the latter could explain the former, because when a system reaches the critical point, it shows a maximum peak in the so-called correlation length. This is an indicator of how sensitive the system as a whole is to the activity of any one of its components. What happens is that the system behaves in a way that allows fluctuations to occur on the scale of the whole system. So, going back to the magnet example, if the temperature is cold, there would be no correlations because all the arrows would not move because they would always point in the same direction. And when it is very hot, the arrows or the spins of the atoms move much more, but not in a correlated way, they are chaotic movements. But right in the middle between order and chaos, these interactions are generated at all scales, which is very surprising because the distances over which these spins could interact are technically infinite. In other words, it is a holistic or global phenomenon. The spin motion of one atom could cause the spin of another atom in another distant part of the magnet. So, a single atom can set off a cascade of events that would propagate through the entire system. Again, something very difficult to understand under the current paradigm, but which would be coherent in Darwinian Physics, since holistic presentism has exactly that behavior. A present that emerges from the smaller or lower scales to the larger or higher ones, instantaneously manifesting in the whole holon as its now.

In 1987, the Danish theoretical physicist Per Bak (1948-2002) wondered whether many different types of complex systems in nature could self-organize around critical points. To illustrate his theory of self-organized criticality, Bak used another simple example: a pile of sand. As a pile of sand gains mass, as happens in hourglasses, friction can no longer hold the grains in place, and a single extra grain of sand falling on top can trigger a mass effect that sends avalanches cascading down the sides. And it turns out that if you look at the size distribution of the avalanches (the big ones, the small ones, and the ones in between), they follow the power law. So, Per Bak's idea was that here is a natural system that organizes itself until the critical point is reached. We don't need to tune it there, as we did with the magnet, by relying on us to get the temperature just right in the Ising model. Instead, the pile of sand evolves into that state on its own. This gave rise to the concept of self-organized criticality. A concept that also applies to the first pilot model of Darwinian Physics, since existence emerging from nothingness tends by itself to holistic equilibrium due to fluctuations of nothingness and causal inertia. If the entire cosmos functions in this way, it should not be surprising that many complex natural subsystems fall into this category, such as earthquakes, stock market crashes, superconductivity, flocks of starlings, or social interactions. It would even apply to our brains through our neural networks, because the power law that occurs right at the critical point would allow small inputs of information (neurochemical signals) to generate large and complex behaviors throughout the neurological system. A process that would have the potential to explain consciousness, intelligence, or free will.

This intimate relationship between holistic equilibrium and complexity is further reinforced if we use another important branch of physics: thermodynamics. This is because complexity in physical systems arises in the middle of the entropy increase process, between minimum and maximum entropy. Again, at the critical point, because at either extreme, when there is very little entropy, like a crystal, or a lot of entropy, like dust scattered in the air, the system is simple, with very little complexity. A simple example of this would be a latte macchiato, which has a low entropy state when the coffee is clearly separated from the milk. And if we wait a while, we will see the result of the second law of thermodynamics, a high entropy state, where the coffee and milk are completely mixed and have a uniform color (light brown). In both cases we have simplicity, either when the two liquids are separated (minimum entropy) or when we have a uniform liquid (maximum entropy). In the middle of this process, however, complexity arises when an entropic equilibrium occurs between its minimum and maximum levels. It is there, when milk and coffee mix, where the interesting unique and chaotic patterns of beautiful and mesmerizing spirals and swirls are formed.

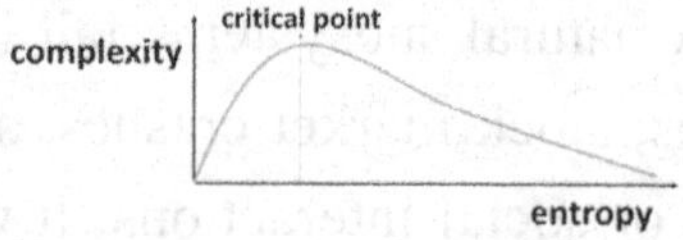

This is the key to understanding complexity, for the evolution of the cosmos in Darwinian Physics describes a similar process in which causal inertia is constantly increasing. In fact, in Chapter XII we will see that causal inertia has much more in common with thermodynamic entropy than it may seem to us now, because both seek the same end, a state where there is no more change or variation. In other words, thermal death is equivalent to absolute being in Darwinian Physics, when "t = g." So this equivalence between entropy and causal inertia makes it so that when a holistic equilibrium or critical point occurs, not only does a new holon emerge, but complexity also emerges, since it is the product of a balance between cosmos and nothingness, order and chaos. All because complex things are a combination of both. A mixture between the dynamics and mutability of nothingness and the stability and conservation of being. A middle state between low entropy (like milk and coffee separated in a latte macchiato, the equivalent of two states of nothingness) and a state of high entropy (where milk and coffee are completely mixed, the equivalent of a single state of nothingness that always appears). It is there, in the middle between these two extremes, that complexity arises. A balance between variability and monotony, randomness and causality, instability and permanence, nothingness and being.

In conclusion, the mechanism of holistic equilibrium provides an explanation not only for the expansion of space, but also for the emergence of increasingly complex realities, since it would describe the creation of systems that are out of equilibrium. The definition of complexity in physics. This is because complex things have a lower level of entropy compared to their surroundings. A clear example is life. Biological organisms are systems that try to maintain a significantly lower level of entropy compared to their environment. Only when these organisms die and reach thermodynamic equilibrium do they equalize their temperature with that of the environment (maximum entropy). And although, strictly speaking, life does not violate the second law of thermodynamics, because it applies only to closed or isolated systems (and the entropy of the solar system is always increasing), it does not explain how it is possible for some physical systems not only to reduce their entropy locally to an incredible extent (to become more complex), but also to maintain this state in such a stable way, as has happened with life on this planet for the last 4 billion years. Especially considering that there is no law of physics that requires or dictates this. This is very important, because many physicists keep repeating that life does not contradict thermodynamics, because the second law does not prevent the appearance of very complex things with very low entropy, as long as it is local, because this law only applies to closed systems or globally. This is technically true, but although it does not prevent it, the truth is that it does not help or promote it very much. In fact, it complicates it quite a bit. So, the fact that life has arisen, and that it has continued to evolve and increase in complexity for billions of years in a solid and constant way, is a great anomaly. It is like winning the lottery 1,000 times in a row. Does it violate the laws of probability? No. But it is undeniably an impressive statistical anomaly that is very difficult to justify without a mechanism behind it. Well, in the case of life, it is exactly the same. The second law of thermodynamics simply does not cooperate for biological organisms to appear and evolve, on the contrary, it tries to destroy them by increasing their entropy.

Therefore, life is a dilemma for thermodynamics for two main reasons. First, while it is true that a complex configuration of organic molecules could appear to form the first protocell simply by chance, it would be an extremely unlikely process. And yet, once the conditions for life were present on the planet (that is, it had cooled sufficiently), life flourished on it. This is the equivalent of shuffling a deck of cards and having them appear in perfect order, of the same color and in numerical order. We could be shuffling cards for the entire history of the universe and still not find that combination, and yet in the case of the appearance of life on Earth, that seems to be exactly what happened, and in the first attempts. And secondly, although thermodynamics does not prevent the above, the creation of that incredible stroke of luck called the protocell, thermodynamics does not promote its survival and evolution. On the contrary, it tries to extinguish it, to kill it, to raise its entropy, not to lower it even more. In other words, statistical anomalies are not conserved, or else they would not be anomalies by definition, since they are transformed into the opposite: patterns or regularities. Well, the evolution of life would be just that for thermodynamics, a series of lucky breaks, like constantly winning the lottery 4 billion years ago. This is not statistically impossible, but it is not justifiable. Clearly, current physics cannot explain the emergence and development of life by chance alone, because it requires a mechanism, a process that learns from itself. In other words, evolution. Something obvious in biology, but very difficult to integrate into physics, because it is totally foreign to what its classical or Laplacean world view preaches.

A completely different scenario from that of Darwinian Physics, since far from being something exotic, the emergence and conservation of increasingly complex phenomena is a process that is not only natural, but inevitable. The product of this emergence and conservation of new holons is the accumulation of holistic equilibrium. The same mechanism that allows the previous holon (the underlying existence) to continue to increase its causal inertia (entropy). So, it is a physical mechanism that had to be added, because thermodynamics alone makes the emergence and especially the development of any kind of complexity, including obviously the biological one, practically impossible. On the other hand, Darwinian Physics offers this natural mechanism of emergence and evolution, not only of life, but of any kind of complexity. This is without contradicting thermodynamics, because the increase in the degree of complexity at the local level is a by-product of the central purpose of the cosmos, which is to exist more and more and to increase its stability as it moves further and further away from nothingness. A purpose that is equivalent to the increase of entropy at the global level, because both mean less and less change. An equivalence that we will explore in much more detail in the third part of the book, especially in the chapter on time. But to return to the previous topic, under this new paradigm we would achieve what is impossible at present, to unify not only two of the most important sciences: biology and physics; but all sciences, by having a universal mechanism of emergence and evolution that can integrate them all. The central topic of the next chapter.

CHAPTER X

THE REVENGE OF HERACLITUS

We have a very simple pilot model of Darwinian Physics that, although it needs much more development, is capable of dealing with the emergence not only of time and space, but also of complexity. This great feat is made possible by a process we will call holistic evolution, which essentially works in two ways. The first is Horizontal Holistic Evolution (HHE), which is responsible for the fact that this insignificant existence arising directly from nothingness can stabilize and become more and more certain by means of causal inertia or $\exists(t)$. The second process is Vertical Holistic Evolution (VHE) and is based on the second type of existence, that which arises thanks to holistic equilibrium. This second evolution is responsible for the emergence of new holons, which allow the cosmos to expand or grow spatially and at the same time increase the degree of complexity of existence. A very important evolution, as we saw in the previous chapter, since it would finally justify not only the increase of complexity in all of nature, but also the apparently strong emergence or appearance of phenomena that are truly novel in the sense that they are not determined, as new holons are created and literally emerge from nothingness. Thus, holons appear spontaneously, with a true degree of independence or freedom, thanks to the random component in the holistic equilibrium.

In this sense, in order to understand well how authentic new things appear and how they can change and evolve over time, we must first solve an ancient enigma, the paradox of Theseus' ship. This is a Greek myth, recorded by the philosopher Plutarch, that tells of the ship in which Theseus returned from Crete after confronting the Minotaur. The story goes that the Athenians kept this ship for centuries to commemorate such a feat. They did this by replacing the old and rotten planks with new ones as they deteriorated. The problem arises when, after a long time, the time comes when all the original parts of the ship should have been replaced. The question then arises: is it still the same ship of Theseus? And if not, at what point did it cease to be? It is a metaphysical question that asks us what identity is. This is important because, like Theseus' ship, our cells are constantly being replaced, with virtually all of them being replaced within eleven years or so. So, are we still the same person? There are several answers to this conundrum. For example, we have mereological essentialism, which says that any composite object is integrally formed by all its parts. Therefore, it does not allow for any substitution, no matter how minimal. This means that Theseus' ship ceased to be a ship the moment the first plank was replaced; or you ceased to be you just a moment ago, because millions of your cells have been replaced by new ones. Not a very reasonable solution. Another solution would be gradual identity, which says that something is according to the percentage of elements that remain. Therefore, Theseus' ship would only be 99.99% authentic after the first board was replaced. A percentage that would decrease as more and more planks were replaced, until there would be nothing left of the authentic ship when the last plank was replaced. However, this is also problematic because we know that living things constantly replace their cells. Therefore, no living being would be the same as when it was born, because those cells from that time are not preserved. Since this still sounds absurd, we have a third option, which is called mereological nihilism, which says that we should not complicate ourselves so much by fixing a degree at which things are, because simple composite things essentially do not exist. There are essentially only the parts, the components. A solution that is very compatible with the radical reductionism of the current foundations of physics, which declares that essentially only the fundamental particles (quantum fields) exist. The building blocks that are neither replaced nor degraded, but remain forever due to the principles of conservation (absolute law). Therefore, it does not matter which things they form: ships, dogs, or people; because they all assemble and disassemble, but what will always last are those fundamental particles. In this way, macroscopic things have no essence in themselves, since they would be simple aggregates, epiphenomena of what is really essential: particles or quantum fields. The problem with this solution is that it seems that such constituents turn out to be smaller and smaller objects. First it was elements or atoms. Then it was subatomic particles like the proton. Now it is the quarks, but even more fundamental elements are being postulated, such as the famous strings, which would be billions of times smaller. In this way, only the strings as such would exist, while all the rest would essentially not exist, since even an electron would be an epiphenomenon. This is the result of extreme reductionism.

So, to stop this reductio ad absurdum of postulating something smaller and smaller, while denying all the substance and essence of what is above (atoms, molecules, dogs, planets and galaxies), Darwinian Physics proposes another solution, one we will call causal continuity, which states that things are as long as they continue to be, in the sense that they continue to repeat the same pattern without interruption. Thus, a thing can continue to be what it is, even as it changes, evolves, and replaces its components, as long as there is no cessation or interruption in the essential pattern of such an object. And since this capacity exists in all things at all scales, from quantum particles to stars and clusters of galaxies, there is true substance and essence in all of them. In this way, we are still ourselves, even though we are very different from our "baby self" and our "adolescent self," because our essential pattern, our main sequence or process, has not been interrupted. We have just evolved. Something that, by the way, is not at all incompatible with permanence, on the contrary, because in order for something to develop or evolve, it is necessary for such a thing to remain so that development, evolution or simple change can take place in it. For example, how can a plant grow and develop if such a plant is not preserved?

However, we can ask another question: what exactly is the essential regularity or pattern that allows an object to change, what is it that remains? To find out, we must distinguish between two types of properties: essential and accidental.

The latter is what can change without the object ceasing to be what it is. For example, a cat sits on the couch. However, it is now walking on the floor. The fact that it is no longer on the couch does not make the cat cease to be a cat, since "being on the couch" is not an essential property of the thing cat. Thus, the essential pattern in Darwinian Physics would be that which is most representative of that particular object. Therefore, if this information is preserved over time, without interruption, the object would remain the same. In this sense, the ship that has been continuously preserved and restored would be authentically that of Theseus, because the essence of the ship has been preserved, its origin, form, material, history, and meaning. A continuation that has not had an abrupt interruption, such as a fire that has completely destroyed it, forcing it to be rebuilt from scratch. In the case of living beings, this fact is much simpler, since we are as long as we live. If our essential biological functions and activities are interrupted, we cease to be, because the moment they cease, we die. A fact that is also completely irreversible. So that regularity or information has been destroyed, it has returned to nothingness. Obviously, to make it more realistic, we have to include the holarchy, which in our case includes the cultural, family and personality layers, in addition to the biological. This is because we would be the same entity biologically because we have not died, but an accident or a disease such as dementia could kill our personality. This would cause us to cease to be ourselves behaviorally, although we would still be the same biologically. So, the definition becomes more complex because reality is complex. This will be elaborated on later, but for now, to simplify the definition, objects are until they cease to be. A conclusion that follows from the very definition of existence in Darwinian Physics as regularity or successive repetition of a state of nothingness. And as mentioned in Chapter VIII, the interruption of that pattern means the death of that existence. The end of that being and its return to nothingness.

This is a worldview that is radically different from what we find in modern physics, because according to its determinism, the real cannot die, because it has never been born or created, since it is eternal. As the famous mantra of the conservation of energy says (which we already know is not fulfilled on cosmological scales): "Energy is neither created nor destroyed, it is only transformed." Thus, in Parmenian physics, everything that exists is part of the same pattern, the same history, the same divine plan, determined by the laws of physics. Therefore, there can be no surprises, no novelties, no spontaneity in an ontological sense, but only in an illusory or epistemic sense. The product only of our ignorance of this plan, of not knowing what the real laws are and what the exact initial conditions were. A radical position, which is due to the exclusion of nothingness, because if there is no nothing, there is only eternity. Or, equivalently, if there is no randomness, there is only absolute determinism. On the other hand, Darwinian Physics has a much more realistic or moderate position, since, thanks to the recognition of nothingness, one can naturally accept the existence of a being, essence or substance in all things; because they always appear, to a certain extent, in an authentically spontaneous (non-determined) way. That is why they are true novelties, because, thanks to the nothingness, new regularities are created, which are always, to some extent, different from any previous pattern or regularity. That is why not everything is written in the universe, because chance in the "now" can interrupt or create new patterns or things. Similar to how random mutations create truly new, unique and unrepeatable genetic codes from pre-existing DNA or RNA sequences. A scientific fact in biology, but one that Darwinian Physics seeks to apply to physics as well. A natural mechanism that would finally provide an explanation for the strong emergence discussed in Chapter VI; because by recognizing the existence of a real randomness or nothingness, it can allow the emergence of a genuine novelty, an authentic essence that cannot be reduced, much less predicted in its totality. Thus, even if the degree of determinism increases (causal inertia) as we move up to higher scales and as the universe ages with time (HHE), nothingness has not completely disappeared because existence has not yet evolved to absoluteness ($t = g$). Therefore, there is still room for spontaneity and novelty, because not everything has been planned from eternity, as is preached in the current foundations of physics.

So Darwinian Physics allows us to explain the emergence of new things. But just as nothingness gives life, it also takes it away; because there is always the probability that such a regularity will be interrupted at some point, causing the things that have emerged to return to nothingness. Something that can also be observed in nature. Its cycle of life and death, of creation and destruction. However, despite this exuberant dynamic of emergence and extinction of essences, the cosmos remains an integral whole, a universe, because a fundamental continuity is maintained thanks to holarchy. Thus, although there are holons that are created and others that disappear, this happens mainly in the higher, more complex holons (more random characteristics of the cosmos). Therefore, the lower holons (more essential characteristics of the cosmos) that support the rest of the higher holons are still preserved. This makes this structure or holoarchy very important; for it allows the cosmos and the things within it to change, vary, evolve, and yet remain one and the same. In the same way that one remains the same person, even if there have been many changes since our childhood (hairstyle, clothes, friends, tastes, residence, work, and so on). So, the cosmos is still one, just made up of multiple layers (holons) like an onion (paraphrasing Shrek). This forces us to consider the whole holarchy when studying anything, because its location among the other holons matters, because the superposition of different holons with different holistic equilibrium creates this complexity. For example, the atomic holon would be the result of the superposition of several electronic holons (electron levels or orbitals) with nuclear holons (quark levels or orbitals). Or the holon of a multicellular organism would be the fruit of several cellular holons and these of molecular holons. Thus, the increase of complexity would be an interholonomic phenomenon that emerges transversely from the set of holons. This is why the vertical dimension is so relevant, because this is where all holons are located. A direction that we already explained in the previous chapter, which would correspond to up-down, and which would be different from the horizontal dimensions (left-right and front-back) because it is fractal in nature due to this accumulation of holons of different sizes. Therefore, the vertical dimension is a matter of scale, from the smallest (nothingness) to the largest (the cosmos). To use an example, if the horizontal dimensions correspond to the movement of a ball to the left or backward, the vertical dimension would correspond to the shrinking or enlarging of that ball.

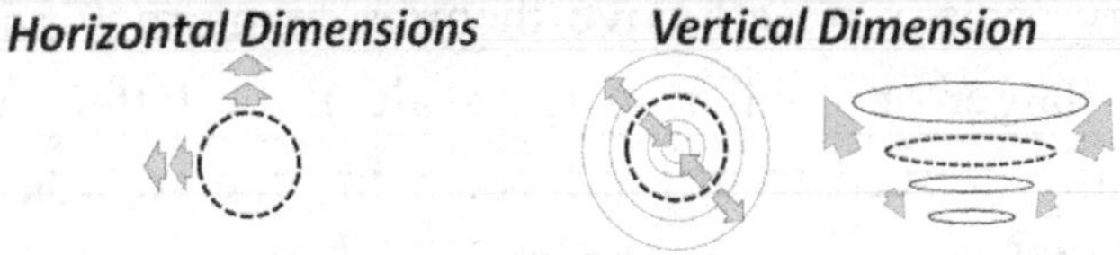

Therefore, the vertical dimension tells us what happens or how existence changes as we go down or up the scale. Or as we would say in Darwinian Physics, when we go from lower holons (closer to nothingness) or higher holons (further from nothingness). This is very important for modern physics, because reality behaves very differently depending on the scale at which it is found. A characteristic of our universe that was recognized and even awarded a Nobel Prize in 1982, thanks to the work of the American theoretical physicist Ken Wilson (1936 - 2013), who formalized the fact that the old theories of physics became effective theories of the newest theories, which became the new fundamental theories. However, he emphasized the important point that these new fundamental theories were more valid at higher energy scales or smaller distances (both equivalent), where the previous theories ceased to be valid. In this way, he was able to show how different field theories relate to each other, formulating the theory of theories, because each theory has its typical distance or energy scale. The higher the energy or the smaller the distance (lower holon), the more fundamental the theory. This means that Darwinian Physics would be the ultimate theory, the most fundamental of all, because it starts from nothing. The largest possible energy scale, because it is the smallest possible space-time scale, where time itself comes from.

But most importantly, Ken Wilson's theory supports holarchy and vertical holistic evolution in the sense that it recognizes that while less fundamental theories (smaller or larger energy scales) are effective theories of more fundamental theories (larger or smaller energy scales), the latter cannot be analytically derived from the former. In other words, a more fundamental theory cannot be derived from an effective theory. Something obvious, since with any of the old theories of physics we could have mathematically deduced all the subsequent theories up to the theory of everything. Something that obviously has not happened and does not happen. That is why physicists have not been able to derive quantum gravity from general relativity and quantum field mechanics, even though we now have impressive mathematical development and computing power through supercomputers. In contrast, it has always been incredibly difficult to obtain a more fundamental theory through a paradigm shift, which requires a scientific revolution. Ken Wilson explains this by noting that many of the details of a parent (more fundamental) theory are forgotten by its daughter (effective) theory. In other words, it is not necessary to know the foundations of physics to understand the rules of an effective theory that applies on a larger space-time scale. A fact that applies not only to physics, but to science in general. Thus, it is possible to divide nature into different sciences that are pragmatically independent of each other, because there is a real discontinuity between scales. This is precisely what allows an extreme specialization in each science and in its different subfields, being a very independent system that can be studied without the need to know the underlying laws.

This discretization between scales is again a scientific fact that is impossible to understand under the current Parmenian or Laplacean paradigm, which conceives the evolution of nature in a fundamentally continuous and predictable way. A totally determined unfolding of the universe guided by absolute laws. In Darwinian Physics, on the other hand, different laws or regularities between scales, which cannot be derived analytically from each other, are obligatory, since here strong emergence is recognized. A real complexity based not on deterministic chaos (weak determinism, as we saw in Chapter VI), but on non-deterministic or truly random chaos. In other words, the emergence of new, larger and more complex holons is an irreversible process. Therefore, laws at different scales cannot be directly derived because of the ontological randomness component present in the holistic equilibrium. This should be evident from the random collapse of the wave function and the Heisenberg uncertainty principle of quantum mechanics, which in principle make perfect precision and prediction impossible, destroying the possibility of weak determinism or exclusively epistemic randomness.

Therefore, the fact that effective theories give no hint as to what their fundamental theories would be is a fact that supports the existence of strong emergence. The spontaneous appearance of authentic new regularities in nature at every scale (micro, meso and macro), because nature is being created in the now, in every instant (holistic presentism). New patterns that, in order not to be completely derived from each other, have to emerge with some degree of real randomness. There is no other way. In fact, this is exactly what we can see in the current applications of artificial intelligence, for example, in the creation of authentic new images, photographs or paintings created by this new technology, which, moreover, are unrepeatable. And how do they do it? Exactly with randomness. For example, the famous DALL-E starts its deep learning work with an image of randomly distributed pixels. Thus, the images are never repeated because the AI starts with a single randomly arranged background. In fact, this would be the same way our creativity would work in the brain, but we will leave that for the next book, when the defense of the existence of free will is addressed. Another natural phenomenon that Parmenian physics denies of its existence ideologically, just like time, quantum randomness and strong emergence; because the only way to explain them is by recognizing ontological randomness or nothingness.

But returning to the subject of this chapter, we now understand why it has been so costly to achieve a new paradigm in each epoch. Something that took hundreds and even thousands of years, because it required a very disruptive innovation, because the fundamental theory that was being sought had no direct relation to the effective theory that was available. A more fundamental theory that also turned out to be simpler, with fewer elements; against the common sense that something better, that explains more things, should be more complicated, with more components. Exactly as the foundations of physics still thinks today, insisting on adding more particles, more fields, more dimensions, and even more universes. This was the key that many theoretical physicists omit or ignore, which explains the lack of progress on the main problem of the foundations of physics: the unification between quantum mechanics and general relativity. Very different laws, because the small one has a much more discrete and stochastic behavior, while the large one is much more continuous and deterministic. A big difference that has confused and frustrated theoretical physicists, but which is naturally explained within holarchy, because the smaller the holon, the closer it will be to nothingness, to the point of space-time; therefore, the more notorious will be the stratification of existence in holons (quanta and fields), and the more indeterminate will be reality (Heisenberg's Uncertainty Principle). This is why, for example, on the small scales of nature we find spontaneous and random quantum jumps, whereas on the higher scales, far from nothing, nature looks smooth and continuous because there is a much larger accumulation of holons, together with a more determined behavior (greater causal inertia). So clearly the vertical dimension is an essential ingredient, as it is the key to correctly interpreting the two pillars of modern physics. Two very different theories because they describe very different scales, but which can harmonize perfectly under this new paradigm. This is the central theme of the third part of the book.

But not only the vertical dimension would be important to unify the foundations of physics, but it would also be very important to relate all sciences in a coherent way, since the vertical holistic evolution does not only bring forth physical holons, but all the holons that constitute nature. To understand this, and to avoid being confused by so much new terminology, let us consider the word reality as a synonym for holon; while the cosmos will represent the set of all these realities or holons, the "whole." And in addition, we will integrate a new concept, vertical distance, which refers to the distance in scale between two particular holons or realities. Thus, since complexity arises from the accumulation of different holons with their own holistic equilibrium, the difference in the degree of complexity between two continuous or successive holons is very small. Therefore, the difference in complexity is only perceptible when a large number of holons are considered, especially when large differences in scale (orders of magnitude) are observed. Thus, the greater the vertical distance, the greater the difference in complexity. This is characteristic not only of Darwinian Physics, but of any kind of Darwinian evolution, since all such processes are characterized by transitions or gradual changes. For example, in historical evolution, no one went to sleep one night in the Middle Ages and woke up the next day in the Renaissance. However, over time horizons of tens of years, we can clearly observe a distinction between these two historical periods. The same is obviously true of Darwin's own evolution, since, for example, no female Homo erectus gave birth to a baby Homo heidelbergensis, but over time horizons of hundreds of thousands of years there was a transition from one species to another. The same thing happens with existence, but in its vertical dimension; so, we can take large sets of holons that are similar (in complexity) to each other, but significantly different from each other.

Thus, we can distinguish four major clusters of realities, which we will call main holons. The first would be the space-time main holon, which emerged directly from nothingness and is the lowest main holon of all. From it emerges, as its name implies, space and time, the fundamental reality that supports all other realities that emerge after it. The second is the physical main holon, the reality of matter and energy, which supports the living main holon, which is the biological reality or living beings. This in turn supports the symbolic main holon, the highest main holon, which corresponds to cultural reality or the world of ideas. Contrary to what Plato believed, however, this last reality would be the least fundamental, the most dependent of all. Each of these principal realities is more complex than the previous one because their information overlaps or accumulates. Therefore, in order to understand the symbolic main holon, we must also understand the biological reality, its mother reality; but also, more indirectly, the physical and space-time reality. To understand the living main holon, we must understand the physical reality, its mother reality; but also, more indirectly, the space-time reality. And to understand the physical main holon, we must understand the space-time reality, its mother reality, but also nothingness itself, from which the space-time reality and everything else originated. This creates a hierarchical relationship (holoarchy) in which the higher holons depend on the lower holons to be and to exist.

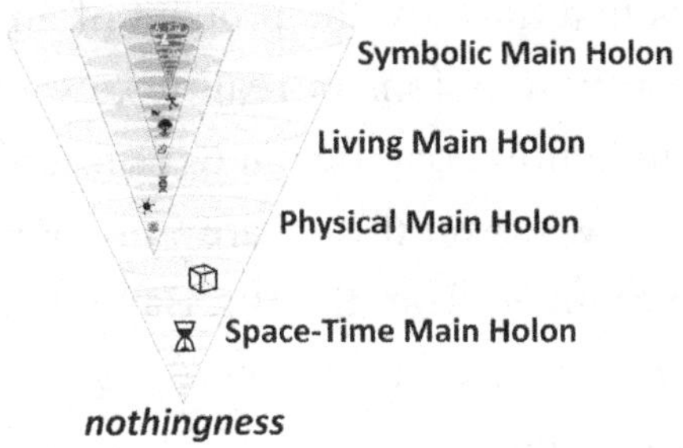

In this way, Darwinian Physics, through its holistic evolution, manages to explain not only the emergence of time and space from nothingness, but also the emergence of all kinds of complexity from simplicity. A feat that would explain the reason why physics has not been able to recognize transformational change, especially disruptive emergence, because it lacks an essential piece that is indispensable to understanding this phenomenon: nothingness; because the ontological recognition of randomness was not only necessary to understand and make sense of quantum mechanics, but it is also indispensable to understand the emergence of truly new, disruptive realities, as was the origin of the main holons. Realities that cannot be fully understood by reducing them to their parts through analysis, since the existence of synergy is omitted or denied. That is why we cannot reduce cultural reality (sociology) completely to people, because it is more than the sum of people. Nor can we completely reduce a person's personality (psychology) to a set of neurons. Nor can we completely reduce living beings (biology) to a mere conglomerate of organic molecules (chemistry), nor molecules to atoms, nor atoms to fundamental particles (physics), because we deny that these realities are unique in themselves, that they are holons. Therefore, the inability of the present reductionist-analytic approach to fully explain the principal realities supports the thesis that they are the product of an authentic strong emergence; that these holons or realities, especially the principal ones, are authentic novelties in the sense that they have an essence of their own. And if they are truly new, they cannot be completely determined from the Big Bang, because that would eliminate all spontaneity, since everything is predetermined. A spontaneity that is clearly observed in quantum systems such as radiation or the decay of more massive particles. Something that cannot be precisely predicted because there is no law that causes it, and it does not exist because it is not the product of anything existential, of any regularity. Therefore, they are intrinsically random phenomena because they are not caused by anything. Therefore, if we simply recognize randomness and spontaneity, as Darwinian Physics does, then we can explain the strong emergence and the existence of an authentic novelty, because there is a real causal separation, because a real randomness intervenes in the process. Then the parts (analysis) are not enough, not even the connections between them (synthesis).

It was necessary to consider randomness, because novelty, as complexity, is not only the fruit of the parts and their relations, but also the product of nothingness. A combination of causality and randomness that we called holistic equilibrium. In this way, the VHE would be the process by which new essences are created, authentic new things; because nature is simply creating itself in the now, evolving in every instant, as it cannot be otherwise in a presentist theory. On the other hand, if the present were an illusion, as current Parmenian physics declares, under an absolutist determinism, then there would only be aggregates of things. We would be just a very complicated collection of electrons and quarks that have always existed and will always exist in eternity. But if we simply recognize nothingness, there can be real novelty, because there would be real discontinuity. A before and an after by existing in the now. Returning the essence to complex things, especially to the principal realities, while saving the infinite instants that eternity requires and the infinite universes that absolute law requires.

All of this may seem very revolutionary, too innovative, but as we saw in Chapter IV, it is in a sense a very old proposal. In fact, Darwinian Physics would be a continuation of an old Western philosophical tradition that has unfortunately been silenced for much of history. A few brave souls have also challenged this dominant absolutist view that has dominated much of philosophy and science. For example, the pre-Socratic Anaximander of Miletus and his arché apeiron (the indeterminate). Something akin to nothingness, in the sense that there could be nothing known, no element. However, the one who comes closest to the proposal of Darwinian Physics in general was another important pre-Socratic philosopher who has been unjustly denigrated by history. We are talking about Heraclitus of Ephesus (540-480 BCE). A revolutionary philosopher of ancient Greece who postulated that the arché, the source or origin of all things, was fire. Thus, he declared that everything is "change," that everything flows (panta rei), because everything originates from the chaotic dynamism characteristic of fire. An origin very similar to the current cosmological origin, since the Big Bang was an incredibly hot event. At the moment, the temperature of the universe is 2.7° K (-270.45°C), the energy of the CMB; and it is evolving towards 0° K (-273.15°C) or absolute zero (absolute order). However, if we go back in time, the universe was not only more compressed or dense, it was also hotter, reaching billions of degrees Kelvin in the first moments of the Big Bang. Therefore, it is quite realistic to say that the cosmos was born of fire. However, the most similar theory of Heraclitus with the present approach was his logos (reason, sense or logic) of the struggle between opposites. A conflict between love and hate, cold and heat, light and darkness, and so on. Something very similar to the struggle between nothingness and cosmos, randomness and causality, the mutable and the permanent, uncertainty and certainty in Darwinian Physics. Heraclitus said that there is a constant struggle between these opposites. A war that produces becoming, the change of existence. However, contrary to what his critics interpreted, or rather misinterpreted, he did not deny any kind of permanence. In fact, Heraclitus said that permanence is the product of precisely this constant struggle between opposites. That is, what produces harmony, stability, and equilibrium is precisely the permanent conflict. For example, the harmony of the bow is the fruit of the constant tension between its two components. The frame, which tries to open, and the string, which tries to close.

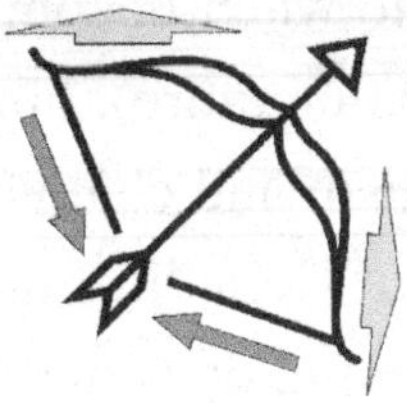

If the conflict were to disappear, if one of the two forces were to win, because the string broke, for example, the bow would cease to exist. The same thing happens with all the examples of homeostasis mentioned above, and in fact with all physical things; because all of them are the product of some equilibrium, some tension between forces, whether fundamental (such as the strong nuclear and electromagnetic forces in the nucleus of the atom) or between compound forces, such as pressure, chemical reactions, hydraulic forces, friction, and so on. Just as we saw in the previous chapter regarding the critical point. Therefore, complexity is nothing more than the balance between the existential, the permanent and stable, and the nothingness, the ephemeral, the changing and the random. Both are equally indispensable, since nothingness provides the dynamism characteristic of complex things. Its degree of freedom thanks to chance. While the existential allows for the continuity and permanence of such complexity. A delicate balance, because if any complex thing were more permanent or stable, it would be dead or frozen; whereas if it were more nothing or chaotic, it would decompose or disintegrate. This is because the properties of things require a certain balance in order to be, and when this balance is broken, because one of the two forces has overcome the other, that thing is extinguished and gives way to another thing with different properties, because it is in a different holistic equilibrium and therefore in a different holon or reality. For example, liquid water becomes ice when it acquires greater stability (lower temperature), but becomes steam when it acquires greater dynamism and change (higher temperature). For this reason, the logos of Heraclitus turns out to be quite similar to the holistic equilibrium, or rather the latter to the former, because from the balance between the opposites (randomness and causality), from the constant struggle between the power of order and chaos, new and complex things emerge. It could be said, then, that Darwinian Physics turns out to be a much more sophisticated and up-to-date defense of the ancient philosophy of Heraclitus of Ephesus; so, it is not surprising that this philosopher is considered the natural antagonist of the philosophy of Parmenides of Elea. For if the latter only recognized Being in his metaphysics, Heraclitus of Ephesus gave priority to Becoming. Thus, Heraclitus was a presentist because he believed that the only thing that exists is what we observe and experience, the now. In contrast to the philosophy of Parmenides and his current disciples, the theoretical physicists, who are eternalists; because they recognize only the being, immutable and absolute, where all past and future has always existed, because change and becoming are mere illusions.

But however absurd the foundations of the current paradigm may seem, the proposal of this book will start from a disadvantage, simply because of the lack of familiarity with it. For this reason, it would be pertinent to compare it directly with the current state of the foundations of physics and thus find out which of the two options is really less reasonable. To do this, we must use as our method of evaluation the new criterion of scientificity that we defined in Chapter VII: holistic consistency and its implicit principle of sufficient reason. From this point of view, contemporary Parmenian physics leaves much to be desired, not only because it is completely disconnected from the other sciences and practically incompatible with them, denying the emergence of novelty and free will, which are essential in biology, psychology, and sociology, but also because it denies fundamental and obvious elements of the foundations of physics itself, such as the becoming of time, because of its eternity, and quantum randomness, because of its determinism. Without forgetting the terrible consequence of infinite parallel universes. A situation that we already know does not improve under the principle of sufficient reason, since modern physics is fundamentally dogmatic, due to its absolute laws. Laws that at the beginning were literally the word of God, but today, fortunately, thanks to the brilliant mathematician Emmy Noether, we know that they are the manifestations of symmetries in nature. So, the laws of physics cannot be the most fundamental entities, because they are the product of the regularities of space and time, of their symmetries. Therefore, if the latter change, the laws also change. Exactly what has already been proven by the expansion of the universe. A phenomenon that still surprises physicists, but which is completely natural in Darwinian Physics, because in this new paradigm, what emerges and evolves from nothing are precisely the symmetries. Let us remember that what emerges from nothing is an invariance in one of its two states, repeating or conserving itself despite its random fluctuation. A conservation that represents a pattern, a regularity, a law. Exactly the definition of existence or being in Darwinian Physics. Therefore, this new theory explains the emergence of the laws of physics, because it explains the origin of the patterns and regularities of nature, that is, the symmetries. Thus, if the cosmos is evolving in Darwinian Physics, through the expansion of space and time due to the emergence of holons, then the laws must also evolve with them, since the latter would simply be the properties of such holons.

This allows Darwinian Physics to really explain everything, even the laws; which is absolute and inexplicable for current physics. In this sense, the fundamental difference between a law of physics and a dog, for example, would be that the former is a much simpler and more stable pattern than the latter. A much simpler regularity due to the fact that it evolved vertically (VHE) much earlier. Therefore, it has less accumulated holistic equilibrium than the dog, so it has a low degree of complexity. This allows it to be a much easier pattern to describe by a mathematical equation. And at the same time, the laws would be a much more stable or constant regularity than that of the dog, since belonging to a much lower holon means that it has evolved much more horizontally (HHE), acquiring a high degree of causal inertia, which gives it this characteristic stability and certainty. So much so that it gives physicists the impression that these are absolute regularities that have always existed. To all this we must add holarchy, for it is this that gives the laws of nature their legal character, for the higher realities must necessarily obey the patterns or regularities of the lower realities, for it is simply the latter that literally support the former. Therefore, the more complex things cannot violate the simpler and more stable patterns of the more fundamental realities, for they would be violating what rightly allows them to be. That is why every science can discover its own laws, because every major reality has more minor realities, and therefore more fundamental laws or patterns, which it cannot violate. Hence there can be no culture without biology, no biology without chemistry, no chemistry without particle physics, no matter and energy without space and time. And now, according to Darwinian Physics, there can be no space-time without nothingness.

Thus, through this chain of interdependence, which we have called holoarchy, Darwinian Physics can not only explain the lawfulness of nature, but also satisfactorily fulfill the new criterion of scientificity proposed in the first part of the book, the principle of holistic consistency. For by explaining the origin of the laws of nature at all its scales, it can also unify the sciences studied at each of these scales. This ultimately allows science not only to describe nature very rigorously, but also to understand it in depth. As we saw in Chapter VII, understanding means coherently relating the facts to the rest of scientific knowledge; for when we have understood new information, it is because we have coherently integrated it into our worldview. On the other hand, if we have not understood something, it is because that something does not fit in with the rest of our knowledge, it is left hanging, loose, which causes us to ask ourselves why? Precisely the cognitive dissonance that occurs with quantum mechanics in the current classical or Parmenian paradigm. Instead, what Darwinian Physics seeks is not only to coherently relate quantum mechanics to gravity and time, but also to unify all major realities through the same mechanism of emergence and evolution. For if we want greater understanding, it will be with greater integration, which cannot be limited to one specialty, or even to an entire science. If we really want to understand the origin and meaning of a phenomenon, it must be related to all the sciences. A process that some have called consilience, the integration of all sciences into a single theoretical framework. Only in this way will we be able to recover a cosmovision of a cosmos as a unique, logical, and integral whole. Just as the universe is. Something that many consider utopian or pseudo-scientific, but which in the new paradigm would not only be reasonable again, but also obligatory, since it would be the only way for us to truly understand nature.

Clearly, Darwinian Physics has a very high degree of holistic consistency, but it must also satisfy the principle of sufficient reason. Something it fully achieves in the only way it could, by explaining "everything" from " nothingness." Something unimaginable for Parmenian physics, because, as we have known since Chapter V, current physics is fundamentally dogmatic, and dogma violates the principle of sufficient reason. For this reason, it can only avoid regression ad infinitum by committing another logical error, that of presupposition. That is, in order not to declare something more and more fundamental infinitely, it is decided to stop this process of asking why? to everything (as if it were a child) and to postulate something as the most fundamental. Something of which it is no longer allowed to ask why? In the case of modern physics, this would be the supreme symmetry, the law of laws, the theory of everything. But even if it were possible to achieve this (and everything indicates that it is not), this is completely arbitrary, since we can still ask the question when we ask about the origin or cause of this fundamental law or symmetry. This is the error of presupposition, the assumption of something as absolute truth because it is so, and that it would be illegitimate to ask what other law is derived from that fundamental law; even though it was perfectly valid to do so with the rest of the non-fundamental laws. A situation very similar to that of theologians, where they explain everything with God, but are forbidden to ask about the origin or cause of God, because it is heresy. So, the big problem with this pseudo-solution is that if such a fundamental premise is left without justification, then all the explanations derived from it (all the phenomena and non-fundamental laws it supports) would also be left without justification, by transferring this absurdity to everything else, leaving all existence without sense or reason, as we saw in Chapter V.

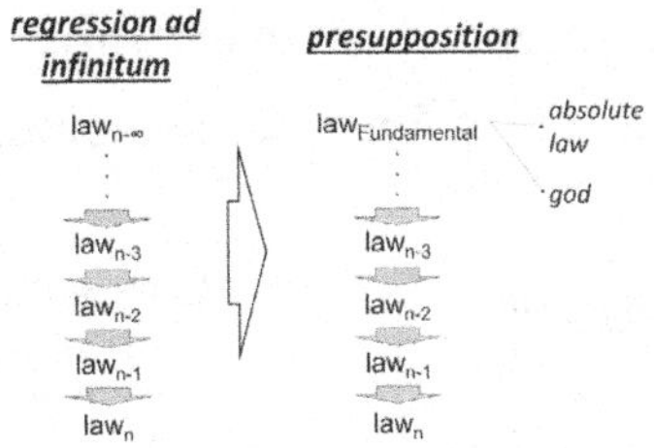

Then the only way to comply with the principle of sufficient reason, without falling into the absurdity of presupposition, is through nothingness; since it is the only arché that can legitimately avoid regression ad infinitum, since it makes no sense to ask for the cause or reason of nothing. Therefore, nothingness would be the only valid solution to the problem of regression ad infinitum, because it also does not fall into the other error of reasoning called the fallacy of circularity, which tries to justify the fundamental premise with the conclusion itself. A classic example is the following syllogism: "God exists, why?, because the Bible says so, and why do you believe in the Bible?, because God wrote it." Or a more modern example, committed in today's physics: "What is time?, what does a clock measure, and what is a clock?, what does time measure" In both cases, the syllogism is inconsistent because it tries to justify a conclusion with the same conclusion. Darwinian Physics does not fall into this error of reasoning either, because nothingness is rigorously justified by its own definition, which is the negation, the opposite of being. And a negation is not a circular argument, especially if its premises are empirically justified. Not only because of the evident certain characteristics of existence, but also its negation is supported by the probabilistic and uncertainty properties discovered in quantum mechanics. Thanks to these definitions, we can derive a darwinian mechanism that allows us to explain the emergence of everything. From space-time, matter and energy, living beings, to consciousness and culture.

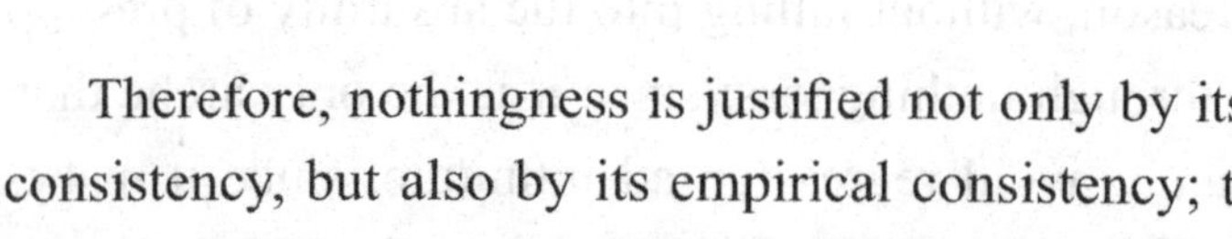

Therefore, nothingness is justified not only by its internal consistency, but also by its empirical consistency; turning nothingness into a fundamental self-consistent premise, since more than reason, what sustains it is the explanation of two natural phenomena: time and chance. A very different situation from Parmenian physics, which is based on assumptions. Let us remember what was said in chapter IV, that absolute laws are impossible to observe and confirm, due to the problem of induction. On the other hand, nothingness would be the true "uncreated creator", being the analog to the simple random genetic mutations that have created the unimaginable abundance and diversity of living species that have existed on this planet in these four billion years. An irrefutable scientific fact that now applies to the rest of nature. In fact, since Darwinian Physics explains ontological randomness and the becoming of time, it would be the first authentic quantum theory; since it arises directly from its principles. This is very different from what happens in current physics, where all current quantum theories are essentially classical, because they start from a classical element (a classical field or classical strings), and then are quantized by, generally, a Hamiltonian. In contrast, Darwinian Physics is truly quantum, since it is by definition anticlassical (this will be explained in much more detail in Chapter XII). In summary, Darwinian Physics is far from being dogmatic, since its nothingness is justified by strong scientific evidence coming not only from physics (time and chance), but also from biology, through its mechanism of evolution by Natural Selection.

Ironically, it is the very absolutism of current physics, which excludes nothingness, that makes physics truly nihilistic by depriving it of meaning; by leaving its fundamental premise (the laws) without justification, by falling into the fallacy of presupposition. Transferring this nonsense to all the other laws that would support this supposed fundamental law. A strategy that has not even worked in reality, since it has been practically impossible to unify general relativity and quantum mechanics in a Parmenian or absolutist way. Darwinian Physics, on the other hand, actually follows the principle of sufficient reason by actually explaining and justifying everything, starting from nothing. An arché that can perfectly explain not only traditional natural phenomena, but also truly spontaneous phenomena such as radioactivity. Something that current physics simply cannot do, because if radioactivity had a Parmenian or deterministic explanation, it would be caused by a law. Therefore, it would be possible to predict exactly when, for example, the nucleus of a heavy atom will emit an alpha particle or gamma rays. However, this is impossible because it is completely random. In fact, alpha radioactivity should not even exist if reality were truly classical or Parmenian, because the energy of the alpha particle, for example, is not enough to leave the nucleus of the atom, and yet it does. All thanks to a quantum mechanism called tunneling, because it seems that there was a tunnel that allowed it to leave, because it lacked enough energy to overcome this energy barrier. Therefore, in classical or Parmenian terms, radiation is impossible; not only because it is random, but also because there is no absolute or classical law that allows it. An impossibility that Darwinian Physics, of course, explains by saying that it is caused by literally nothing. And not only does it explain this, it also explains the rest of quantum phenomena, which are inherently random. Something we will see especially in Chapter XII. Thus, Darwinian Physics not only explains all of existence, but also reveals its meaning by discovering its purpose, its end, to be the opposite of nothingness (certainty and regularity). That is why Darwinian Physics, unlike modern physics, is teleological, because what gives meaning to existence is nothingness itself. Just as darkness gives meaning to light, or cold gives meaning to heat, or death gives meaning to life.

To all this we can add that it is very easy to criticize Parmenian physics because, like any absolutist or radical position, be it philosophical, religious or political, it is easy to show that it is false because of the absurdities it generates. For example, the current paradigm denies the existence of novelty, change, time, complexity, free will, the present, randomness, and generates far-fetched conclusions such as eternalism and the multiverse. Absurdities that are not reached in Darwinian Physics because it does not go to the other extreme and postulate that everything is randomness and change. What is really postulated is that the universe is in an intermediate state because it is evolving. We are a combination between permanence and change, between certainty and uncertainty, between absoluteness and nothingness. Something that also varies according to the scale (vertical dimension). Therefore, the position of Darwinian Physics is really moderate, ontologically, however radical it may seem to theoretical physicists; because we are in full transition between two extremes. So, the universe is obviously lawful, and it strictly follows patterns and rules, but at the same time, not everything is completely determined. Not everything is a slave to fate, because the future is not written in stone, because it does not exist. So, there is still some nothingness in existence, although it is quite suppressed in the face of the long holistic evolution. However, even though nothingness is rather weak at the current level of evolution of the cosmos and on our scale of reality, it is strong enough to be detected by our measuring devices, for example in the intrinsic randomness of quantum systems such as radioactivity. Therefore, although the amount of nothingness is small, it is measurable; and it would be responsible for providing that small dose of freedom or surprise to existence. As a scientific fact with the potential to explain not only microscopic physical systems, but even the free will that very complex beings like us would have. An indeterminacy that would extend vertically through the holoarchy, from nothingness to the atoms and molecules that make up our neurons. Because if the physicists are right, not only would the whole universe lose its meaning, but also symbolic reality, since it would render psychology, sociology, economics, politics or justice completely absurd; because none of this matters if the will is a mere illusion, completely determined as a stone. A nonsense that extends to time, since the now disappears, because it would also be an illusion, since in eternalism only the future and the past exist.

If it is still hard to believe that Darwinian Physics can be any more reasonable, compare it with the best that today's the foundations of physics have to offer: the famous string theory or superstring theory. The pinnacle of Parmenian physics and the perfect example of all its vices. First of all, this theory was not born to be the theory of everything, as it is currently sold, in the sense that it unifies the four fundamental interactions (gravity, electromagnetism, weak and strong nuclear forces), but was postulated to explain only one, the strong nuclear interaction. However, after being overwhelmingly surpassed by quantum chromodynamics, physicists apparently could not devote themselves to researching anything else, so it was recycled for the purpose for which it is known today. All thanks to the fact that a supposed advantage was found, the possibility of representing bosons. These are a type of particles whose main characteristic is that they have an integral spin. This property makes them responsible for transmitting the fundamental forces or for being intermediaries of interactions, such as the photon, which is responsible for transmitting electromagnetism. A theoretical possibility that lies in the fact that strings naturally oscillate through all their harmonics (standing waves that fit on the string): 1, 2, 3, 4, 5... ∞^+. A property that all strings have, such as those of a guitar. Just shake it harder and harder to create more and more standing waves.

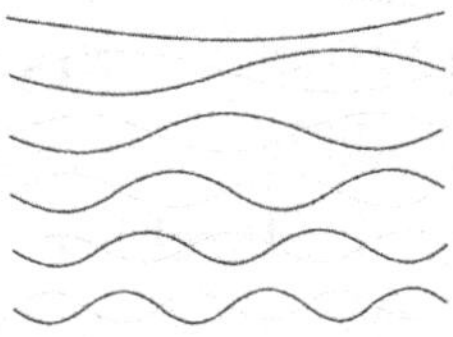

So these harmonics would represent the spin of bosonic particles. Like the famous Higgs boson of spin 0, and the photon and gluon of spin 1. But it would also include the hypothetical bosonic particle of spin 2, the famous graviton. The particle that should carry the force of gravity. Since strings naturally represent all possible bosons, they allow us to unify all the fundamental interactions mediated by these bosons under the same mathematical object. However, we know that the universe does not only consist of bosons or forces, but also of matter, particles called fermions. These particles are characterized by having a half-integer spin, such as the ½ spin of the electron and quarks. Since strings do not vibrate to produce this kind of spin, an extra degree of freedom along the string called supersymmetry was introduced to include it. This is a very exotic symmetry which, represents an invariance under a transformation between bosons and fermions. This means that if we replaced all matter particles with force-carrying particles and vice versa, the universe would remain the same. This is obviously not true. Therefore, it would be necessary to duplicate all existing particles in order to make this transformation. In other words, every particle would have to have a supersymmetric partner. This is something that no experiment has observed, and it is quite worrying for string theory, since without this exotic symmetry not even a single electron can be modeled. In fact, the problem of supersymmetry is so severe that one cannot even take all the particles of the Standard Model and simply duplicate them to obtain it, because the Standard Model of Particles itself cannot be supersymmetric. You would have to add new normal or non-supersymmetric particles to the Standard Model in order to duplicate them through supersymmetry. Therefore, the simplest supersymmetric model is the Minimal Supersymmetric Standard Model of Particles (MSSM), which is created by introducing an extra Higgs field, which has apparently not been observed. And if that wasn't catastrophic enough for string theory, you don't even have a mechanism to break supersymmetry; because even if it exists, it has to be broken at low energies, because the universe we observe is obviously not supersymmetric. So, it is not just an experimental problem, as string physicists like to pretend; it is also a theoretical problem, of internal consistency, because to this day, despite all their efforts, they are not even close to generating a mechanism that breaks the supersymmetry they so desperately need.

In addition to the above, the theory has been sold, and continues to be sold, as a simple and elegant theory. Over the years, however, even string physicists themselves have had to admit that it has turned out to be much more complicated and ugly than they would have liked. For example, they started out naively believing that they could describe all of physics with a single fundamental element, the string. A string that at its inception was classical, that is, it had a thickness. Again, like the string of a guitar. But when they saw that this did not work, they invented a one-dimensional string, without width. A leap in abstraction that is difficult to explain. But since they could not explain the curiosities of quantum mechanics with this exotic element either, they simply attributed such properties to it by quantifying the vibrations of this one-dimensional string with the rules of quantum mechanics. In other words, they simply exchanged one strange physics of quantum point particles for another strange physics of quantum one-dimensional strings, without clarifying anything in the process. So, it does not resolve any doubts about quantum mechanics, neither the measurement problem, nor the intrinsic randomness, nor its quantization, nor its non-locality, the problem of time, and so on. Therefore, instead of solving mysteries, it adds them, through its supersymmetric one-dimensional quantum strings: what is it, where did it come from, who put it there?

But if you think that at least these are the only leaps of faith that have to be made, you have to assume the existence of 6 extra spatial dimensions. Exactly, in addition to left-right, forward-backward, and up-down, we have to assume the existence of 6 spatial directions that nobody can imagine. Dimensions that apparently nobody has ever observed, that exist only in the world of mathematics, but that are essential for these mysterious one-dimensional quantum strings to yield something minimally similar to the Standard Model of Particles. In the face of the obvious absence of extra dimensions, string theorists try to justify it by saying that these are very small spatial dimensions, curled up at every point in space. Conveniently tiny dimensions to hide the fact that they have not been observed by any experiment. Nor do they explain why there are only 6 ridiculously small spatial dimensions, while the 3 spatial dimensions we know are as big as the universe itself: what was the mechanism that decided that, who built the universe that way? Nobody knows, and no string theorist cares. But even if we accept this cosmological deus ex machina, the integration of extra dimensions literally creates infinitely more problems than solutions. This is because there is no single way to roll up these extra dimensions, so all ways of compacting or rolling up would be valid. This is no small matter, because there would be at least 10^{500} (a 1 followed by 500 zeros) ways to compact these extra dimensions, although they could be infinite. A huge problem, which string physicists call a landscape, because each compactification yields a different solution, a different physics. A monstrosity of ways to do physics that cannot even assure us that any one of them describes anything like our universe. This is very important because, as so often in this theory, string physicists are experts at turning a terrible flaw into a virtue, so they see something positive in this ridiculous number of solutions, saying that since there are so many solutions, the right one must be there, among all of them. As if it were guaranteed. The only problem is to find it. But this is completely false, because you can have an infinite set and still not find what you need, even if you look forever, because it is simply not there. For example, the set of even numbers is infinite, but if you are looking for the number 7 or 23 as a solution, you will never find it in that set, even if you have infinite candidates. So, the landscape is a brutal defect of the theory, not only because it does not give a unique solution, but because everything indicates that its innumerable solutions have nothing to do with a universe with three dimensions, without supersymmetry and in accelerated expansion. That is, with the real world. So, we may be looking for an imaginary needle in an infinite haystack.

But not only are string theorists unable to provide a single solution, they are also unable to provide a single theory. So far, there are five different superstring theories: Type I, Type II A, Type II B, Type SO (32), and Type E8 x E8. Theories that could be related by dualities, forming what they call theory M. However, it is only a conjecture, one of many in these theories. Its ambiguity is so great that string physicists do not even know what M means. Its own author, the American physicist, but above all the mathematician Edward Witten (born in 1951), a messiah for many physicists, said that he would explain the meaning of this M when he knew what the theory meant, because to this day no one can say exactly what string theory is, not even in principle. In fact, you hear the word string less and less at string conferences, because they are not even sure if string theory is really about strings, or something else even more abstract and ambiguous. In conclusion, string theory is just a theoretical framework, one in which theories and solutions can be developed, in fact too many theories and solutions, but which has not been able to produce a single one that is really useful. Instead, it can only offer conjectures.

But the absurdity does not end there, because string physicists promised that all physics would be explained with a single element: these one-dimensional quantum superstrings; however, current string theory includes, in addition to supersymmetry and extra dimensions, branes. Another element that would be as real and fundamental as the strings themselves, since it would be a generalization of the latter. Since all particles in the Standard Model are represented by open strings, physicists realized that the ends of these strings must end somewhere in order to vibrate. Again, much like the ends of the strings on a guitar must be anchored and taut in order to sound properly. So, they invented branes, fundamental objects of all possible dimensions. For example, there would be 0-branes or dimensionless points where the ends of the strings would be anchored. Objects similar to the point particles of the Standard Model, which they wanted to replace. But in addition to the above, there would also be 1-branes, which would be the strings; 2-branes, which would be two-dimensional membranes; 3-branes, or three-dimensional membranes; 4-branes, or four-dimensional membranes; and so on. The novelty is that these branes would also vibrate, just like the strings, imparting different properties to the strings anchored to them. For example, the string representing a quark would be anchored at one end to a brane that would give it the electric charge, while the other end would be anchored to another brane that would give it the color charge (strong nuclear force). The obvious problem is that it violates the original idea: to explain all the diversity and complexity of nature with a single and simple element, the strings. A claim that disappeared completely when, in order to get something vaguely reasonable, they had to assume more and more things: supersymmetry, extra dimensions, and now a zoo of branes for all tastes. Something that becomes even more disturbing when you realize that some of these branes can be the size of the entire universe. In fact, our universe would be a 3-brane. Something completely absurd, because these branes are as fundamental as strings, so they would not be composed of anything more fundamental. So, it is hard to believe how something that should be fundamental, an atom of existence, can be as big as the whole cosmos. In conclusion, string theory has long ceased to be the simple and elegant theory it claims to be, and has become a convoluted theory with many assumptions that nobody understands.

The greatest failure of string theory, however, is its inability to deliver its eternal promise, the holy grail of physics, quantum gravity, or the unification of general relativity with quantum mechanics. Its best attempt was the famous AdS = CFT conjecture of the Argentinian string theorist Juan Martin Maldacena (b. 1968). As its name suggests, it is just an unproven hypothesis, like everything in this theory, relating a kind of AdS gravity theory to a kind of quantum field theory called conformal (CFT) with one dimension less (holographic principle). This is a conjecture that string physicists are very proud of, but the big problem is that it has nothing to do with the real world. First of all, this conjecture is what physicists call a toy model, consisting of an extremely simplified model with as many symmetries as possible, many of which they already know to be false. However, they include them out of necessity in order to calculate something. Furthermore, CFT means that the Standard Model of Particles should be scale invariant. In other words, that the physics must be the same no matter what the size. Which is patently false. The physics of the small is clearly different from the physics of the large, something that applies to quantum field theory as well, since masses and interactions are very different depending on the scale at which they are found. Something that Darwinian Physics clearly recognizes through holoarchy. On the other hand, on the gravitational side, our universe is clearly de Sitter, not anti de Sitter (AdS); because the cosmological constant (Λ), which is measured and for which a Nobel Prize has already been awarded, is positive ($\Lambda > 0$). That is why our universe is expanding at an accelerated rate. However, string theory is completely incompatible with de Sitter solutions, because its essential requirement, supersymmetry, prevents it. Therefore, string theorists only work with AdS (anti de Sitter) geometry, i.e. with a negative cosmological constant ($\Lambda < 0$). So not only does the conjecture have no connection to the real world, but one could also say that the theory itself has been falsified again, because string theory has always been able to yield solutions only with negative cosmological constants. Something that was never believed to be true, but which string theorists expected to find a solution with a zero cosmological constant. That this would be the correct string theory. But unfortunately for them, it was already discovered that the universe is of the de Sitter type. And since then it has been impossible to get a de Sitter solution in string theory, leaving only the hope of getting it in the future, but in a very convoluted and very unnatural way. And yet the string theorists go on as if nothing had happened, completely against the Poppenian spirit of science.

Continuing with the conjectures, another very popular one, especially in recent times, is ER = EPR, also invented by Maldacena together with Leonard Susskind (American theoretical physicist born in 1940). This conjecture relates wormholes (Einstein-Rosen bridges or ER), which are completely fictitious mathematical objects, impossible for physics because they require "negative or anti-gravitational energy," to a real quantum phenomenon that has nothing to do with the above-mentioned quantum entanglement (Einstein, Podolsky and Rosen paradox or EPR). This conjecture is famous because it is the reason why from time to time headlines appear in the press claiming that a wormhole has been created in a laboratory or in a quantum computer, because every time quantum objects are entangled, which is quite normal in experimental physics institutes and especially in quantum computers, it is said that a wormhole has been created. A bizarre connection, frankly ridiculous, which is justified by the fact that they share two characteristics: the first is that both two entangled particles and two wormholes connected inside them are always connected, even if they are far apart; and the second is that both are not useful for transmitting information, since the entangled particles are randomly correlated, while the two wormholes are not passable, since you can only enter them, but not leave them, just like black holes. From these two similarities, string theorists infer that they are the same thing. This makes the same sense as saying: horse = table, because they both have 4 legs, if the horse is brown and so is a wooden table, and furthermore if the table is strong enough to sit on, just as we can ride a horse. If these three conditions are met, then we can say that both the horse and the table are the same thing. The ER = EPR conjecture is one of the best examples of the desperation of string theorists to publish anything due to the great scarcity of real results.

And like the previous one, there are many other outlandish conjectures based on the same erroneous concept that Einstein's space-time is not fundamental because it is also the product of the interaction between strings, like everything else in this theory. Therefore, space-time is on the same level as electrons and protons, except that this time they would be closed strings forming a kind of superfluid or Bose-Einstein condensate of gravitons. This means that space-time does not exist at the level of strings, just as moisture does not exist at the level of H^2O molecules. Therefore, space and time would be emergent phenomena that arise at higher scales. This creates a huge logical problem, because if space-time is the product of interactions between strings, then: where are the strings if they inhabit nowhere, how can they be strings with length if there is no space, where do the extra dimensions come from, how can the strings vibrate and interact to create time if there is no time to vibrate and interact? In other words, the whole theory is based on the fundamental logical error of confusing the content with the container. Simply put, space and time cannot arise from physical things, because physical things need space and time to exist in the first place. This easily explains why string physicists have found it impossible to make their theory of quantum gravity concrete, because the idea is essentially meaningless.

A small detail that it is incredible that the great majority of theoretical physicists forget, but which is again explained by the extreme specialization, this time in the specialization in quantum field mechanics or Standard Model of Particles, to the detriment of general relativity. This is because in this second pillar of the foundations of physics Einstein had made it very clear: gravity is no longer a force, as it was in Newton's theory. Now gravity is the product of the curvature of space-time itself. In other words, gravity is inertia itself, force-free motion, but in a non-Euclidean or curved geometry. This is exactly what is meant by the principle of strong equivalence. The most important pillar of general relativity and, according to Einstein himself, the happiest idea of his life. And yet virtually all theoretical physicists, including apparently string physicists, continue to stubbornly try to describe gravity as if it were just another force, equivalent to an interaction like electromagnetism and the strong and weak nuclear forces. Since the latter are the product of the exchange of bosons, like the photon in the case of electromagnetism, gravity should be the product of the exchange of gravitons. In the case of string theory, represented by closed strings. This is why physicists speak of the quantization of gravity, because they are convinced that spacetime itself is just another quantum field, equivalent to the field of the electron. Thus, as in particle physics, interactions are described by quantum fields placed in space-time (flat or without gravity); then they continue with the same strategy, interpreting gravity as yet another interaction, the fourth, by a quantum field. The problem is, where to place the quantum field of gravity in space-time, if gravity is space-time itself? A strategy that obviously has not worked because it is absurd; but they keep trying because they do not know how else to do it. This is an excellent example of the phrase popularized by Abraham Maslow (1908-1970) in his book The Psychology of Science (1966): "I suppose it is tempting, when the only tool you have is a hammer, to treat everything as if it were a nail."

And in spite of all this, string theory continues to enjoy great popularity, because sadly it is the best that the foundations of physics have to offer today. A pseudo-theory that represents everything that should not be done to generate a new paradigm, because it is essentially the ultimate consequence of continuing with the strategy of the rigid paradigm. Therefore, string theory does not solve any fundamental problem. It does not solve the problem of measurement in quantum mechanics, it does not answer what time is, it does not reveal where quantum randomness comes from, what caused the big bang, where laws come from, why the universe is expanding, why the speed of light is invariant, and so on. Questions that need to be answered, at least one of them, in order to build a new paradigm. In fact, it does the opposite, because instead of solving mysteries, it adds even more questions, such as: what are these strings, where did they come from, what are these branes, how did these extra dimensions get compactified, how did supersymmetry break, how do you get a "$\Lambda > 0$," where did this multiverse come from, why 6 tiny dimensions and not more or less, and so on. In conclusion, it is not an economic theory at all, since it depends on a huge number of assumptions. Moreover, it works only with conjectures, because it is incapable of producing any exact demonstration or prediction. Therefore, it is an incredibly speculative theory that cannot be falsified. Let alone trying to integrate it with the other sciences to satisfy the principle of holistic consistency, since it is not even capable of being coherent with itself and with physics itself. It should come as no surprise that the legendary theoretical physicist Richard Feynman did not like this theory, especially the attitude of string physicists. In an interview with P.C.W. Davies and J. Brown, he said: "I think the problem in theoretical physics is to prove yourself wrong as quickly as possible! What the string theorists are doing now is that they are not proving themselves wrong, because they are taking liberties with their own equations, saying, 'Okay, maybe six of the ten dimensions are curled up, and that leaves us with four,' without proving that six are curled up, or considering the reason that seven are curled up. They are not sufficiently testing their ideas with experiment because it is difficult to calculate anything. That is, they're in the air, so they don't have to pay much attention to it!" What Feynman is criticizing is the anti-Popperian spirit of string physicists, because being a scientist means being critical of your own theory. Trying to prove that you are wrong. Exactly the opposite of what string theorists do, who are dedicated to fanatically defending their theory, making it more and more convoluted in order to forcefully fit the data. Something that has been more than proven in the case of the absence of supersymmetry and the accelerated expansion of the universe.

Instead, what Darwinian Physics seeks is not only to defend the spirit of falsification, but also to add another additional condition, holistic consistency, in order to make science not only more rigorous, but also more complete, recognizing all its stages of development (normal science, crisis, and revolution). Therefore, Darwinian Physics is not just a small modification of the foundations of physics, but a true scientific revolution, repeating the same process that has overcome every crisis in physics. A semi-disruptive evolution, as we have learned since Chapter II, happens through a modification in the hard core of the current paradigm. An update that occurs, for the most part, through the rejection of an absolutist postulate or principle, through a natural constraint. In fact, the last revolution in physics is a very good example of the latter, because the birth of Einstein's relativistic one happens when a speed limit was applied to what can be caused or affected, the speed of light or "c" (299.792.458 m/s). Something that directly clashed with Newton's theory, which included a gravitational action that was transmitted at an infinite or instantaneous speed. Therefore, thanks to Einstein, we know that gravitational waves travel at the speed of light or "c." Another example is the elimination of the postulate of the infinite continuity of matter and energy. This was also thanks to Einstein when he demonstrated the existence of atoms through Brownian motion, because until then atoms were practically considered pseudoscience. However, the most important contribution in this sense was made by the German physicist Marx Planck, who, to solve the problem of the black body (ultraviolet catastrophe), took a desperate measure, in his own words, postulating that energy is not continuous, but is transmitted through discrete units that he called quanta, thanks to the introduction of a constant that we now call "$\hbar$" (6.63×10^{-34} Js).

These two simple restrictions: "c" and "$\hbar$," changed everything. It is the difference between the physics of the 20th century and the physics of the 19th century, because when we remove these restrictions, we recover the physics of the 19th century, because we have absolute values again. For example, when $c = \infty$, the relativistic effects disappear; and when $\hbar = 0$, the quantum effects disappear. So, we can say that 19th century physics is only a special case of modern physics when we have absolutist postulates where $\hbar = 0$ and $c = \infty$. But when such absolutist postulates are eliminated and we put a natural limit, physics becomes more realistic, which avoids getting absurdities like an instantaneous cause that breaks causality in the case where $c = \infty$, or that all things radiate light of extremely high energy due to an infinite frequency of energy when $\hbar = 0$. So, in this context, Darwinian Physics simply tries to continue this process of making physics more realistic by continuing to eliminate absolutist postulates. And this time it does so by imposing a natural limit on the mother of all absolutes, the laws of physics themselves. More specifically, the principle of conservation of information or also known in quantum mechanics as unitarity. This point will be better explained in chapter XII, but for now we need only know that unitarity seeks to conserve all the quantum states of the wave function, so that the sum of all its probabilities always yields 1 ($U = 1$). Well, unitarity is the quantum version of the classical Parmenian principle that confers the eternalist (that each instant has exactly the same information as any other instant) and deterministic (a single history already written forever) character to physics. So, now it is the turn to put a natural limit to the origin of parmenism, declaring that the unitarity cannot be 1, but less than one ($U < 1$); because the difference or the remainder corresponds to nothingness, the probability of not being or existing. Something that has not yet been discovered, because the degree of existence of the universe is so high, so close to 1 (it is very likely that the cosmos continues to exist), that it gives the impression to physicists that U = 1, which supports their absolutist belief.

However, to suppose that the laws of physics are immutable for all eternity has been the great mistake, not only because it will always be a conjecture (problem of induction), but because we have already discovered that the cosmos has had a beginning, a Big Bang, and that everything evolves in nature. For this reason, Darwinian physics simply seeks to continue the tradition of naturalizing physics, moving it away from its Pythagorean-Platonic past, by adding to the list of restrictions ($c \neq \infty$ and $\hbar = 0$) that the laws are neither absolute nor eternal ($U < 1$). Thus, current theoretical physics would be transformed into a new special case, where U = 1 is just an approximation, as "$c = \infty$" and "$\hbar = 0$" were in the past. Again, this will all seem very radical at first, but it is a much simpler, more conservative and realistic solution than that offered by the current foundations of physics. And, above all, it is a more satisfactory alternative because it provides concrete answers to the great fundamental puzzles of modern physics. This is the subject of the next and last part of the book.

Third Part

Answers to the Great Unknowns

CHAPTER XI

THE PROBLEM OF TIME: THERMODYNAMICS

The time has come to justify all the big words that were said in the first two parts of the book, and to show how Darwinian Physics can be better than Parmenian physics in offering concrete explanations and solutions to the great puzzles that the current foundations of physics is unable to address. And what better problem to start with than the most difficult and important of all, time. Because incredibly few physicists are interested in it. And the two main theories of physics, interpreted in the Parmenian paradigm, don't have much to say about it. For example, time is completely absent in quantum mechanics because of the measurement problem or the collapse of the wave function. This will be discussed in much more detail in the next chapter, but for now we just need to know that time is an external element of the theory that is introduced by the order in which measurements are made on a quantum system. This is because before the measurement, the quantum system is indeterminate, in a quantum superposition. Therefore, the determination of the system is fixed by the experimenter when he decides to make an observation. Thus, the history of the system depends on the order in which the measurements are made. Therefore, it is impossible to assign an intrinsic time to the quantum system, since it depends on the experimenter. In fact, this is the reason why quantum mechanics is a timeless theory in the current paradigm, since the measurement problem has not yet been solved. Fortunately, the other pillar of physics, general relativity, can tell us a bit more about time. However, it treats time (as well as space) in terms of metrics. This means that the theory deals exclusively with intervals or distances between two points in spacetime. Therefore, the theory cannot tell us anything about what a space-time point, also known as an "event," is; and therefore cannot tell us what an "instant" is. This is why we do not know what space-time is in Einstein's theory. Its fundamental element.

Then, faced with this ambiguity in the two pillars of modern physics, the theorists were able to interpret them according to their paradigm, so they applied the Parmenian solution, which consists in denying the existence of what is not understood. Thus, physicists replicated the same formula invented by Parmenides more than 2500 years ago, solving (in quotation marks) the problem of time by denying the existence of such a problem. This philosophical position is called Theory B, or eternalism, as mentioned several times in this book. In this view of time, the present does not exist in the sense that there is any moment or point in time that is special or different from the rest. For this reason, all instants, all moments, must exist simultaneously, and none of them can be distinguished from the others. As we saw at the end of the second part, this is a direct consequence of the dogma of absolute laws, because if information is conserved, every moment has the same importance, so there cannot be a privileged moment that functions like now, making time symmetrical or reversible. The modern Parmenides who popularized this position was Einstein's former teacher, the Lithuanian-born German mathematician and physicist Hermann Minkowski (1864-1909). One of the professors who believed that Einstein was a complete failure. Something that must have changed his mind when he saw the recognition his former student was receiving. Then, out of curiosity or perhaps remorse, he began to work on his former student's theory: special relativity, giving it a more geometric interpretation by conceiving of time as an additional dimension, the fourth dimension. The result was a view of the universe as a tetra-dimensional block in which all the past and all the future exist simultaneously, indistinguishably and immutably in space-time.

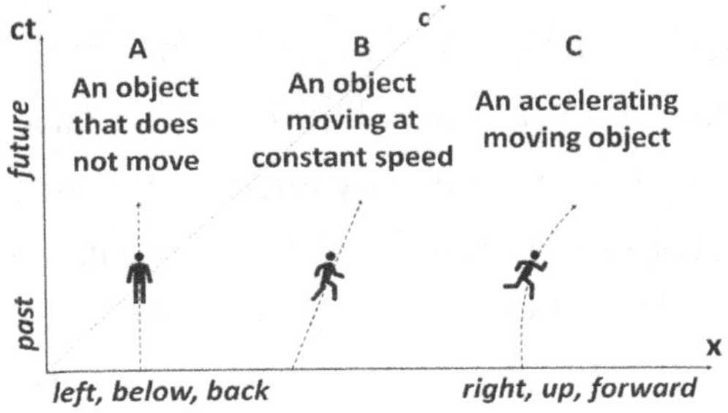

In the figure above, we have the space-time diagram, which shows the different displacements that can be performed. On the horizontal axis, the three dimensions of space are simplified to just one spatial dimension. The vertical axis represents time, with "down" representing the past and "up" the future. The diagonal line shows the maximum possible speed, which is the speed of light c. Therefore, all possible motions must be above this line, since they correspond to speeds less than the speed of light. In this sense, we have that case "A" corresponds to a person standing still in place. Since he is not moving through space, all his displacement in space-time is temporal, he is simply moving forward into the future. In other words, he is aging. In case "B" we have a person who, in addition to moving in time (upward), also moves in space (to the right) at a constant speed, walking relative to the first. On the other hand, in case "C," we have a person who is accelerating, running faster and faster relative to the first, because in each time interval his displacement in space is greater. In this way, the movements of objects, their dynamics, can be geometrized by straight lines and parabolas in a Cartesian plane. The consequence of this is that, as in geometry, the points on the lines, the equivalent of "events" in relativity, are indistinguishable from each other. Therefore, there is no particular point or event that we can identify as "now." An instant that performs the vital task of dividing time into past and future. An absence that literally freezes time because there is only one possible history, the complete line or world line, as physicists say. That is why this view of time is called a block universe, because all the events that have happened and will happen already exist. They are fossilized in eternity, unchangeable, describing a universe where our destiny has been literally written in stone since forever.

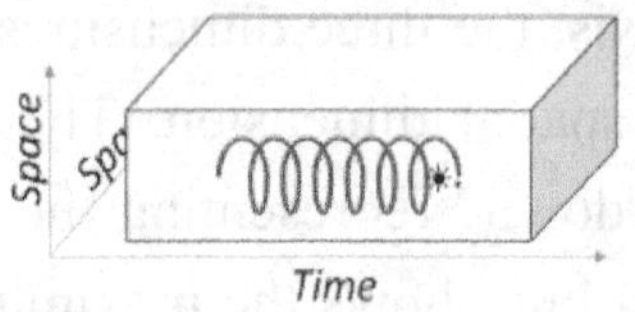

In the figure above we see how the universe would be for a physicist, where the spiral line would represent the orbital motion of our planet around the sun, for example. An existence as a compact block, where all instants are indistinguishable, inseparable. Therefore, it is wrong to think of the block universe as a set formed by sheets or layers, each of which would correspond to a moment; because there is no moment shared by all in the universe. This is because special relativity has destroyed the concept of simultaneity by discovering that the passage of time is different for everyone. In other words, time is relative. Your "now" is not the same as my "now" or your neighbor's "now" (this will be discussed in more detail in Chapter XIII). Therefore, since no universal moment or lamina can be defined, the block universe must consist of a single piece in which all moments are merged. Thus, not only the present disappears, but also the very concept of time in modern physics, because everything is literally frozen in one continuous block, eliminating its flow or becoming. Let us recall that in mathematics a straight line, no matter how long or short, is always made up of infinite points. Therefore, it is impossible to distinguish any instant.

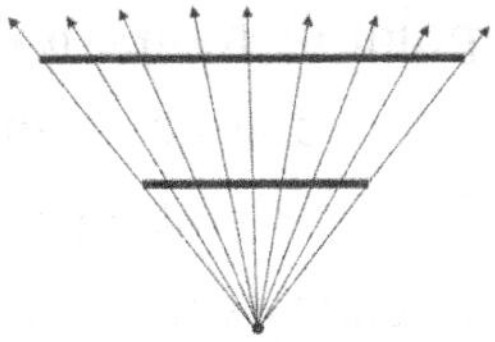

As you can see in the figure above, the same number of straight lines or arrows (vertical lines) can be traversed by the two parallel segments (horizontal lines), even though one is significantly longer than the other. This means that both segments are made up of the same number of points, infinity. The same would happen with Einstein's space-time block, which would contain infinite instants, all coexisting simultaneously in eternity. This view of the block universe helps us better understand why eternalism is so attractive to physicists, even though it completely contradicts our experience, because it fits perfectly with their dogma of absolute laws. Mathematical equations that are also continuous, and that draw a single, immovable, deterministic line or history made up of infinite points or moments. Thus, Parmenian physics is eternalist, because no "now" can be distinguished in the absolute laws. If we return to the Cartesian plane and use the laws of motion to describe the trajectory of a ball when it is thrown, for example, it will draw a parabola. In that curve is all the history of that phenomenon. All of its past and future. No moment or point within that line is special with respect to the other points; because each point, among the infinite points that make up that parabola, contains all the information (position, velocity, mass, spin, and so on) necessary to reconstruct the entire curve or trajectory of that ball. In other words, each point serves to know the complete history of this physical system, since all these properties are conserved in time. Therefore, no point or moment of the parabola serves to represent the present, since none is different enough to perform the important task of separating time into what we can "remember" and what we can "expect." This causes the concept of "now" to disappear as it is diluted in the concepts of past and future, which in turn are merged in eternity.

Some physicists might argue that this interpretation is outdated because it is based on classical physics and not quantum physics. So how can we update it, what would Nobel laureate Richard Feynman say? Well, he would say that the ball follows not only the classical trajectory, but all imaginable trajectories. Thus, it follows infinite paths, each with its own probability of occurrence, but existing simultaneously with all the others because they can interfere with each other. But what seems to be radically different from the classical case is not very different from the problem at hand. In fact, it makes it much, much worse, because not only does the present continue to be non-existent, but now infinite "presents" cease to exist, because no point of these infinite curves or timelines is special either. Only the curves or histories are, with different probabilities. But the points within these curves remain irrelevant. That is why the "now" also disappears in quantum mechanics, according to the Parmenian interpretation.

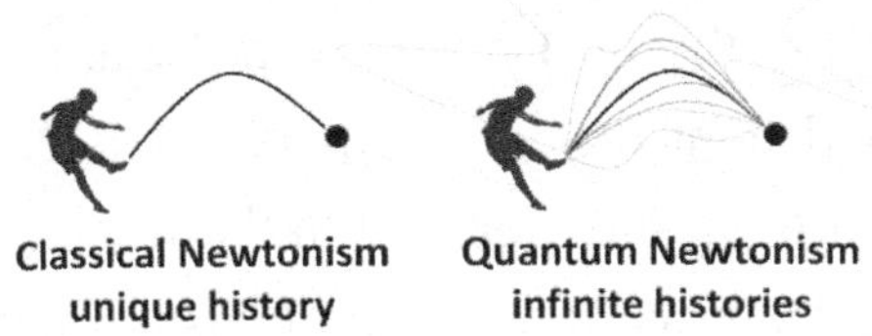

This is the consequence of the forced interpretation of quantum mechanics under the Laplacean or Parmenian paradigm, since under the dogma of conservation of information, renamed unitarity in quantum mechanics, Ψ never collapses because it evolves deterministically forever under the absolute quantum law: the Schrödinger equation. The only difference from the classical non-quantum case is that instead of a single block universe, we now have infinite block universes, each with its own probability of occurrence. This is the origin of the many-worlds interpretation, and the reason why it is increasingly popular among physicists, because it is the inevitable consequence of its hallowed classical principle (which will be explained in much more detail in the next chapter).

In conclusion, in both pillars of modern physics, under the Parmenian paradigm, time loses its meaning and is reduced to the absolute laws of physics. It does not matter that experience says otherwise. Mathematics rules. In this way, time is equated with another spatial dimension, because if "here" and "there" exist simultaneously in space, what "was" and what "will be" must also coexist in time. This obviously freezes the flow of time and eliminates its becoming. In fact, the Parmenian paradigm is so incompatible with time that it even kills causality itself. This is made quite clear by an authority on time, astrophysicist Sean Carroll, in a talk at the Royal Institution in October 2016 called The Big Picture (the same name as his book he was promoting there). Here is what Sean says: "This solid surface we used to walk on is our old Aristotelian notion of cause-effect, purpose and meaning in the world; which we have left behind thanks to the fundamental laws of nature, according to our best scientific knowledge {...}. Aristotle would have said that if things move, it is because there is something moving them. There is a cause or a reason why they move. Laplace, on the other hand, says that they simply do so because they obey equations. Instead of cause and effect in the universe, we just have patterns. Think of the integers, write the numbers 0, 1, 2, 3; and also, the negatives, -1, -2, -3.... There is a pattern. If you tell me any number, for example: 3, I know what the previous number was, 2. I would also know what the next number is, 4. But I'm not saying that three is the cause of four or vice versa, there's just a pattern that relates all these numbers to each other.

Laplace says that the laws of physics are like that. It's not that there's an impulse or an incantation or a guiding force. There is just a pattern that says if this exists, then that follows, and vice versa. This way of thinking about how the world really works, at a fundamental level, is something we still have a hard time coming to grips with." To emphasize this point, Sean Carroll quotes the famous British philosopher, mathematician, logician, and writer Bertrand Russell (1872-1970), who said, "The law of causality is, I think, like much of what goes on among philosophers, a relic of a bygone age, which, like the monarchy, survives only because it is wrongly assumed to do no harm." Then Sean Carroll goes on to say, "It's not that there's no such thing as cause and effect, it's that it's not a fundamental principle of how reality works. It's a very useful way of thinking about things at the macroscopic level, which we call the emergent level of reality, but we don't find it anywhere in the most fundamental laws of physics."

Thus, in current physics, not only do becoming and the present disappear, but even causality itself, because at the fundamental level there are only mathematical equations (laws). Thus, the Parmenian paradigm strips away everything that makes time unique and special and reduces it to just another dimension of space. Something that is formalized in theoretical physics under time-reversal symmetry, which means that the laws of physics are completely time-reversal, since the equations work exactly the same whether the time parameter "t" is positive or negative. One way to understand this would be to record someone juggling three balls. A person would not be able to tell whether the video was running normally or in reverse, because it looks the same in both cases. Therefore, time would behave the same as space, since space is also symmetric (space translational symmetry). However, time is obviously different from space, because a point here does not necessarily affect another point there. On the other hand, the points or moments of time are interdependent, since an earlier moment obviously affects another later moment, because it causes it, it affects it. Today depends on yesterday, and tomorrow depends on today. Therefore, when causality is eliminated, everything that makes time special is eliminated: the becoming and the now. It is reduced to another dimension of space. That is why they cannot distinguish a direction in time, in the same way that a person floating in space, without a planet or a moon as a reference, could not tell where "up" or "down" is (as happens to astronauts floating inside the International Space Station). Neither would it make sense to ask where the future or the past is, because according to the laws of physics we should rejuvenate as normally as we age, just as we can move to the left as easily as to the right in space.

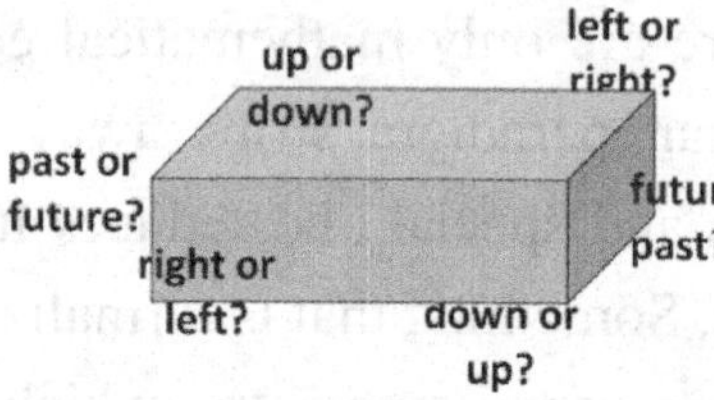

The figure above shows the consequence of time absolutism. A block universe where we cannot distinguish not only between up and down, but also between past and future, because if we flip or rotate the block, it would look exactly the same. In other words, a completely symmetrical universe, just like the laws of physics. Obviously, this does not make sense, because obviously everything is moving towards a future that is also very different from the past. In order to explain this obvious temporal asymmetry, while maintaining the absoluteness of the laws of physics, we had to resort to the second ingredient of the Parmenian paradigm, the initial conditions. This is where the second law of thermodynamics of the Austrian physicist Ludwig Boltzmann (1844-1906), the father of statistical physics, comes in. Theory in which a temporal direction to the universe is generated, but in a circumstantial way. This means that the flow into the future is not a fundamental phenomenon, but a purely statistical one. The product of the interaction of many elements, such as a gas or a liquid, made up of millions and millions of molecules. Since the universe consists of billions of galaxies with billions of stars and planets, the passage of time in the cosmos is also a macro-phenomenon, the result of the interaction of all these elements. However, each element separately is governed by the laws of physics, and therefore they do not distinguish the past from the future. But as a whole they do. For example, a gas always expands, not because each molecule of the gas has that tendency. Each molecule can go in any direction because it has no preferred direction. However, it is this very indifference (symmetry) of each molecule that causes the whole (the gas) to have a strong tendency to go from a concentrated state to an expanded state (asymmetry). In other words, the absence of a preference in the particular creates a preference in the general. This makes the forward direction very practical, since it would be very tedious to apply Newton's equations of motion to, say, every molecule in the gas. Instead, it is much easier to approach the system as a whole and obtain an average by using the tools of statistical mechanics.

To understand how this works, we need to distinguish between two aspects of a thermodynamic system: the microstate, which represents the particular configuration or combination of the components of the system; and the macrostate, which would be a global state of the system that can be compatible with multiple microstates. In this sense, the microstate is what we cannot know about the thermodynamic system because it is not possible to measure each of its elements to know its exact configuration, for example, the position and velocity of each molecule of a gas. While the macrostate would be what we can know, since it is what can be measured globally in the system, such as pressure, temperature, volume, density, and so on. Boltzmann's great discovery was to realize that the macrostate is the one that undergoes the direction of time, for example, when a gas cools and expands, and that, in addition, this colder or expanded macrostate coincides with the fact that there are more microstates compatible with it. In other words, thermodynamic systems tend to have macrostates with more and more microstates. And this is easily explained by statistics. To see this, let us use a simple example. Imagine a small system consisting of four electrically charged particles, which can have either positive or negative charge. We cannot see the charge of each individual particle (they are very small), but we can measure the electric charge of the whole system as a whole. This is a very natural situation, since all the objects we see every day are made up of many charged particles that we cannot see, such as the electrons (negative charge) and protons (positive charge) that make up atoms. So, going back to the example, if we measure the system and find that it has an electric charge of +4, this means that all four particles have a positive charge. Or, if the system is electrically neutral (0), it has two positively charged particles and two negatively charged particles (-2 +2 = 0). Then the global charge of the system, which we can measure, is the macrostate, while the charge of each particle that creates it, which we cannot see, is the microstate. The interesting thing about all this is that the number of microstates or existing configurations is not the same for each possible macrostate. For example, for the macrostate +4 and -4 there is only one possible microstate (the situation where the four particles have the same charge). On the other hand, for the neutral macrostate we have six microstates or configurations compatible with it.

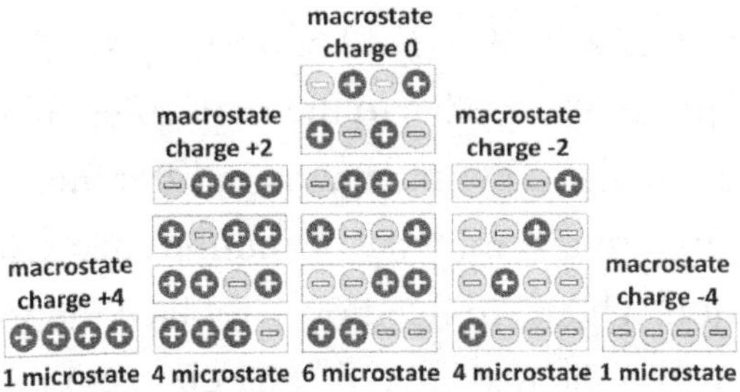

In the figure above, we can see all the existing combinations or microstates for this system, which are clearly not evenly distributed among the possible macrostates. The more neutral the system as a whole, the greater the number of possible configurations (microstates), making these macrostates more probable, simply because they have more chances. For example, the system above has 16 possible microstates (1+4+6+4+1). The +4 or -4 macrostates have only one configuration or microstate assigned to them, so the probability of them being chosen by chance is 1/16 or 6.25%. On the other hand, the neutral macrostate has 6 possible microstates, so its probability is 6/16 or 37.5%. This means that each microstate has the same importance or relevance, so each has the same probability of occurring, 1/16 or 6.25%. Therefore, it means that there is a symmetry with respect to electric charge, since the particles are completely indifferent to having positive (+1) or negative (-1) charge. It is as if each particle flipped a coin to choose its electric charge. For example, to get a macrostate of +4 or -4 in the above system, each particle would have to agree on the result of its coin flip. That is, each should only get heads or tails. The calculation is 0.5 x 0.5 x 0.5 x 0.5 x 0.5 = 6.25%, equal to 1/16. While in the case of the neutral macrostate it is this probability multiplied by 6, because there are 6 possible combinations where there are two positive and two negative particles: 6 x 6.25% = 37.5%. The same probability as 6/16. So, the same principle applies to the molecules of a gas. All we have to do is change a symmetry or indifference with respect to electric charge to a symmetry or indifference with respect to spatial direction. Thus, we need only replace the toss of a coin (two choices) with the toss of a die (six choices) for each molecule of the gas to decide where to go (1 = up, 2 = down, 3 = left, 4 = right, 5 = front, and 6 = back). So, if each molecule were to do this, the inevitable result is that the gas expands, because that is the macrostate with the most possible configurations or microstates. This is because if the gas were to contract, with all the molecules meeting at the same point, it would be equivalent to all the molecules getting the same result in a dice roll. This is extremely unlikely. For example, if the gas consisted of only 10 molecules (an extremely small gas), the probability would be $(1/6)^{10}$ or 0.000001654%. An extremely small probability. Now imagine the probability of a gas consisting of hundreds $(1/6)^{100}$, thousands $(1/6)^{1,000}$ or millions $(1/6)^{1,000,000}$ of molecules contracting. It is not surprising that no one has ever seen it happen.

In conclusion, this distinction between past and future, or asymmetry of time, would not exist at a particular level, but only globally; because it is a purely statistical phenomenon. Therefore, the future does not fundamentally exist because it is indistinguishable from the past. What we call the future is simply a direction that appears on macroscopic scales, a phenomenon that physicists call the arrow of time. However, at the fundamental level, at the scale of individual particles and atoms, this difference between future and past does not exist due to the time symmetry of absolute laws. Now we can understand where the famous entropy (S) in the second law of thermodynamics comes from, since it corresponds to the number of microstates or possible configurations for each macrostate. Therefore, the fact that physical systems tend toward a more entropic state simply means that they evolve toward a macrostate with more compatible microstates. Simply because it is the most likely. That's why we don't rejuvenate or things don't warm up or repair themselves. Not because there really is a future, but because this tendency is statistically evident on the largest or macroscopic scales of the cosmos. In this way, physicists have found a way to explain this temporal asymmetry of nature, despite having absolute or timeless laws, and have managed to provide a worldview with a little more sense of time, abandoning this perfectly uniform and symmetrical block universe, product of its absolute laws, and finally obtaining something more realistic. A universe that is irrevocably heading towards an increasingly entropic future.

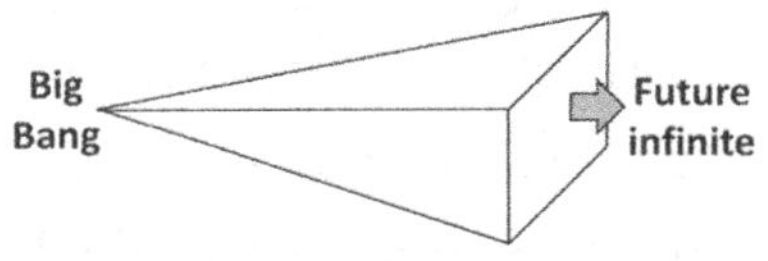

Despite this considerable improvement, the truth is that the second law of thermodynamics is far from answering all questions about time. In fact, it generates an additional question that is even more difficult to answer, again because of the absolute laws. For if time were really symmetric, then the second law of thermodynamics should apply to the future as well as to the past, being indifferent to both directions. This means that entropy should not only increase forward in time, but also backward, because there is no preference in one direction (symmetry) under absolute laws. So, we should not only age into the future, but also into our childhood. Something totally absurd. Therefore, it is wrong to believe that the second law of thermodynamics alone solves the mystery of the arrow of time, because it needs another ingredient. A fundamental assumption or principle called the past hypothesis, which says that the reason there is more entropy today than there was yesterday is because there was less entropy before yesterday, and less entropy before that, and less entropy before that... and so on and so forth. In other words, entropy does not increase in both directions in symmetric or absolute time, because there was always less entropy before than after. So, the second law of thermodynamics only works because it applies to a universe that requires a very special condition of entropic asymmetry. Only by accepting this initial hypothesis can thermodynamics explain this great difference between past and future that we observe in the universe. This makes the time of the Big Bang an incredibly rare, unique event. If the laws of physics are symmetric in time, and all past and future already exists in the block universe, then entropy should not only increase towards our future, but also towards the past, in the time before the Big Bang. Therefore, the arrow of time must run in both temporal directions because of this absoluteness of the laws of physics.

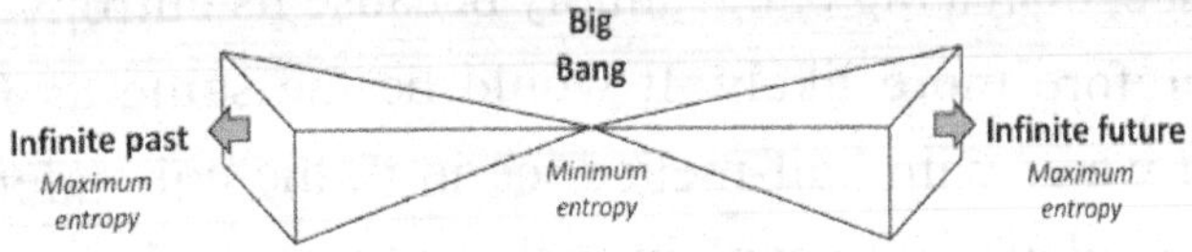

Then we would have a mirror universe identical to ours, governed by the same deterministic laws that apply to the same initial conditions: the Big Bang. Two identical universes for the price of one, but one evolving temporally into the past of the other, which would be the future for the other. An absurdity that not only adds yet another parallel universe to the long list in Parmenian physics, but also makes the Big Bang an incredibly special moment, because in both eternity into the future and eternity into the past we encounter an entropy that increases toward infinity, whereas right in the middle, at the Big Bang, we have the minimum possible entropy. Therefore, compared to the infinite instants of eternity, the Big Bang is a very, very exceptional instant. All because the symmetry of the absolute laws of physics itself forces the universe to expand and cool in both directions. And yet, at the Big Bang, the universe was in an incredibly dense and hot state. How did this happen? It is so paradoxical that other, even more absurd scenarios become more likely by comparison. For example, it would be more reasonable to believe that all the particles in the universe came together in a configuration similar to when the universe was supposedly a billion years old, a thousand years old, or even yesterday. This is because all of these possibilities are more likely than all the particles in the universe coming together in a big bang 13.8 billion years ago. This is because any configuration of the recent universe is easier to recreate spontaneously, by chance, than a big bang, simply because its entropy is higher and therefore more likely. It would be the same as finding a glass of water with half-melted ice in it, the equivalent of our universe today. So, to explain such a situation, obviously without a refrigerator, there would be two possibilities; the first is that each of the water molecules in the glass, all of them, spontaneously rearranged themselves into a very specific configuration to form a complete ice cube, and that it has been melting ever since until we find it half melted. Or the second possibility would be that only half of the water molecules spontaneously rearranged themselves to form that "half ice cube" a few seconds before we found it, and that we believe it has been melting from a full ice cube for longer than that. In terms of probability, this second option is more feasible, in the same way that it is far more likely to get a million heads in a row in a coin toss than it is to get two million heads in a row. Similarly, it would be much more likely that the universe we observe today formed recently, giving us the impression that it is much older, than that a real big bang happened by chance 13.8 billion years ago.

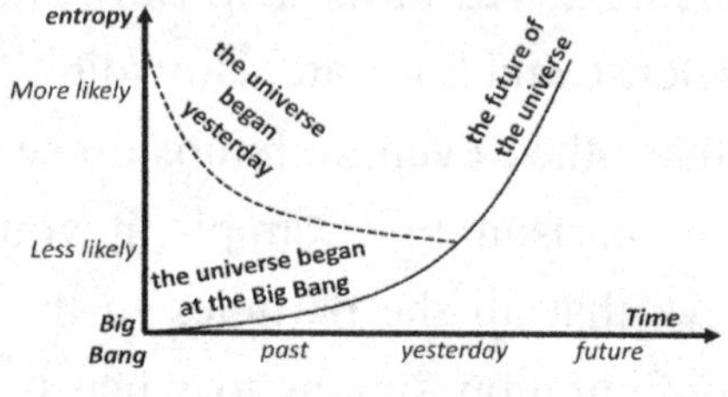

Another more absurd possibility, but even more likely than the previous one, is that instead of an entire universe having recently formed by chance, looking 13.8 billion years old, only a single brain has formed, with all the memories that make you believe you live in a universe that emerged from a big bang.

In other words, it is more likely that you are the only thing in existence, in the form of a floating brain that spontaneously formed from a completely thermodynamically dead universe (maximum entropy, the most likely state), than that there was actually a Big Bang. Simply because the number of atoms that must come together to form a brain with all the memories it contains in a universe like ours is much, much smaller than all the atoms in the universe coming together in a big bang. This criticism is known as Boltzmann's Brain, since he is credited with the original idea that the universe could have started from a random fluctuation. And as we have just seen, this is incredibly unlikely. So, it is quite clear that Parmenian thermodynamics is not only incapable of solving the problem of the origin, but makes it much worse. A fact that is not improved by the introduction of cosmic inflation, as many physicists think. In fact, it makes it worse, incredibly so, because it makes the initial conditions of the universe even less likely, as we will see in Chapter XV when we analyze the Big Bang theory in depth.

Then, clearly, current thermodynamics is very far from solving the problem of time, not only because it exacerbates the problem of origin, but also because it says nothing about the other relevant problem of time, the problem of the present. This is because it continues to be interpreted in an absolutist or eternalist way. Therefore, the probability or randomness used by thermodynamics is purely epistemic. Randomness is not recognized as something real or ontological, as we saw in the first part of the book. Then, Boltzmann's thermodynamics, interpreted under the Parmenian paradigm, uses randomness as a mere effective tool, as something fictitious, which is only a product of our ignorance, due to the impossibility of measuring with perfect precision each one of those innumerable parts (particles, atoms or molecules) of the physical system. So, it is wrong to think that all those microstates of a given macrostate, its entropy, is something real, because there is really only one microstate at a time. It is like throwing a dice. To calculate the result of a throw, we have to consider all possible outcomes, from one to six. But when the die is thrown, we know that only one result will come out. Similarly, thermodynamic systems have only a single microstate, a single configuration, but since we cannot know it, we must consider all possible microstates. To return to the example of the system of four electrically charged particles, if we measure that it is neutral, this does not mean that it has all 6 microstates simultaneously, as if there were a quantum superposition of all of them. It simply means that it has one of the 6 possible states, any one of them, but since we cannot know or measure it, we must consider all 6. Therefore, 5 of the 6 configurations or microstates are completely hypothetical, they do not exist. It is just a statistical tool that helps us compensate for the lack of knowledge about what the real configuration or microstates of the system are.

Therefore, current thermodynamics is fundamentally epistemic, not ontological; because entropy measures the degree of our ignorance with respect to the particular (microscopic) state of thermodynamic systems, due to our unavoidable metrological limitations. Therefore, the more entropic a system is, the more ignorant we will be with respect to what particular microstate that system would have, since there would be many possible configurations that would be compatible with that entropic macrostate that we can measure and know. Returning to the example of the system with four electric charges, if it has entropy 6 or is neutral, this means that we have an uncertainty of 16.6% or 1/6 chance of guessing what microstates the system really has. On the other hand, the system with entropy 1, or when the global charge is +4 or -4, means that we have no uncertainty about the configuration, since we know with 100% certainty that each of the particles of the system has the same charge. So, in this sense, the supposed random choice (the electric particles flipping a coin to choose their charge, or the molecules of a gas flipping a die to choose the direction to go) among all those possible configurations is not real. It is only a computational tool. This means that thermodynamic systems are ontologically just as determinate and certain as non-thermodynamic systems, such as throwing a ball. The only difference is that in thermodynamics, physical systems are so complicated that it is technically impossible to know exactly what their past was and what their future will be. This is why statistics is used. However, ontologically speaking, there has always been only one possible history, only one sequence of successive microstates. Therefore, this uncertainty and probability is very similar to the nonlinear systems we saw in the first part of the book, in the sense that both have a weak determinism, with a purely epistemic uncertainty, again the product of our ignorance, and not the fruit of any real or ontological randomness. In this way, stones and apples, as well as gases and climate, would always have a single history. A single past and future determined from eternity. That is why thermodynamics, under the present paradigm, still does not answer what the present is, because just as in a parabola describing the trajectory of a ball there is no point inside it that is special, in thermodynamics the microstates are not special either. Therefore, all microstates have exactly the same probability, regardless of their configuration or whether they belong to a high or low entropy macrostate. Therefore, it is also impossible to identify a microstate that can serve as the present to divide eternity into past and future. The only difference that thermodynamics brings is that at least this time we have a clear distinction between past and future. A past where the entropy is lower and lower, while the future has a higher and higher entropy. But it is a temporal asymmetry that still belongs to a block universe, where there is only a single history or timeline petrified in eternity, without a becoming or even a "now."

And as if it were not enough to solve the problem of origin and present, thermodynamics also fails to explain how eternalism is consistent with quantum mechanics; by saying nothing about the problem of measurement or collapse of the wave function, which occurs in a truly random, not epistemic, manner. This obviously destroys the fundamental determinacy postulated by the eternalism of current physics. Nor does it say anything about the reason for the violation of time reversal symmetry in quantum field theory. Something very important, because if the asymmetry or arrow of time is an exclusively macroscopic or thermodynamic phenomenon, then there should be no microphenomenon, especially at the level of quantum particles, that is capable of violating such temporal symmetry. In other words, no particle should be able to distinguish between past and future. However, this is exactly what has been observed in experiments with particles under the influence of the weak interaction. As we saw in Chapter IV, the decay of a particle called the B meson was used, which can oscillate and decay into another similar particle ($B^0 \rightarrow B^+$), which can decay back into the same previous meson ($B^+ \rightarrow B^0$). Now, it has been observed that these two decays, to and from the B+ meson, take place in different time intervals.

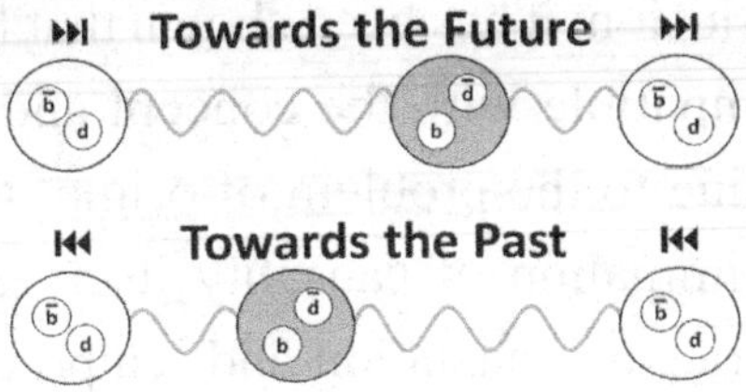

This difference makes it possible to clearly distinguish where the future is from where the past is, which should be impossible according to the Parmenian paradigm. Thus, empirically, the direction of time is not only a thermodynamic phenomenon, but also a fundamental one. A temporal asymmetry that is also mandatory in the Standard Model of Particles in order to maintain its internal consistency while respecting its principal symmetry or CPT (Parity, Charge and Time Reversal Symmetry). An invented symmetry, which, as was said in Chapter IV, requires the violation of each of these symmetries in order to be valid. Symmetries that the physicists were convinced were fundamental, but when they saw that each of them was violated by the weak nuclear interaction, the physicists were forced to merge the three in order to save the consistency of their paradigm. Therefore, the fact that time is not fundamentally symmetric is not only an experimental fact, but also a theoretical requirement of the two pillars of physics, because we find it not only in the expansion of the universe itself (general relativity), but also in quantum field theory, due to its CPT symmetry (violation of T-symmetry). A fact that still surprises and shocks physicists, and which they apparently still cannot explain, because it simply does not fit into their absolutist temporal worldview, where time must be completely symmetrical, homogeneous, unchanging, in order to maintain the absoluteness of their physical laws.

In conclusion, it has been shown that Parmenian physics is totally incompatible with the concept and empirical experience of time, due to the problem of origin, the problem of the present, the elimination of causality itself, and the violations of the symmetries of translation and temporal inversion in observations and experiments. These are all facts that theoretical physicists are reluctant to accept, so they delude themselves by ignoring them and taking refuge in their imaginary Platonic world. If you think this is an exaggeration, that physicists are being taken out of context, let's see what some of the most influential physicists think. Let's start with Julian Barbour, a British physicist born in 1937 who is considered an authority on the cosmology of time. He is also the author of several popular books, including The End of Time, in which he explicitly states that time is an illusion because there is no motion or change whatsoever. He also states that such an illusion is a product of our brains, saying in his book: "The mature brain is a time capsule. History is in its structure." He also says that the great problem of modern physics, the reason why it has not been possible to unify general relativity with quantum mechanics, is due to the fact that time is still considered to be real. An opinion very similar to that of another authority on the subject, Italian physicist Carlo Rovelli (born in 1956), best-selling author of several books that have sold millions of copies worldwide. One such book is The Order of Time, in which he says: "In the elementary grammar of the world, there is neither space nor time. [...] At the most fundamental level we know today, then, there is little that resembles the time of our experience. There is no special 'time' variable, no difference between past and future, no space-time." He also stated in an interview with Penguin Books UK 2018 about the same book that "time is about losing things or opening up new spaces, in both cases it is emotional [...] at the microscopic level, the fundamental equations of the world do not distinguish the future from the past, so it seems that time does not exist at the fundamental level." Another British physicist and writer, Paul Davies (born 1946), in an interview with FQXi (Foundational Questions Institute) in 2013, said, "The flow of time is an illusion, and I don't know many scientists and philosophers who would disagree with that, to be perfectly honest [...] and presumably the explanation for this illusion has to do with something that is here (in the head); and it's related to memory, I suppose, the formation of memories and so on. So, it's a feeling that we have, but it's not a property of time itself [...] Time doesn't flow. That's part of psychology." Or the case of the famous American physicist and best-selling author Sean Carroll, who wrote in his book The Most Fundamental Ideas of the Universe: "The arrow of time is not written into the fundamental laws of physics. It is merely an epiphenomenon, the product of very special initial conditions. The directionality or flow of time is not part of the fundamental architecture of reality, which makes no distinction between past and future."

Then, since the dogma of absolute laws is obviously more important than our obvious experience of the becoming of time, many physicists have decided to throw such a problem to another science. In this case, they blame neurologists or psychologists for our temporal hallucination, as we saw in the previous paragraph. This absurd pseudo-solution is further proof of the serious consequences of extreme specialization, for they simply make fools of themselves when they go outside their specialty. This is because all neurological activity, experience, and memory depend on information received from the senses.

Information that comes from outside, from nature. Therefore, if the future exists out there, it should be possible to perceive that information, for example, by seeing and remembering the future. Moreover, if the brain voluntarily chooses not to perceive or remember the future, it would be something that would not make the slightest sense in Natural Selection; because seeing the future would be an extraordinary strategy to be selected for. How would it not be an evolutionary advantage to be able to see the future in order to hunt or avoid being hunted? And even if we believe that there is a gigantic conspiracy in the animal kingdom to hide the future from us, our brainless machines should easily measure or observe a different moment than now. In other words, all our technology would essentially be time machines, since they would interact directly with the block universe and be able to measure other instants of eternity. Therefore, when we take a photograph, we should be careful not to end up photographing the person 10 years younger. In short, the debate about whether time exists or not makes as much sense as the debate that some fish would have who think they are very intelligent and revolutionary because they question the existence of the water in which they swim.

In contrast, Darwinian Physics offers a much simpler and more reasonable explanation for this alleged dilemma. It is simply trying to solve a real dynamic problem (time) through a purely kinematic or static approximation, such as the equations of the laws of physics. And since mathematics is the only tool theoretical physicists use to think, they end up confusing the tool, the representation, with reality itself or what they want to represent. Completely forgetting that mathematics is only a tool, perhaps the most powerful of all, but still a tool. A tool to describe, to represent, to make models of reality, but it is not the real thing itself. It is only the best method we have to describe the regularities that exist in nature, which are also very different from mathematical patterns, since the former emerge, change and evolve. In short, they flow with time. Mathematical patterns, on the other hand, are static, unchanging, absolute; because they are symbolic, not physical entities. To make a mistake on this basic point would be like confusing a real person with his or her photograph, just because the latter is a very accurate representation of the former. However, no one would marry a photograph (I hope), knowing that it is a representation, but not a real person. One that moves, changes and evolves over time. So, the idea of the block universe, like the Minkowski diagrams, are simply models, incredibly useful and necessary representations in the work of a theoretical physicist, but mathematical tools. Like the Feynman diagrams, which are mathematical equations made into drawings, which are incredibly useful in facilitating calculations in quantum field theory, but physicists know, or should know, that they do not represent anything real. That they are fictitious things, because what would really exist at the fundamental level would be quantum fields (second quantization), not particles (first quantization). Therefore, the whole discussion about whether time is real or not is absurd, because it is based on the nonsense of taking mathematical models too seriously, to the point of considering them more real than the real thing.

And we may ask ourselves why, to what is this fundamental confusion due? Part of the answer has already been given in the first part of the book, since the extreme specialization and monopoly of mathematics in theoretical physics would have a great responsibility, having excluded philosophy, making theoretical physics extremely instrumental. Therefore, only what can be calculated counts, nothing more, remaining almost exclusively in the epistemic field. It no longer matters what nature itself is, what these equations and models mean, but only whether it can be calculated or not. This completely abandons the original purpose of science, which was to discover and understand the ontological, what nature really is and how it works. All because it has been forgotten that mathematics is a tool, perhaps the most powerful tool ever invented, but it is not the real thing. We have already seen this confusion between epistemology and ontology in Chapter IX, with regard to the principle of non-contradiction and quantum mechanics. And just as it was expressed there, this inability to distinguish between the thing itself and what is used to represent it is not only very recurrent, but also much more dangerous than theoretical physicists imagine, since it leads back to fundamental idealistic physics by disconnecting from the real, by getting lost in the Platonic world of ideas. A philosophical position very similar to that of the famous German philosopher Immanuel Kant (1724-1804), who was also not interested in knowing what things are in themselves, what he called the noumenon, because he assumed that it was impossible to know it, since we have no direct experience of the real. We only know the things that our brain interprets from the information that our senses receive. Therefore, what we know, what he called the phenomenon, is not the same as the noumenon or the real world.

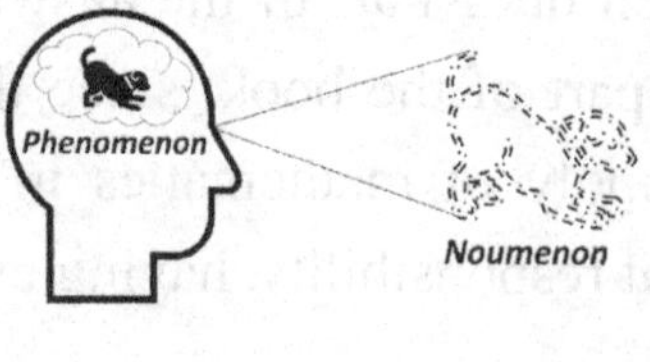

So for Kant, as for many physicists, time is a property of our cognitive box, of our brain. Not a real property of nature. A subjectivity that for Kant did not prevent him from doing science because we all have the same cognitive apparatus, the same kind of brain; and therefore the same kind of subjectivity. Therefore, universal knowledge can be acquired in the sense that everyone can understand it, reproduce it, or experience it. However, this position completely eliminates the main characteristic of science, realism, because it completely abandons the main aspiration of science, which is to know and understand what the real world is like. This is why most physicists despise philosophy, especially metaphysics, because they have completely abandoned the task of knowing the noumenon, of understanding what the world itself is like. Hence they are not interested in questions such as: what is energy, spin, a quantum field, space, the random collapse of the wave function, and, of course, time. A position that may even be respectable if one suspends judgment with respect to the noumenon, as happens in Kant's philosophy. The problem is that by eliminating philosophical training in physics, one ends up doing bad philosophy; because one does not suspend judgment on the noumenon, but ends up interpreting the real world anyway, making statements about what is out there, from its mathematical models; attributing the absolute and eternal character of its equations to an intrinsically dynamic, mutable, and chaotic nature. Hence, many physicists openly preach that time does not exist, as we have seen recently. But even with the correct Kantian attitude, by consciously adopting its instrumentalist stance, as in the case of the German physicist Sabine Hossenfelder, such an attitude is effective only in the stage of normal science, because in the stage of crisis it is not only useless, but can even hinder the solution (revolution), by discouraging the addressing of ontological questions or fundamental premises in the hard core, considering them as metaphysical acts that serve no purpose.

Therefore, as just said, the foundations of physics becomes more and more idealistic. And not even in the good sense as Kantian (instrumentalist), but in the end they end up making metaphysical statements about the real world, imposing a particular reason (rationalism) over the obvious evidence (empiricism) of quantum randomness and the becoming of time, by denying that these exist. This ends up completely reversing the hierarchy of existence, because first there was a Big Bang, with an expansion of the universe, the formation of stars and galaxies, the appearance of life on any rocky planet, the evolution of multicellular life, the emergence of a fish from the water, the descent of a primate from the trees, and then a hominid appeared and said: cogito, ergo sum. In other words, nature is completely independent of our beliefs and convictions. And it is exactly the opposite that is supported by current theoretical physics, due to its devotion to the sacred equations or laws of physics, because it confuses a certain mathematical model (the map) with the real (the territory).

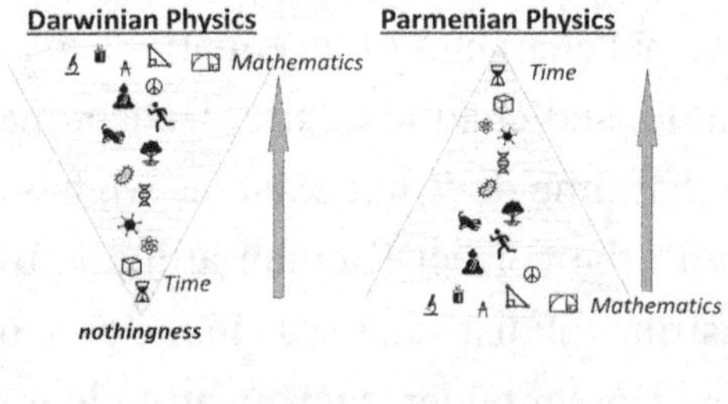

In other words, the foundations of physics ends up seeing the world upside down. A serious attack against realism in science, which is even more evident in Parmenian quantum physics and its "shut up and calculate" motto, where it explicitly recognizes and defends non-realistic positions such as the subjective collapse of the wave function, or the increasingly popular qbism or quantum Bayesianism. Without going any further, this Kantian position is the official one under the Copenhagen interpretation. For example, one of the main intellectual fathers of this interpretation, Neils Bohr, said: "It is wrong to think that the purpose of physics is to find out what nature is like... Physics is concerned with what we can say about nature." A philosophical stance confirmed by one of his students and closest colleagues, Werner Heisenberg, who said, "The laws of nature that we formulate mathematically in quantum theory no longer deal with the particles themselves, but with our knowledge of these elementary particles. Therefore, we should not be surprised that the existence of time is denied in the foundations of physics, because nature is conceived through the filter of mathematics, forgetting that mathematical laws are only a tool to describe the regularities in nature. An error that has promoted these idealistic positions, making physics abandon the main purpose of science, which is to tell us how nature really is, to know the noumenon. Precisely where time is, outside our heads.

So, having discovered the epistemic error of the foundations of physics, what would a realistic or ontological account of time look like? One possibility would be Possibilism or the Growing Block Universe. This temporal worldview assumes the existence of the past, because it has already happened, and the present, because it is the instant in which we exist; but there would be no future, because it has not yet happened. One of its main proponents is the British physicist Fay Dowker, Professor of Theoretical Physics at Imperial College London, and she says that instead of having an infinite block universe in the future (eternalism), it ends in the present. Therefore, the block universe grows in every moment, in every "now." This temporal worldview seems more reasonable to us because it is more reasonable to believe that the past exists because it has already happened, while the future does not exist because we have not yet experienced it. But although it makes much more sense than eternalism, it suffers from the same problems, because it still does not solve the problem of the present: how does a new now emerge, where does it come from, what is the mechanism? Let us remember that Einstein's special theory of relativity eliminated simultaneity, so what would this present be, where is it, and how can it be defined? Fay Dowker would answer by saying that it would be similar to how a pile of sand grows in an hourglass, where the pile would correspond to the block universe that grows thanks to the grains of sand that would be atoms of space-time. However, it does not explain how these space-time atoms are created or appear, what they are, why there were none before but started to accumulate since the Big Bang? This is a problem that all models of cellular automata or graphs suffer from, such as Wolfram's theory, which assumes the existence of these hypergraphs, which are the equivalent of space-time atoms, but does not explain where they come from, nor why they are like that, nor how they are created and appear. Well, the typical problem of the origin of always. They simply assume that things are like that, and under this assumption they try to come up with a mathematical model that can be reconciled with observations and experiments. However, despite this omission, none of these attempts have been able to come close to the official models of physics, for example, being able to derive the equations of quantum mechanics, let alone general relativity. This is why the eternalist worldview is more popular, because it saves all these problems by applying the Parmenian strategy of denying the problem. Thus, by making the creation of nothing impossible, there is no need to explain how the "now" comes into being, since all these instants of time have always existed, since eternity, thanks to the fact that the laws of physics are absolute.

But in addition to the problem of the present, possibilityism or the growing block universe shares the other problem of eternalism: explaining where all that past is, those 13.8 billion years of history and eternity that should come before. Because they avoid having to explain the whole infinite future of eternalism, which is a great saving, but they still assume the existence of all the past. So where is it, how is it preserved, why don't we have access to it? Parmenian physics answers that it is out there, but our brains ignore it for some reason. So would the growing block universe solve it in the same way, are we being fooled by our fat ball in our heads? Also, assuming they only exist from the past and not the future sounds more reasonable, but physically it is much easier to "supposedly" travel to the future due to relativistic effects like time dilation (this will be explained in Chapter XIII), but so far it is impossible to travel to the past due to causal paradoxes, like the classic time travel to kill your grandfather before your father was born. So, who traveled back in time if you were never born? Paradox. So, if the growing block universe recognizes the past, why does it seem impossible to travel to it, doesn't it seem like an unnecessary assumption, why would one have to assume the existence of a past that would be logically impossible to visit? Then, can something exist that is impossible to see? In other words, can there be something somewhere that is impossible to go to or visit?

For all of the above, it seems the most economical (Ockham's razor) would be to assume the existence of the only thing that we are certain exists because it is the only thing that we have observed with our senses and measured with our instruments, the "now." And spare us the assumption that there are infinite instants in the past and in the future that we can neither access nor observe. This is exactly what Darwinian Physics does, because the real illusions would be what we have never experienced or measured, the past and the future. This extremely radical position for physicists, but actually the most conservative, is called the A-theory of time, or presentism, as already advanced in Chapter IX. And it is so because it automatically resolves the greatest discrepancy between theory and experience in the history of science. A denial of the obvious that began with Parmenides, who denied the possibility of change simply because he could not understand it. The same monumental error that physicists make today in denying time, simply because they cannot understand it according to their religious absolute laws. Quite unscientific behavior, since the real difference between philosophy and science is precisely to always privilege experience and evidence (empiricism) over any particular reason (rationalism). It does not matter how logical, reasonable, or beautiful it may seem to us, because a particular logic or reason can always be wrong, because it is based on false assumptions; as has been demonstrated throughout the history of science, and as will continue to happen in the future. Thus, physicists have reverted to the old bad habits of the Greek philosophers by privileging a particular reason, the Parmenian paradigm, over the obvious experience. Thus, at least in this sense, Darwinian Physics turns out to be much more scientific, since it recognizes only the existence of the present. This is completely in line with our experience, but also with the empirical evidence, since no apparatus or instrument has measured any other moment than the present, since no one has created a time machine outside of science fiction movies and novels.

However, the presentist theory of Darwinian Physics not only recognizes the evident, the present, ignoring what has never been observed, the past and the future, but also does so by providing a mechanism that explains how it arises and evolves. And it does so not by an assumption to be believed on faith, but by another empirical scientific fact, ontological chance. A natural phenomenon that we have recognized and called nothingness. A solution that allows us not only to comply with the principle of sufficient reason, but also to obtain a high degree of empirical consistency by explaining quantum mechanics by solving its main problem, measurement. This is because the problem of the present and the problem of measurement are essentially the same problem, since what causes the collapse of the wave function is the present itself. Therefore, the collapse is random because it is the product of nothingness itself. From which emerge the presents that ascend into existence. Therefore, what causes the collapse of the wave function is neither consciousness nor other parallel universes, but time itself, its becoming. Specifically, the materialization of presents arising out of nothing. This will be explained in much more detail in the next chapter, but as an introduction, to prove this intimate relationship between the present and randomness, the determination of the properties of quantum particles, such as the spin of an electron ("up" or "down"), is random, as has been mentioned several times in this book. In this way, we can use the same analogy as always, that of a coin toss. It is a very good analogy because in both cases the probability is independent at each instant. For example, if we get 100 heads in a row, that lucky chance does not make it more likely that the next flip will be tails, because the probability is still 50%. Each roll is completely independent of the previous results. The same applies to electrons. If you randomly measure the spin "up" of the same electron 100 or 1000 times in a row, this will not at all change the probability that the spin "down" will be observed in the next measurement. It will still be 50%, because the history does not affect the present action. The same is true for the rest of the properties of quantum systems. This peculiarity of quantum objects is not only a proof of ontological randomness, but also of presentism, since what happens "now" is fundamentally independent of the past, at least at the quantum level. What must be added is the most obvious, that randomness is independent of the future, because the outcome is not determined. Just what randomness means. Between these two facts, quantum mechanics would prove that nature, at its most fundamental level, is presentist, that the "now" is fundamentally independent of what has happened before and what will happen after. This renders both the past and the future redundant, as they become unnecessary assumptions given the inherent randomness of quantum mechanics. Therefore, holistic presentism not only solves the mystery of time empirically, but it would also be the most satisfactory and, above all, the most economical explanation, since it saves us from having to assume not only the existence of infinite parallel universes, but also of infinite instants of the past and the future. All this thanks to the recognition of what we already know to exist: the now and quantum or ontological randomness.

So in order to see how exactly this presentism works, we must first explain what the present is, and how this definition is compatible not only with quantum mechanics, but also with the rest of the current foundations of physics. To do this, we have to do exactly the opposite of the Parmenian paradigm, and recognize change ontologically, because there is no other way. In fact, this omission sums up the whole problem of time in physics, because one cannot understand what time is without change. Therefore, in Parmenian philosophy, time is eliminated because change is eliminated first. Fortunately, in Darwinian Physics, change is not only possible, but indispensable, because nothingness is pure change. Therefore, time can emerge from nothingness because time is created from these changes. Remember that nothingness is chaotic change, as stated in the previous chapter. This means that it oscillates between two states in a completely random manner. Fortunately, through holistic evolution, orderly changes can emerge from this nothingness that correspond to regularities or patterns that can be repeated and conserved. It is this definition of existence that explains the properties of time, such as its continuity and predictability, which are based on the conservation of a state. The tendency to repeat itself, to remain the same. In this way, the evolution of time is the evolution of existence itself, because everything begins in a state of very low entropy, where everything changes very quickly, nothing lasts very long (very chaotic changes), and progresses towards an increasingly entropic or stable state, where everything changes more slowly (more ordered changes), until we reach a point where nothing seems to change, everything remains the same, because there is no difference between before and after, because everything is completely frozen. Thermal death or thermodynamic equilibrium of the universe, where time comes to a standstill because existence has reached its maximum degree of order, being totally invariant changes. Therefore, time as we know it exists only in this transition, in evolution, because in pure nothingness there is no time, because there is only chance. While in absolute existence there is no time either, because nothing changes, everything remains, everything continues exactly the same, as in the eternalistic world view of the block universe of current physics. Thus, time as we know and measure it is a product of complexity, of the mixture of chaos and order, because in both nothingness and absoluteness, time as we know it finds its end.

This temporal worldview would be compatible not only with the two pillars of physics discussed in the following chapters, but also with the theory we are analyzing in this chapter, Boltzmann's thermodynamics. In fact, it would be even more compatible because the use of probability would not only be instrumental or epistemic, but also real or ontological. In other words, the particles would really be tossing a coin or dice to know what to do at any given instant. In fact, Darwinian Physics is so compatible with randomness that it recognizes both types, epistemic and ontological. And both coexist harmoniously in the new paradigm because they are on different scales in the holoarchy. In other words, there is no contradiction between these two definitions of randomness in Darwinian Physics, since the epistemic or thermodynamic type (weak determinism), due to our metrological limitations, dwells more in the macroscopic scales of nature (higher holons). On the other hand, in the microscopic terrain (lower holons) we find ontological randomness due to quantum mechanics. But here not only is quantum randomness recognized, but it would also explain the main feature of this pillar of physics, which owes its name to the famous quanta, or the discreteness of the behavior of systems. Something that also applies to thermodynamics, since the "changes" or random selection between the microstates of a thermodynamic system would have a more literal interpretation here, since they would be the product of the materialization of the present coming from nothing. Thus, what entropy would really measure would not be how many microstates there are (remember, they are hypothetical), but the degree of stability of the physical system in a given macrostate. Its degree of probability of changing globally, due to these constant random fluctuations at the microscale, is the product of time itself, which comes from nothingness. For this reason, Boltzmann's thermodynamics assigns the same probability, value or importance to each microstate, because the macrostate, the system as a whole, has always been relevant. Specifically, whether it is about to change or whether it is in a more stable or permanent state.

This solves a big mystery about what entropy really is, because although it is one of the most famous and used concepts in physics, there is a lot of confusion about what it represents, because again, due to lack of philosophical skills, many physicists forget that entropy is a purely epistemic concept. As has been repeated several times, the rest of the microstates determined by entropy are hypothetical, since obviously a thermodynamic system can only have one microstate at a time. So, it is wrong to relate entropy to something real in the system, such as its spatial configuration, and say that it measures the degree of chaos or disorder; because the spatial configuration, how the parts are arranged in the system, is completely irrelevant. Again, all microstates have exactly the same probability because they have the same relevance. It does not matter what they look like.

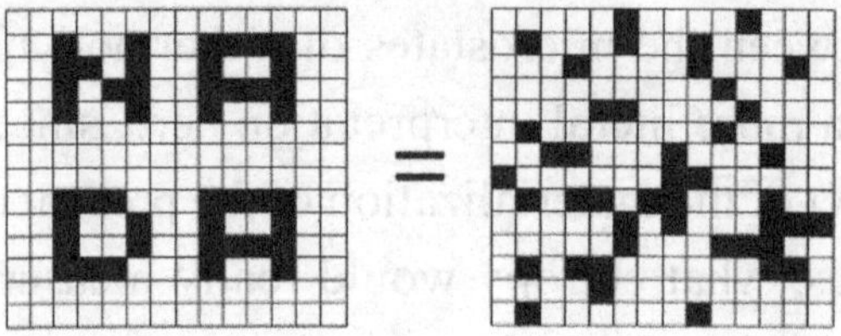

For example, in the figure above, it seems to us that the first configuration is more special, more ordered than the second configuration, but thermodynamically both are identical. The first is not less probable than the second because both have exactly the same meaning as any other possible combination; because all microstates are identical in thermodynamics. What is special or relevant are the macrostates, because they differ from each other according to the number of compatible microstates. For this reason, it is officially incorrect to use the concepts of chaos or disorder to refer to entropy, although many physicists continue to use them, especially in the science communication, because the configuration (microstates) does not matter. Thus, if we want to interpret entropy ontologically, we must do so temporally, not spatially, since the true referent of entropy is the degree of stability of the system as a whole. Its tolerance to change or variation in the macrostate. It is therefore a fundamentally dynamic (temporal) concept, not a static (spatial) one. This is quite consistent with the arrow of time. Therefore, if one wants to relate entropy to something real, one must relate it to the macrostate, to what can be measured; and not to something hypothetical, such as the number of microstates. Then, contrary to popular belief, entropy has more to do with order than with disorder or chaos. Only it would be a temporary order, in the sense of less change. This is why all systems naturally tend toward thermodynamic equilibrium, because the most entropic state is also the most stable state. Where the system as a whole remains more constant, without major changes. In this way, entropy as a temporal order is much more consistent to represent the arrow of time, as it measures the degree of durability, the ability of a macrostate to "remain," to "endure." Something that makes much more sense than associating entropy with disorder, or even worse, with "chaos. Because to relate the scientific fact that the universe and all things naturally tend to become increasingly cold, uniform, static, simple and boring (thermal death or thermodynamic equilibrium) to a more chaotic state does not sound very consistent. It is much more consistent, thermodynamically speaking, to associate entropy with order, with not changing.

This ontological interpretation of entropy confirms this great compatibility between Boltzmann's thermodynamics and Darwinian Physics; because it would indicate that there is an intimate relationship between entropy S and the degree of existence or "P(H)," because they both grow with time and because they basically represent or measure the same thing, the degree of stability or immutability of a closed system. The ability to stay the same or not to change. Therefore, contrary to popular belief, the universe is evolving toward an increasingly ordered state, since entropy, as ontologically defined, can refer to what it actually represents, time (not space). That is why the universe is becoming slower, duller, more uniform, and colder, because it is trying to "change" less and less, to become more ordered in time.

$$S \equiv P(H)$$

Having shown that thermodynamics is compatible with Darwinian Physics, we will now see how thermodynamics can solve the major problems of time under the new paradigm. Let us start with the problem of the present, which would be solved immediately by being in a presentist paradigm. Something that can be achieved in a simple way, since it is only necessary to take the random selection of microstates, which was only instrumental in the Parmenian paradigm, more literally, since it would correspond to the materialization of the present. Which is also entirely consistent with reality, since, as has been said above, thermodynamic systems have only one microstate at a time, one possible configuration, just as there is always one instant at a time, one "now." Thus, the present in thermodynamics would simply be the last microstate that has occurred. While the past or all previous microstates or instants would be represented in the "now" in the form of "memory" or "information." Something that would correspond to what we have called causal inertia; but that would be easier to visualize in thermodynamics. This is because the memory or information in the last moment or microstate would be encoded in the form of the macrostate in which it is. If the current microstate is in a low entropy macrostate, it is because it has had little past (little causal inertia); or if it is in a high entropy macrostate, it is because it has had much past (much causal inertia). Therefore, this information or memory is already in the "now," so the presence of a past with an independent existence inhabiting a fourth spatial dimension is simply redundant, because this information of the history of the system can be derived from the "now" itself, in the current microstate. A knowledge of the past that is not exact or perfect, but that does not matter to Darwinian Physics because it is not deterministic or absolutist. Moreover, perfect knowledge of the past has always been hypothetical, because it is technically impossible, since no measurement technology is perfect. Not to mention the indeterminacy principle of quantum mechanics itself, which makes absolute certainty impossible, this time in principle. Thus, uncertainty is not only technical or epistemic, but also fundamental or ontological. Therefore, it is not necessary to know exactly each of the microstates that existed before, but only enough to understand how much the present has evolved. And this is exactly what thermodynamics, as interpreted by Darwinian Physics, offers. In short, thermodynamics is easily interpreted by holistic presentism, since thermodynamic time is discrete by dividing the history of macroscopic systems into microstates that would correspond to instants or "nows." And since these instants are fundamentally independent of each other and have the same probability, then the last microstate or the present configuration of the system corresponds to the present. This saves us the necessity of assuming the existence of all the infinite instants, this time in the form of all the microstates of the past and of the future, which would exist simultaneously, forming a block universe, because the past is simply transformed into information or memory of the new now (causal inertia). On the other hand, the future or subsequent microstates do not exist either, because they simply do not yet exist, since the "now" has not yet evolved to be located in those even more stable macrostates or with a higher degree of entropy or order.

This makes thermodynamics infinitely more economical, since it requires only one instant, one microstate at a time.

But the ontological recognition of randomness allows us to solve not only the mystery of the present, but also the most difficult, indeed impossible, problem for Parmenian physics: the origin. This is because the explanation of the very low entropy at the Big Bang, the most improbable, unimaginable event, is simply due to the fact that the evolution of the cosmos began from nothingness. A very unstable macrostate, since it was literally changing at every instant. In other words, for every change in the microstate (present), the whole macrostate changed as well; because there was no distinction between micro and macrostate at the space-time point of nothingness, because there was simply no scale (vertical dimension). But from there, through holistic evolution, the cosmos tends toward more and more durable macrostates, because larger scales or holons emerge, which are also more and more resistant to changes at the level of microstates. This allows the cosmos to better withstand (with greater probability) the random changes of nothingness, making nature more and more stable, in the same way that particles flipping a coin are more stable in a more entropic or homogeneous macrostate, as we saw in the example of the four charged particles. Only here we would have to change the flipping of the coin for the present particles coming from nowhere. This is the advantage of equating the thermodynamic entropy S with the degree of order or P(H) of existence, since both are fundamentally probabilistic concepts and both tend to increase with time. In other words, the second law of thermodynamics is equivalent to horizontal holistic evolution itself in Darwinian Physics, because both seek the same goal, an increasingly stable, more durable, less changing global state of the system. This solves all the problems of thermodynamics by abandoning its epistemic and eternalist version and interpreting it in an ontological and presentist way.

We have seen how the presentism of Darwinian Physics is consistent with quantum mechanics and now with thermodynamics. However, all of its criticisms must be addressed, the most common being the duration of the present. This is because it seems incommensurable, since the now can always be segmented into smaller and smaller pieces. Moreover, if we are aware of it, it has already passed, it is already past; and if we are waiting for it, it is still future. Then it seems that the present is really an illusion, because it cannot be touched, like a mirage. In Darwinian Physics, however, this criticism is easily resolved by holarchy, because there is not just one present, but many of different sizes. Thus, there would be a fundamental presence, the oscillations of nothingness, which would explain the invariant speed of light (this will be explained in much more detail in Chapter XIII); while the other major presences would be simple resonances propagating toward the higher holons. Then the different "nows" that are measured would not refute holistic presentism, on the contrary, because they would be the different presents of the different holons, as we saw in the previous chapter. This is why the present seems incommensurable, because everything is present.

Another criticism against presentism comes from the need to assume change in order to explain time. In this sense, to deny change would be to deny time. And an argument in this sense would be that if there is change, there must be something to change with, because one always changes with respect to something else. Therefore, change is always relative. So, if time really exists, it must flow in relation to another time, for example, in relation to a supertime. So, when we move through time, it would be like moving one second for one super-second. Something unrealistic, since there is no evidence that there is anything beyond the space-time of our universe. So, if there is no supertime to change with, our time must be just an illusion. Darwinian Physics solves this by eliminating this supertime, precisely because of the presentist holoarchy. In this way, the present of the higher holons changes with respect to the present of the lower holons, and all of them (the cosmos) can change because they do so with respect to nothingness. The latter is pure change, not just because it is the least change or the fastest possible change, but because it changes with respect to itself. In fact, this argument is another justification for the necessity of the two states in nothingness, because in addition to the argument of randomness, brevity, and uncertainty, we also have the argument of mutability. Thus, for nothingness to be pure change, it must change with respect to itself; and for this, at least two states are needed. A condition which, as we saw in Chapter VIII, is unavoidable in order to fulfill the definition of nothingness. In conclusion, time exists not because it changes with respect to a supertime, but because it changes with respect to nothingness. That is why existence tries to distance itself from it as much as possible, to change less and less, to evolve holistically, to obtain longer and longer or orderly changes, to become more existential, more stable, more predictable, more permanent. So obviously becoming makes no sense under the Parmenian paradigm, but the problem is obviously not nature, but this Parmenian logic. Therefore, if we want to explain time and the present, it is essential to recognize nothingness or ontological randomness; because as long as the latter continues to be considered a mere statistical (epistemic) tool, instead of something real, "becoming" and "now" will disappear. An inevitable conclusion, because without chance, without nothingness, there remains only absoluteness, eternity.

It has become quite clear that holistic presentism is well compatible with current physics and is internally consistent, so why is it so radically different from today's dominant eternalist worldview, which denies the very existence of becoming and causality? How can Parmenian physics work so well if reality is presentist? This can be better understood if we use an analogy again. In this case, we can imagine the fundamental laws of nature, the two pillars of modern physics, as the equivalent of a movie reel. In this context, for the movie of the universe to exist as such, the real thing that we experience as spectators, we also need the movie projector. The apparatus that sets the tape in motion and projects it frame by frame, generating the becoming and the cinematic experience of existence. So here is the key question: what is the mechanism that projects or makes the laws of physics work in the Parmenian paradigm? Where is the projector in modern physics? Nowhere. They have completely forgotten about it. They are convinced that the film tape, without any change or movement, where the first frame exists simultaneously with the last on the reel, is actually the cinematic experience of existence, not the projected film. This is essentially the time problem of the foundations of physics. The "forgetting of the projector." The omission of the mechanism that generates the present, the change, the becoming. That is why it has been said that this is the most important problem of physics, since it is practically 50% of it. Because while physicists have concentrated exclusively on the software of our universe (the laws of physics), they have completely ignored the hardware, the computer that executes such a program. In other words, what gives time to these static laws. This is very ironic, because today theoretical physicists routinely work by simulating their mathematical models on supercomputers to see, for example, if their theoretical predictions replicate the evolution of galaxies. This is where physicists recognize the need to run their equations, to have a computer. So where is the computer that executes the laws of nature? They have not even asked themselves this question because they have completely forgotten the problem of change and the present; they are lost in the abstract, timeless and absolute world of mathematics, constantly confusing the map with the territory. So, it is not surprising that the foundations of physics have stagnated in recent decades, when it seems to have completely forgotten the other half of physical reality.

A very different situation from that of Darwinian Physics, because here the projector is recognized from the beginning and in a fundamental way, by not denying the existence of time. But what would this projector be, what is the mechanism that executes the software of the laws of physics in every present? The answer: nothingness. It would be a very special projector, because it would not only project the tape to create the "now," but it would also write the script of the film. Let us remember that patterns emerge thanks to vertical holistic evolution and are preserved and stabilized thanks to horizontal holistic evolution. In this way, the first regularities are transformed into the laws of the latest regularities, which must be respected because they belong to lower holons. In this way the vertical dimension becomes relevant not only to space but also to time, for with each new holon not only space grows but also time. But it is not a growth into the past, as the growing block universe assumes, but an expansion of the present itself into the vertical dimension. Then, if nothingness is randomness, only the present can exist; for randomness means an inability to know, to predict the future; therefore, if randomness is ontological, it is due to the fact that the future does not exist, for nature is fundamentally indeterminate, uncertain. Therefore, there can be no future if there is real chance. But how do we explain that chance exists at all? According to Darwinian Physics, because there is no past. Because there is no previous moment that causes, that determines. Therefore, if there is unequivocal evidence that randomness is real, such as that provided by quantum mechanics, then this is strong evidence that neither the future nor the past exists; because real randomness implies the non-existence of the future, which in turn would be the result of the non-existence of the past. This is a much more reasonable and empirical conclusion than the eternalist position of the current Parmenian paradigm, where there is only one possibility, one history, in which all instants of the past and the future coexist simultaneously. Therefore, not only is it necessary to assume that fate has already been written forever, because the future is real, but also to assume the existence of an entire past, which is logically impossible to travel due to causal paradoxes. A huge leap of faith that also contradicts the scientific evidence of the flow of time and, above all, of the randomness inherent in quantum mechanics. The most accurate theory in all of physics. Many assumptions simply to keep alive the dogma of the absoluteness of the laws of physics.

In conclusion, thermodynamics interpreted in the new paradigm can really explain what time is, thanks to holistic presentism, declaring the revolutionary statement (sarcastically speaking) that there is only what we experience, the "now." The past exists only as information that exists in the present. As biochemical information stored in our brains, or as physical remains such as fossils, statues, or photographs. Therefore, if this information is destroyed, the past is also eliminated, as happens with neurological lesions and the destruction of archaeological sites. The future exists only as dreams and expectations. Again, as information that is in the present, and that we can use to make predictions and estimates, under the assumption that the past (what has been said before) will repeat itself. An assumption that very often fails, especially in the case of meteorologists and economists. But it works quite well in simpler and more ordered realities. Hence the extraordinary precision and success of physics; compared to psychology and sociology, which study higher and more complex realities. Therefore, at least in this respect, and despite the many criticisms that can be made of this proposal, Darwinian Physics would be much more empirical than current physics, because it is based on the only thing we can prove, the now and chance. Two closely related empirical elements, because for Darwinian Physics, the now emerges out of chance, out of nothingness, at every instant. Similar to how an entropic macrostate emerges from the random selection of possible microstates. In the current Parmenian paradigm, on the other hand, no evidence has been found for an eternal and absolute Platonic-Pythagorean world beyond our present; no evidence for time travel, nor for the future already being written in stone. On the contrary, irrefutable evidence has been found for a fundamental indeterminacy of nature due to quantum mechanics. The problem of time is thus a very good example of the book's central criticism of the current foundations of physics. That it is only an effective model. Nothing more, nothing less. A very powerful and successful one, but not a realistic explanation of nature, because it is simply incomplete. Perhaps physicists have revealed their plan, their software, through their laws. Something very impressive, no doubt. But they have completely ignored how they came into being in the first place. And above all, they have ignored who breathes life into these mathematical laws, how they are executed and projected. All because they have forgotten, omitted, and even denied time itself.

CHAPTER XII

THE PROBLEM OF INFORMATION: QUANTUM MECHANICS

If the time problem of the previous chapter could be avoided by theoretical physicists, the problem of this chapter cannot, because it is too obvious to be ignored, although this does not mean that they will not try. The equations of quantum mechanics work wonders in predicting atomic and subatomic phenomena with impressive accuracy, but only that. The theory, such as it is, is incapable of explaining why it works. Predictions that have been confirmed in countless experiments, going through all the tests that have been imposed on it, which have not been few. That is why quantum mechanics is still more a discovery than a theoretical creation as such, because nobody expected it and no physicist understands it yet, as Richard Feynman himself assured. That is why the best physicists of the 20th century racked their brains trying to make sense of it. To quote the famous physicist Richard Feynman again: “If you think you understand quantum mechanics, you don’t understand it. In fact, the problem was so difficult that physicists quickly gave up a few years later. Since the theory was effective and worked beautifully, a strong heuristic (positive and negative) was quickly put in place to do normal science with it, with the goal of developing the quantum theory as quickly as possible, and thereby winning awards and recognition in the community, even though it remained a nonsensical theory to everyone. This is the best example of the excessively pragmatic or instrumental character that modern physics has acquired, in the sense that it looks only for models or mathematical descriptions that serve to make predictions. It does not matter if such a model is incomprehensible, which again confirms the Kantian attitude of current physics, since it is essentially epistemic, since it is not very interested in the noumenon, in nature itself, as we saw in the previous chapter.

An attitude that is well summed up in the phrase popularized by physicists themselves, the famous "shut up and calculate." An attitude that has been reinforced over time by the great progress and development of the theory: why bother the physics community with philosophical questions when the theory works wonderfully? Even many physicists justify this disinterest and laziness in finding an explanation by again blaming neurologists and psychologists. Remember what was said in Chapter VII, that many physicists excuse themselves by saying that quantum mechanics is impossible to understand because our brains have not evolved to these microscopic scales. This again shows how ignorant they can be when they step outside of their specialty; for we have not evolved in interstellar environments either, yet we understand the motions of stars and galaxies quite well. Also, classical physics is quite non-intuitive because of all those relativistic effects where time dilates and space contracts. However, physicists don't seem to have much trouble understanding this, because they don't complain too much, even though we haven't evolved to travel at the speed of light either. In fact, even Galileo's physics, where heavier objects fall at the same speed as lighter objects, is not intuitive (in fact, many people still ignore this, which is very embarrassing for the current educational system). In conclusion, physics is not about adaptation or habituation, it is about simple logic. Whether our universe is coherent or not. Because if it is, as all scientific and empirical knowledge indicates, finding no evidence of magic or miracles, then it is because our world is fundamentally comprehensible; because if it were not, it would violate the principle of sufficient reason to have two or more different logics coexisting in parallel in the universe. This would make existence itself inconsistent. Something that is simply absurd, since nature, in all its scales and manifestations, functions in perfect harmony. So let us stop being so narcissistic and presumptuous in blaming Darwin's evolution or nature, because the most obvious and simple thing is that we are the problem, because we have not yet found the right explanation. Just as it has happened countless times in the history of science, when it seemed impossible to understand a natural phenomenon, which later could be solved and understood perfectly.

In this scientific spirit, Darwinian Physics would be one of these candidates to shed light on the meaning of quantum mechanics. To reveal its meaning and unify it with what we already know about nature at large scales (holistic coherence). And as we will see below, it would be the most serious candidate to achieve this. But before we demonstrate this, let us first see what the problem is. The reason why quantum mechanics has been such a headache for the leading physicists of recent times. It all boils down to the seemingly trivial problem of knowing what the heck is going on when you observe a quantum system, say a single electron. This unknown is known as the measurement problem and is due to the famous wave-particle duality of the constituents of matter. To better understand this apparent contradiction, we need to review the famous double-slit experiment. This experiment is a very good way to find out whether something behaves as a wave or as a particle. The procedure is quite simple. You have to throw that something against a barrier with two slits and see what appears on the other side. If it leaves two marks, then that something is a particle. On the other hand, if it leaves several marks, which is called an interference pattern, it is because what we are sending out are waves.

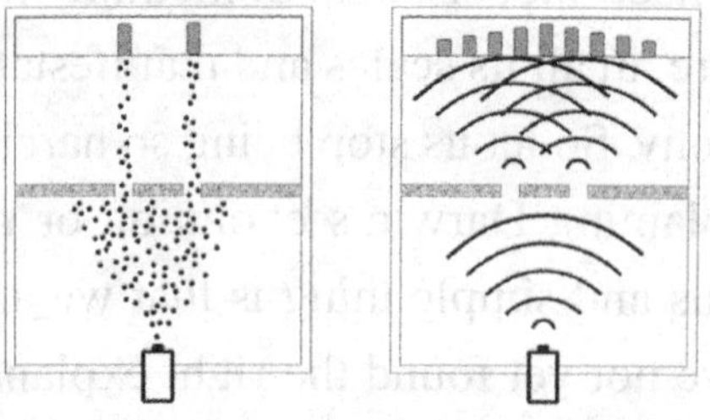

It makes sense, doesn't it? If it's a wave, it will split through the two slits, becoming two waves that interfere with each other, creating those multiple lines. But if it is a particle, it can only pass through one slit or the other, leaving a single mark just on the other side of each slit. This identity test works well for macroscopic objects like water waves and marbles, but not for quantum objects. To see why, let's shoot electrons this time. When an electron is sent through the two slits simultaneously, it leaves only a single point, a single mark, on the detector screen. This shows exactly what we expected, that it behaves like a marble. A single electron, a single mark. So far, no surprise. The problem arises when we keep bombarding it with more single electrons, because after a few hours, all those dots start to form an interference pattern. A completely unexpected result, because it shows that the electron also has a wave nature. This is very strange. It is as if each electron had traveled as a wave, passing through the two slits at the same time and interfering with itself on the other side; but when it reaches the detector screen, it immediately transforms into a particle, leaving only a single trace. This is the only way a series of individual electrons can leave an interference pattern on the detector.

Since no one understood what was going on, after countless heated discussions it was decided not to waste any more time on the subject, so a protocol was established on how to use this new model, but in a purely pragmatic way. This is the origin of the famous Copenhagen interpretation, which is taught in physics departments all over the world and whose main proponent was the Danish physicist Niels Bohr (1885-1962). Here, the behavior of the quantum object, such as the electron, is described by the physical law called the Schrödinger equation, which describes the evolution of the wave function (Ψ). A mathematical object that contains the probability distribution of all the properties of a quantum system, such as its position, momentum, spin, and so on. However, this law only works when the quantum system is not observed, because when it is measured, an additional equation must be introduced, called Born's rule, in honor of the German physicist Max Born (1882-1970) who postulated it. This rule consists in squaring the modulus of the part of Ψ to be measured ($|\Psi|^2$) in order to transform its possible values into probabilities. And to all this we must add a third element, the measurement postulate, better known as the collapse of Ψ, which consists in transforming the probability distribution obtained by the Born rule into a single possibility with 100%. More precisely, the only possibility that is observed, hence its 100%. Thus, through these three steps, we have that before we measure, the quantum object evolves in time with a vibrational behavior, similar to a wave; while when we observe it, this wave is reduced to a single concrete possibility, acquiring a corpuscular behavior, like a particle.

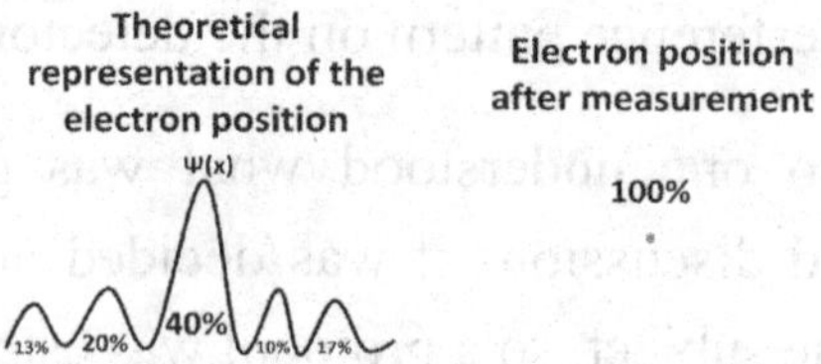

This means that the electron, for example, would actually be a wave (Ψ), but a probability wave, in the sense that the quantum particles travel as a wave, but interact by collapsing into one of its possibilities, randomly. So, where there is more wave amplitude, there would be more probability. The problem with this interpretation is obvious: what is this Ψ really, where does the Born rule come from, why does it collapse randomly, what makes it collapse, what is the physical mechanism? All questions to which the mathematics of quantum mechanics has no answer. Moreover, the collapse of Ψ is completely incompatible with the Schrödinger equation, the quantum law, since the latter is linear. This means that if you have two different states of a system allowed by Ψ, then the sum of these two states is also an allowed solution. This is the basis of superposition, of the same electron passing through both slits, or of the cat being alive and dead at the same time in the famous Schrodinger's cat thought experiment. However, we have already mentioned that Ψ collapses when observed, so the electron is there but not there, and the cat is either alive or dead, but not alive and dead at the same time. So, we see only one possibility of Ψ, one state, not all states at the same time. And the second problem is that the Schrödinger equation (and its relativistic version, the Dirac equation) is reversible, i.e. symmetric by time reversal, like any physical law. This means that it should be possible to know the entire past and future of the quantum system, because it is deterministic. However, the collapse of Ψ is obviously not reversible, because it is random. Therefore, one cannot predict exactly what will happen when Ψ collapses, nor can one know exactly what Ψ was from a measurement already made.

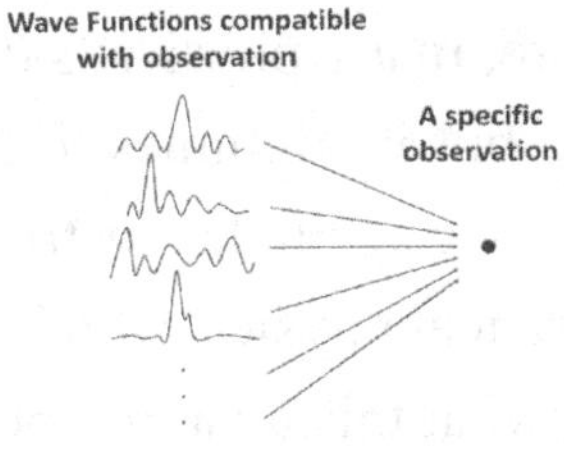

Thus, for any given observation, there would in principle be an infinite number of Ψs that could have produced it. The only difference is that some would be more likely than others, but all could have been. Therefore, we cannot know the past with certainty, because we cannot rescue the original Ψ from its collapse. Nor can we know in advance exactly what outcome a given Ψ will produce, since collapse is random. All this completely destroys the reversibility of the physical law, since it is not possible to go back to the past due to this random collapse. This completely destroys determinism. As said in Chapter IV, it would be like having six identical dice and trying to guess which of them will produce the result we observed, 4. It is not possible. It is the same with quantum mechanics. When you measure a quantum system, you cannot go back and retrieve the Ψ that produced it, nor can you know what value will result. Only this time the randomness is ontological, not epistemic as in the case of dice. Moreover, one has no idea how it works, nor what it means, since the mechanism for the collapse of Ψ is unknown. Remember that Born's rule is only a postulate. Therefore, this second equation of quantum mechanics, the squared modulus of the wave function ($|\Psi|^2$), is not derived from the first equation or from Ψ. It was added only to make the Schrödinger equation useful. In fact, even Ψ cannot be derived from anywhere. Recall that Erwin Schrödinger invented it after a moment of inspiration with his lover in the Swiss Alps, as noted in Chapter IX. For all these reasons and more, many physicists do not agree with this official interpretation, but it continues to be used because it is practical, confirming once again this instrumentalist Kantian attitude of current theoretical physics, which is not very interested in seriously knowing what this quantum noumenon is.

So it is not surprising that Albert Einstein was completely against this official interpretation of quantum mechanics, not only because it is obviously ad hoc, but because it completely destroys all of classical physics, which is based on the deterministic certainty of absolute laws. This is why Einstein postulated that quantum mechanics could only be an effective theory, because in order for it to really be a complete description of nature, which for Einstein meant that it should be Parmenian, an ingredient would have to be added to it that would eliminate this ontological randomness in order to make it epistemic or statistical, as happens in Boltzmann's thermodynamics. In other words, to turn it into a randomness resulting from our ignorance, lacking all the necessary information to make a perfectly accurate prediction. This missing information, which would make quantum mechanics deterministic, is the famous hidden variables. Parameters that we have not been able to measure or observe, hence the hidden variables; but if we knew them, quantum mechanics would automatically become classical or Parmenian, making Schrödinger's equation as absolute as any other law of physics, such as Maxwell's, Newton's and, of course, Einstein's equations. To this end, Louis-Victor de Broglie proposed the first quantum theory with hidden variables, called the pilot wave model, which consisted in keeping the particles as corpuscular, but they would be transported by a wave, similar to how the crest of a wave carries the surfers. In this way, particles and waves would remain conventional things, nothing strange; while randomness would be the fruit of ignorance about the exact position of the particle in the wave. An ignorance of position that would correspond to those hidden variables. However, it did not take long for de Broglie himself to abandon his interpretation, realizing that his idea could work very well with the variable position. The problem is that in quantum mechanics, all properties of a quantum system have this wave-like behavior, since there is a superposition not only of positions, but also of moments (different speeds simultaneously) or energies. Therefore, it is hard to imagine how the spin of a particle would surf in a spin-up and spin-down wave. Moreover, this quantum behavior applies not only to particles, but to a set of particles, an atom, a molecule, or a cat. So, the interpretation was very naive, because this pilot wave was increasingly difficult to conceive in a traditional or classical way.

Despite the rapid failure of the "pilot wave," the assumption of the existence of hidden variables, which made quantum mechanics deterministic, was still very attractive and tempting to many physicists, apparently because it saved their paradigm. Then, in 1935, in order to demonstrate that current quantum mechanics was only an approximation to a classical theory, Einstein, together with Podolsky and Rosen, found an apparent contradiction in the official formalism of the Copenhagen interpretation with special relativity. They predicted a quantum phenomenon known today as "entanglement." Something that sounds very complicated is much simpler than it seems. For example, if a billiard ball hits (interacts) with a similar one (same shape, composition, size and weight) at high speed, the first one would have to move the other one, since the sum of the two resulting speeds is very close to the speed of the ball it hit (assuming there is very little friction or drag), due to the conservation of linear momentum. Logical, because the amount of motion should be the same, being the motion of the striking ball transferred to the ball that was hit. Because if the struck ball did not move, or moved too little, or vice versa, too fast compared to the speed of the ball that hit it, it would be very strange. Where did the missing motion go, or where did the extra motion come from? The same thing happens at the quantum level, except that we can replace the speed of the billiard balls with the spin of the electrons. So, if two electrons have interacted, and one of them we have measured has a spin ½ (up), then the other must have a spin -½ (down), and vice versa. This is because the sum of the spins of both electrons must be zero (-½ + ½ = 0), due to the angular momentum of the entire system must be conserved, if it forms a singlet state. So far so good, except for the small detail that, according to the Copenhagen interpretation, the properties of quantum mechanics are fundamentally indeterminate until the moment of measurement. In other words, before we observe, we only have the Ψ of the electron, where the spin is in quantum superposition up and down. Only when we measure the electron and collapse its Ψ can it happen to have a specific spin: up or down. It would be like not being able to know the specific speed of each billiard ball after its collision if we did not observe it, even though we had all the information about the situation: the strength of the shot, the angle, the distance, the weight, and so on. Therefore, it would be impossible to blindly predict the speeds of the balls, because they would have all possible speeds simultaneously, from the maximum speed at which the first ball was shot to zero. However, regardless of the fact that we do not know the exact velocity of each ball, we would be certain that the sum of the velocities of both balls must be equal to the velocity of the first ball thrown (again, assuming no friction). Therefore, if we observe the velocity of one ball, we will automatically know the velocity of the other, the difference with respect to that maximum possible velocity. And here is the problem that Einstein, Podolsky and Rosen found with the official interpretation of quantum mechanics, because if the determination of the spin of the entangled electrons is random (50% of having spin up or down), and also the angular momentum must be conserved, then every time the spin down of one entangled particle is observed, the other particle must always have spin up, and vice versa. This is despite the fact that these entangled particles can be separated as much as you like, with one here and the other on another planet or in another galaxy. So, according to the Copenhagen interpretation, every time we determine the spin of one entangled particle when we observe it, we automatically determine the spin of the other entangled particle as well. This is very much at odds with special relativity, because it would seem that the information about what spin the other entangled particle should have would be transmitted at infinite speed every time one of the two is measured, obviously violating the maximum speed of relativity, that of light. This is why Einstein called this quantum phenomenon spooky action at a distance.

So, to eliminate this postulate that the measurement of one entangled quantum particle instantly causes the determination of the quantum properties of the other entangled particle, Einstein and company simply postulated that there is no such communication at infinite speed, but a simple correlation already determined from the beginning. To understand this solution, let us, as always, use an analogy. In this case, the classic analogy to spins would be gloves, because they come in pairs (left and right glove) as well as spin (up and down). Then entanglement would be easy to understand, as if we put each glove in a different box and sent them to distant places. In this way, we would not be surprised if we did not know which glove we had before opening the box, but after opening the box and knowing which one we got, there would be no mystery in automatically knowing which glove the other box, the one on the other side, should have. Einstein and company's explanation of entanglement was exactly like this example. It is not that the spin is determined at the time of measurement, but that the spin has always been determined, just like the gloves. So, there is no spooky action at a distance between entangled particles, just a perfect correlation that has already been determined. Something we could, in principle, know and predict if we had all the information (those hidden variables). So, the hidden variables would not only solve this supposed spontaneous and random collapse of the wave function, but also this spooky action at a distance.

For a long time, this discussion between Bohr and Einstein about whether there really are pre-determined hidden variables or whether reality is fundamentally random or not (which is why Einstein said: "God does not play dice") remained on the fringes of physics, due to this Kantian character of "shut up and calculate" that dominates physics. So strong was the negative heuristic of that time that decades later a physicist had to take a sabbatical year to research it. In other words, he had to work for free. That physicist was John Stewart Bell, mentioned in the first part of the book, who in 1964 developed a mathematical theorem that assumed Einstein's position on the existence of hidden variables. Under this assumption, he was able to make a prediction that differed from the Copenhagen interpretation, the famous "Bell inequalities." In this way, the brilliant physicist John Bell was able to turn a dilemma that was thought to be purely philosophical (in a pejorative way, of course) into an experimental problem. Into something that could be solved in a laboratory. This is because the introduction of hidden variables into quantum mechanics, in order to transform it into classical mechanics, imposes restrictions on its predictions. In particular, the correlations between entangled particles cannot be greater than a certain value. Therefore, it was possible to design an experiment to test this inequality and thus prove whether reality is truly quantum or not; because if the result were greater than Bell's inequality, it would prove that reality is fundamentally truly quantum and not classical. An extraordinary result that, incredibly, nobody cared much about, again due to the shut up and calculate! culture. It took several years for this work to be rediscovered and recognized, so that in the 1980s several experiments carried out by John Clauser and refined by Alain Aspect, both winners of the 2022 Nobel Prize in Physics for this research, finally proved the violation of Bell's inequalities. In other words, the prediction made by quantum mechanics under hidden variables was falsified, since it was observed that the correlations between entangled particles could not be classical; because they turned out to be greater than what the various hidden variables allowed. Therefore, at its most fundamental level, the world is authentically quantum, not constrained by any underlying classical theory.

Thanks to Bell, Clauser, Aspect, and many others, we now know that quantum mechanics is neither classical nor Parmenian; so what exactly is quantum reality? Let us remember that the Copenhagen interpretation is only a protocol of use, because it does not provide any explanation of its steps, especially with respect to collapse. Therefore, it does not clarify anything about what quantum mechanics is. In view of this, many physicists have proposed different explanations, called interpretations of quantum mechanics, which can be divided into three categories, according to the classical or Parmenian principle that must be sacrificed in order not to contradict the experiments. These principles are: realism, which means that nature is objective or independent of any observer; locality, which means that things can only influence their immediate environment, therefore there is a maximum speed at which phenomena can propagate or cause, the speed of light; and determinism, which assumes that nature is absolute because there is no randomness or ontological uncertainty. So, depending on which principle is abandoned, we can basically have three groups of interpretations. Let us start with the non-realist interpretations, where the Copenhagen interpretation is no longer seen as a simple protocol of how to use the mathematics of quantum mechanics to make useful predictions, as instrumentalists or Kantians would say, but is interpreted literally. Thus, since the formalism of quantum mechanics says that the wave function collapses when a measurement is made, it is assumed that it is the very act of measuring, of observation, of consciousness itself, that causes the collapse of the wave function and thus determines nature itself. This folly, also known as the subjective collapse interpretation, is the strategy of using one mystery, such as consciousness, to solve another, the collapse of Ψ, which obviously solves nothing. There is no basis and much less evidence to believe that the human mind is a separate and distinct reality from the material. Nor that the human mind is the only thing that can collapse the Ψ. Nor even that it is better than the mind of a mosquito, or that of a plant, or that of a stone, or any other quantum particle, since what really causes the collapse of Ψ is the interaction itself, regardless of whether what is interacting thinks or not. And as was obvious, there is no mechanism proposed by which consciousness can interfere with the world of things, such as telepathy or anything like that. So, it is not only absurd but also cosmologically narcissistic to suggest that reality "is" only when we observe it. As if nature needs us in order to exist. And yet it was proposed and incredibly defended by great 20th century physicists like Neumann, John Wheeler, Eugene Wigner, Fuchs, or Schack.

Another anti-realist or idealist position that has recently gained momentum is Qbism or quantum Bayesianism. This is an interpretation that has recently become very popular, even though, incredibly, it means the same thing as the subjective collapse that so many physicists loathe and shudder at (another proof of their lack of philosophical skills). The only difference now is that this mysticism is cloaked in a more modern and sophisticated jargon. Thus, the randomness of quantum systems is interpreted in a directly epistemic way, through Bayesian probability, which is subjective because it measures our degree of ignorance and our assumptions. Not all observers will agree with the results, because the probabilities reveal the beliefs and expectations of each individual. Thus, each time we make an observation, we update our personal beliefs about the probabilities of all possible outcomes. This means that it does not resolve or clarify anything about the real, about the noumenon, because it completely denies it by being inherently subjective. It is so anti-realist that it is even used to defend the hypothesis that we live in a simulation, mixing it with the jargon of computer science and quantum computation. Thus, it would justify this indeterminacy of the properties of quantum systems, even when measured, as a way for the simulation to save computing power. In other words, exact values are given only when necessary, when someone observes. A hypothesis that would not only be the mother of all conspiracy theories, but would also literally put physics on the same level as religion, since believing in an almighty God who created and controls everything is completely equivalent to assuming that a hacker from a higher universe programmed our entire cosmos in his garage. The religion of the 21st century.

And continuing along this idealistic line, we can add one of the most recent interpretations, called Relational Quantum Mechanics. Its creator is the Italian physicist Carlo Rovelli, the same one who preached in the previous chapter that time does not exist (so it should come as no surprise what follows). Now, in his new book called Helgoland, he tells us the same thing as subjective collapse, only instead of it being only consciousness that collapses the wave function, now it is everything, including animals, plants, rocks, and other particles that the subjective collapse interpretation excluded. So, it is every interaction that constructs and materializes existence. An interaction that would take time to happen, but which he denies exists. But even omitting this small detail, he does not explain how this relationship happens, what the mechanism is, how things really materialize, especially if there is no time. Moreover, the interpretation itself lacks logic, because for there to be a relation, there must first be things that can be related, of course. But Rovelli says just the opposite, that things do not exist in themselves. They exist only when they are related. This makes as much sense as saying that a marriage exists first, and only then do the persons who would be the spouses of such a marriage exist, which is absurd. So, Rovelli ends up saying exactly the same thing as the Copenhagen interpretation without clarifying anything. He just says it in a much more ambiguous and romantic way that gives the impression, at least to him, that it is more revealing. However, not only does it not resolve any fundamental questions about quantum mechanics, it also contributes to this deterioration of confidence in science by supporting these idealistic positions that are later used to justify spiritual or mystical pseudoscientific beliefs. Ideas that are not only irrational, but often prove to be very dangerous, as they are used by charlatans who deceive many people through self-help books and quantum healing treatments, putting the lives of these people at risk by convincing them that such alternative treatments are better than traditional medical treatments. This is one of the many dangers of denying the most essential principle of science, realism.

The second category of interpretations would be the non-local ones. Again, hidden variables are used to save determinism. However, in order to respect the violation of Bell's Inequalities, which has already been experimentally proven, they would have to be non-local hidden variables. In other words, they would have to be variables that communicate at infinite speed, obviously faster than the speed of light, to explain the correlation that is stronger than the classical correlation in entanglement. One of these proposals is a recycling of the Broglie pilot wave of the American physicist David Joseph Bohm (1917-1992), also known as the Broglie-Bohm pilot wave. The problem, in addition to those already identified by de Broglie himself, is that the whole point of the original idea, which was always to save all of classical physics, which obviously includes the locality principle, has obviously been lost. Therefore, they would no longer be classical hidden variables, but strange hidden variables with a spooky action at a distance. Einstein would not be happy at all, especially because it violates his special theory of relativity by completely breaking its most essential principle, that the speed of light is the maximum speed, because here there would be physical objects, these hidden variables, that would communicate with each other at infinite speed. This clearly violates causality. Therefore, in this interpretation, physicists speak of retrocausality; for the future could affect the present because these non-local hidden variables could transmit information instantaneously, allowing effects to occur before causes. But this is a fanciful interpretation, because quantum mechanics cannot be retrocausal, because it is simply timeless, as it is now formulated. Therefore, only the results of some experiments performed with quantum entanglement are interpreted, where the entangled pairs are separated far enough to measure effects faster than the particles can travel below the maximum speed of light. Thus, assuming the existence of non-local hidden variables, one concludes that the future has affected the past. But this is cheating, because measuring later or not does not change what was measured earlier. It just reinterprets the same data one way or the other, whether it was measured later or not. Therefore, empirically speaking, the future does not change anything that happens in the present.

Another type of hidden variable theories would be the superdeterministic ones, first mentioned by John S. Bell himself as one of the possible consequences of violating his inequalities. It is currently advocated mainly by the German physicist Sabine Hossenfelder. Here the hidden variables are not strange or exotic, but quite classical. In other words, they are local hidden variables. So how do you solve the problem of entanglement, the stronger-than-classical correlation? By taking seriously what the Parmenian paradigm means, which is determinism. In other words, superdeterminism is simply the usual determinism, that of Newton or Laplace. The confusion of physicists is that they usually take for granted something that is actually quite incompatible with their cherished Parmenian paradigm: statistical independence or, as many also call it, free will, which means that experimenters really are free to choose how and when to make observations and measurements. Therefore, it would be very much a coincidence that there is such a strong correlation between entangled particles. But in the eternal block universe of the Parmenian paradigm, where all the past as well as all the future already exists, already happened, is real, where does this supposed statistical independence come from, this supposed freedom of the experimenters to make these measurements? From nowhere. So, it is not that the future causes the past, as the previous retrocausal interpretation believes, but that the whole future has already been determined since the Big Bang. A whole that includes not only the particles and the hidden variables that cause this correlation, but also the experimenters who make the measurements, since they are also made up of the same particles with hidden variables. In other words, it is the whole universe that conspires to measure quantum systems at the right time and in the right way to produce such results; for it is not that the information travels instantaneously, nor that it comes from the future, but that it is already contained in all things, determined as early as the Big Bang, including when physicists will make observations. Remember that in the block universe there is only one possible history, already written in stone. Although this is a conclusion very much in line with the paradigm that physicists so strongly advocate, the vast majority of physicists abhor this interpretation. In fact, they find it ridiculous and dangerous. But what is really ridiculous is that it is the same consequence that you get with the absolutism or determinism that they defend with so much faith. So again, it is another proof of the serious lack of philosophical skills in the foundations of physics today, not knowing what the paradigm they so religiously defend really means.

However, although the above proposal would explain the collapse of the wave function and entanglement in a way that is very consistent with the eternalism of the laws of absolute physics, it has a small problem for Parmenideans, namely that it requires the assumption of hidden variables that no experiment has detected or given any evidence for. Moreover, it would be a very conspiratorial way for nature to work, unnecessarily complicating reality with variables that we suspiciously cannot see. It sounds as ad hoc as extra dimensions that are suspiciously small so that we cannot detect them. This brings us to the most faithful interpretation of the formalism of quantum mechanics under the current paradigm, the many-worlds interpretation; first postulated by the American physicist Hugh Everett (1930-1982), later rescued by Bryce DeWitt and Neil Graham, but currently championed mainly by the physicist Sean Carroll. It is so faithful to the Parmenian paradigm that it replicates its old strategy of solving the problem by denying the existence of the problem. In this case, if the problem is the postulate of the collapse of the wave function, then the postulate is simply eliminated. It is as simple as that. So only the quantum law, the Schrödinger equation, exists. Therefore, Ψ continues to evolve intact, without disappearing. In this way, quantum mechanics is interpreted from a simple interaction between waves. In fact, the many-worlds interpretation is so compatible with Parmenian physics that one simply has to take Feynman's path integral literally. The mathematical tool used by particle physicists to perform calculations in QED (quantum electrodynamics) theory, where particles, such as the electron, are assumed to take all possible paths as they travel from point "A" to point "B."

The small problem with this interpretation is that Ψ has never been observed, so it is necessary to make the small assumption (sarcastically speaking) that all the infinite outcomes of this Ψ that we do not see still exist, but this time in other parallel universes that would be just as valid or real as ours.

In other words, every possible path computed in QED would be a real world among Everett's many worlds. And since there are infinitely many paths in the path integral, there would be infinitely many parallel universes. Here, everything that can happen must happen in one of these universes. So, for example, you are the President of the United States, the Pope, Brad Pitt, a Nobel Prize winner, a porn star, a tramp, and so on. All of this is happening right now, but in these other worlds. This madness is the result of interpreting the randomness of the collapse of Ψ in the other possible way within the Parmenian paradigm; because if we have qbism through the Bayesian definition of probability, then the many worlds are born out of the frequentist definition of probability. This is the reason that all the possibilities that we do not see from Ψ must also be real, in the same way that one particular result of a throw of a dice does not negate the existence of the other five possible results.

All thanks to the elimination of the problematic postulate of measurement, just as the problematic becoming of time was eliminated. The latter is very important because there are many physicists who do not like this many-worlds interpretation because they mistakenly believe that it is an idealistic (not realistic) and presentist (not eternalist) interpretation. This is because they think that the universe bifurcates every time someone makes a measurement. This is perhaps confused by the classic example of Schrödinger's cat, where someone opens a box and causes the universe to bifurcate into a universe where the cat is alive and another universe where a dead cat is observed. So, they think it is similar to a subjective collapse interpretation where consciousness is the one that creates the universes by observation. And since this observation is happening in the present, they conclude that the bifurcation and creation of universes is also happening in the present. But this is all wrong. A confusion that is understandable since they are more mathematicians than philosophers. Remember that this interpretation is achieved by eliminating the collapse postulate, so there is no way to cause something that no longer exists. Moreover, we already know that in the Parmenian paradigm there can be no "present," so it is impossible to bifurcate universes in the "now." Therefore, there has always been one person observing a live cat and another observing a dead cat, each in their own eternal block universe, because in this paradigm there is only the past and the future. The only difference now is that all possible pasts and futures must exist, each in its own eternalistic world.

Then the many-worlds interpretation saves all the classical principles: realism, locality, and determinism; hence it is the best Parmenian interpretation of quantum mechanics. In fact, we can reduce this whole dilemma to just two classical or Parmenian principles: realism and determinism, because let us remember that locality would already be included in determinism, as we saw in the explanation of superdeterminism (determinism = superdeterminism). Therefore, we can reduce all the conflict and entanglement in understanding quantum mechanics by trying to make it compatible with these two presuppositions. This is because if you want to eliminate ontological randomness in order to save determinism, but at the same time do it in a realistic way, then you cannot use the subjective interpretation of randomness (Bayesian probability), as qbism does, because it would be idealistic. Then the only option left in the current paradigm is the frequentist interpretation of probability, which requires a statistic, a population of data, because it is impossible to do statistics with a single case. It would be like doing a survey with a single person, or thermodynamics with a single molecule. Therefore, it is necessary to make real each of the infinite possibilities of Ψ that are not observed in order to satisfy the statistical requirement. We inhabit only one of them, without interfering with the other universes. Unfortunately for fans of this interpretation, however, the argument does not hold, because the definition of frequentist probability cannot be strictly applied in this case, due to the fact that one cannot do statistics with infinite values, because there is no probability of anything between infinity. For example, what is the probability of our universe among infinite parallel universes? It makes no sense.

Despite this absurdity, this interpretation continues to be popular among physicists. This should come as no great surprise, since it preserves the very essence of their coveted Parmenian paradigm: the eternalism and determinism of absolute laws. In fact, the interpretation of the many worlds is so faithful to the current paradigm that it makes transparent its true fundamental position, the same desire of Parmenides, Plato, and the classical theists (Neoplatonism); that the most real thing is not what we observe and experience through our senses, but something beyond, the Absolute. This desire is particularly evident in quantum mechanics, because, at least up to now, laws have always represented regularities that we observe in nature. That is where their legitimacy came from. In this sense, the absolute was supposed to be subordinate to the sensible, the empirical. Thus its Parmenian roots were disguised. Now, however, with quantum mechanics, the true principle of the Parmenian paradigm is revealed, since the law is considered to be suprasensible, platonic, or divine; for it does not matter if its consequence is infinite unobservable universes, we must believe them by faith because the law dictates it. That is why Ψ cannot collapse, because the Schrödinger equation is absolute, like any Parmenian law. It does not matter that such laws are differential equations and therefore continuous or infinitesimal functions. Something that was not so problematic before quantum mechanics, because such infinity was within the world line, in a single history. We only had to make the small assumption that all infinite instants of the past and future exist, thereby eliminating the present and the becoming of time. But now, since the physical law of the collapse of Ψ must be saved, these infinities of the calculus have been externalized and must endure not only an eternity, but also an infinity of worlds parallel to our own.

This is the great problem of the Parmenian paradigm, which again confuses the epistemic with the ontological; for in taking the physical law as the most real thing, one forgets that these are differential equations, mathematical tools invented to deal with the problems of dealing with infinity in the Parmenian paradigm. For what are derivatives but a way to make sense of "0/0," and integrals a way to make sense of "∞·0"? So you have to work with limits, with approximations; because in mathematics "0/0" and "∞·0" are meaningless. And if they are meaningless in mathematics, they are even more meaningless in the real world. So, if neither differential equations are realistic, much less the laws that use them, because they are merely tools to represent the real. Something that becomes even more obvious in quantum mechanics, since it is anti-classical. It therefore reveals a finite, discrete and uncertain world. But despite all the overwhelming empirical evidence of quantum mechanics, physicists still try to fill every gap of nothingness or uncertainty in nature with infinities through differential equations. Which in itself is not bad, if it is practical, if it is instrumental, if it works. But if you forget that they are representations, models, and instead consider absolute laws more real than the real, as is currently the case in the Parmenian paradigm, then you end up assuming that matter, energy, space, and time itself must also be continuous, infinitesimal. All because they want a world full of certainties, so they have to fill every possible interval and space with an infinite, which inevitably generates these infinite universes of eternal blocks.

Obviously, Darwinian Physics does not fall into this trap; it simply recognizes that mathematics is only a tool. And although the laws of physics based on differential equations are impressively useful and accurate, they remain mere models, approximations of the real, not the real. Thus, from a Humean point of view, Darwinian Physics recognizes the Schrödinger equation as a faithful representative of the quantum world, but not as the quantum world itself. So, it should not be taken as the word of God itself, especially when it requires another ad hoc equation (the Born rule) and a postulate (the collapse of Ψ). A reality that also applies to the most modern version of quantum mechanics, the Standard Model of Particles. Here, as with all current physical laws, continuous mathematical tools are used, so they do not work. Calculations in quantum field mechanics yield infinities because they again assume an infinitesimal division of spacetime, so you have to add infinitely many terms, which yields infinity in their solutions. This is the reason why we have to use perturbative techniques, which physicists call "renormalization," to extract and hide these infinities from the calculations in order to make the Standard Model of Particles useful and thus to get finite predictions. Something completely ad hoc, because like the introduction of the Born rule and the measurement postulate, renormalization is not derived from the equations of quantum field mechanics. Its only justification is that it is useful. Therefore, quantum field mechanics alone is wrong, because in its essence it is still a classical, continuous, non-quantum theory. So, it should come as no surprise that it leads to infinity.

Now that the distinction between mathematical reality, especially the continuous one, and the real world is clear, Darwinian Physics sets out to create the first truly quantum mechanical model. For, as already mentioned, all quantum theories so far have been developed by proposing a classical model or equations, differential or continuous, and then quantized according to the rules of quantum mechanics. This is because physicists simply do not know how to do anything else in the Parmenian paradigm. That is, they start in the Platonic world of absolute mathematics and try to bring it down to earth from there. Darwinian Physics, on the other hand, starts directly in the mundane and stays there, in a finite, quantized, discrete world, with the ontological randomness and becoming of time that we observe. That is why it solves the great mysteries of quantum mechanics so naturally, because it arises from its non-classical principles. For example, it easily solves the problem of the collapse of Ψ by adding the missing element to quantum mechanics: time. This problem of the absence of time in quantum mechanics is known as Pauli's paradox and refers to the inability to construct a quantum operator for the time variable. It has been possible to construct a quantum operator for the variable "space," "position," "linear momentum," and even for the variable "energy," but it has not been possible to construct it for the variable that determines a duration, precisely because the intrinsic randomness of quantum mechanics prevents it from integrating it in an eternalist or absolutist way, as it has always been. Let us recall that time is introduced into quantum mechanics in an external way, through the sequence of measurements. If the problem of measurement, the reason for the random collapse of Ψ, has not been solved, then the problem of time in quantum mechanics cannot be solved either. Apparently, only Darwinian Physics has taken this fact seriously, because in this theory the Pauli paradox is solved by explaining the collapse of Ψ by time. In other words, both problems would be essentially the same. Therefore, the probability wave referred to by Ψ is actually the present coming from nothingness, while its collapse is the materialization of that now.

This simple explanation solves the problem of measurement, since Ψ collapses when it is measured, not because the act of measuring produces it, but because in a presentist theory it cannot be measured at any other time than in the now. In other words, it is just an inevitable coincidence, because the collapse of Ψ is the very materialization of a new present. Precisely the only thing that can be measured or observed. The collapse, on the other hand, is random because time itself has a nihilistic origin in Darwinian Physics, since the present comes from the chaotic fluctuations of nothingness itself. In other words, the collapse of Ψ is random because literally nothing causes it, because if it were caused by something, a regularity, a law, then the effect or collapse would also be orderly, it would be predictable. But if the collapse is completely random, as all the experiments reveal, then the only way that this could be possible is if nothingness itself is responsible.

This collapse mechanism also solves the puzzle of what Ψ is, because it is wrong to think that Ψ is something concrete like a classical wave, as de Broglie thought in his pilot wave theory. As mentioned above, the Ψ is not only used to predict the position of a particle, but also to predict any attribute of that particle, such as its spin, momentum, energy, and so on. Moreover, Ψ applies not only to fundamental particles, but to composite particles, to atoms, to molecules, to a ball, to a planet, in short, to everything. This is why many theoretical physicists, including Stephen Hawking, have flirted with the idea of a Ψ for the whole universe. Therefore, any information that can be obtained from a quantum system fluctuates as a probability wave. A well abstract and ambiguous nature, which Darwinian Physics can explain quite well, because the real referent of Ψ is a time wave, a potential present. That is why Ψ is omnipresent, because every property of a physical system needs a present to be, to determine itself, because everything that exists is immersed in time. In this way, Darwinian Physics through holistic presentism turns out to be a much more elegant solution than the many-worlds interpretation, because with the same mechanism it explains what the collapse of Ψ is and how it works, solving at the same time the problem of measurement and Pauli's paradox. All this without introducing extraordinary elements like hidden variables or parallel block universes. On the contrary, it was only necessary to add the missing element to physics, time, to make quantum mechanics complete.

Thus, the Darwinian interpretation of quantum mechanics turns out to be realistic, since it belongs to the category of objective collapse interpretations. And it is obviously non-deterministic, since the randomness it recognizes is ontological. And if there is no determinism, then it is not local. But this non-locality is fully compatible with Einstein's special relativity. This will become much clearer in the next chapter, but for now all we need to know is that the speed of light is the maximum speed at which information can be transmitted, that it is a pattern or a regularity. In other words, that it travels something that exists. However, non-local or instantaneous communication between two entangled quantum particles is completely random. This means that you cannot use this quantum phenomenon to build an intergalactic telephone that would allow us to communicate instantaneously at any distance, because only noise would be heard. The only peculiarity is that this noise would be perfectly correlated. In other words, if on one side of the channel we get a completely random sequence of "1" and "0," on the other side of the communication channel we would have exactly the same thing, noise, only that every time on one side there is "1," on the other side there is "0," and vice versa. A correlation that can only be confirmed by taking one of the lists to the other side of the communication channel and comparing them. A journey that, of course, has to be made below the speed of light. Since entanglement is completely random, no information or regularity is transmitted at infinite speed, so it is not possible to violate causality. Therefore, the effects (regularities) will continue to happen after the causes (continuation of regularity), respecting the speed of light.

The previous point is very important because many physicists believe that there cannot be causality without determinism. In fact, they are often considered synonymous, so that a theory that violates determinism would also violate causality. But this is yet another example of the great philosophical shortcomings of physicists, because one does not necessarily imply the other. There may well be non-deterministic or probabilistic causality, because causality strictly means a relationship of dependence between two events in time. For example, if "A" happens, then "B" will happen. Therefore, "A" causes "B." However, this degree of dependence can be absolute or deterministic, in which case every time "A" happens, "B" will happen with 100% certainty; or it can be with a degree of probabilistic dependence, for example, if "A" happens, "B" may happen with 70% probability. Therefore, causality can exist without determinism. In fact, the truth would be the other way around. There can be no causality if there is determinism. Remember what Sean Carroll (the greatest advocate of the many-worlds interpretation today) said in the previous chapter, when he quoted the great British philosopher Bertrand Russell, that it is Laplace's determinism itself that eliminates the need for causality; because causality is an inherently dynamic concept, an active process, whereas absolute determinism is total paralysis, the freezing of time in eternity, where nothing acts because everything has already happened forever. Therefore, far from being a theory incompatible with causality, Darwinian Physics rescues it by reviving time, by restoring to it its active, generative role, it's becoming.

In conclusion, Darwinian Physics is realistic and causal, but neither deterministic nor local. It is precisely these last two principles that have been experimentally proven to be violated in quantum mechanics due to the collapse of Ψ (indeterminism) and entanglement (non-locality). An authentic quantum theory will be able to reveal the true meaning of quantum mechanics.

As already mentioned in Chapter VIII, the true representation of nothingness is the Bloch sphere. A space-time point whose two states are in constant superposition, fluctuating and annihilating each other to remain nothing. Therefore, despite having these two states, it remains something fundamental, irreducible, unique and indivisible in both time and space. That is, a space-time point. A representation that finally explains, within Darwinian Physics, what the three steps of the Copenhagen interpretation mean. For if Ψ = time, in presentist terms, then it is consistent to represent Ψ mathematically by a probability amplitude containing a complex component. A component that is indispensable to calculate the time evolution of Ψ in the Schrödinger equation, but that must be eliminated to make a prediction, leaving only real values in Ψ, by Born's rule (squaring the modulus of the wave function or $|\Psi|^2$). Here we move from a probability amplitude that includes complex numbers (numbers with imaginary parts or "i = √-1") to a traditional probability distribution that includes only real numbers (decimal numbers between 0 and 1). These two steps are as controversial as the Ψ-collapse itself, but compared to the latter they are practically not discussed.

In Darwinian Physics, these three steps are obviously addressed, because this probability amplitude would correspond to the present potential (Ψ) that emerges from nothingness by means of the vertical dimension, from below. Hence the need for complex numbers, since it would correspond to the axis perpendicular (vertical) to the real axis (horizontal dimension). But when a prediction is to be made, this complex component must be eliminated, since this vertical dimension is no longer needed, since the facts occur in a concrete holon, that is, in the horizontal dimensions. Therefore, in our everyday world, to represent all the phenomena we observe, we only need real numbers to predict and model them. That is why Ψ is unobservable, because it refers to the potential present that emerges from nothingness in the vertical dimension, while we inhabit a concrete holon, in the horizontal dimensions. But we cannot stop there, because Ψ still remains a probability distribution ($|\Psi|^2$), although it does not contain complex numbers. In other words, we do not see imaginary numbers, but neither do we see probabilities in everyday life. For this reason, the measurement postulate must be applied. That is, we must collapse the classical probability distribution that we obtained with the Born rule in order to eliminate the superposition of possibilities. Here, the present ceases to be a potential and becomes a de facto present, so we must pass these infinite decimal numbers between 0 and 1 and convert them into a single value: 1 or 100%. Precisely the possibility or the fact in which it ends up being observed or measured, because now "it is," it exists. It is no longer potential or nothingness. In short, Born's rule is to take the potential present that travels in the vertical dimension from nothingness (Ψ), to place it horizontally in a concrete holon ($|\Psi|^2$), and then to actualize or materialize such a present into a concrete possibility by means of the so-called reduction or collapse of Ψ.

In this way, Darwinian Physics is able to explain the entire Copenhagen Protocol in a very simple and coherent way; because it is able to explain all the peculiarities of quantum mechanics, such as the fact that it is irreversible, instantaneous, and non-local. In other words, what physicists hate and dislike the most. Features that not only violate the absoluteness of the Parmenian paradigm, but also destroy the linearity of the

Schrödinger equation, the quantum law. So, it is not surprising that they had to be eliminated in the many-worlds interpretation.

So, let's start with an explanation of instantaneity. This will be discussed further in the next chapter, but briefly, Darwinian Physics explains instantaneity because it is presentist. Therefore, it makes sense that quantum phenomena occur precisely in the "now," since the collapse of Ψ would be evidence of holistic presentism, since it is the product of the "present" that materializes in each instant. A present that occurs in a random way, since the presents of quantum systems are more striking in their nihilistic origin, being presents of lower holons, closer to nothingness. A present that also marks a before (because it is not reversible) and an after (because it cannot be accurately predicted) in a physical process that is also fundamental. In other words, the collapse of Ψ breaks with eternity, separating past and future, because it is simply now. A presentism that is reinforced by another very important feature of quantum mechanics, the indistinguishability of quantum particles. For example, if two electrons are detected and named electron "A" and electron "B," one might think that it is possible to continue to identify them as such in subsequent observations. But this is impossible. Since their behavior is random, they do not follow a pattern, a regular path, so they cross and mix in such a way that it is not possible to know which electron was which in the next detections.

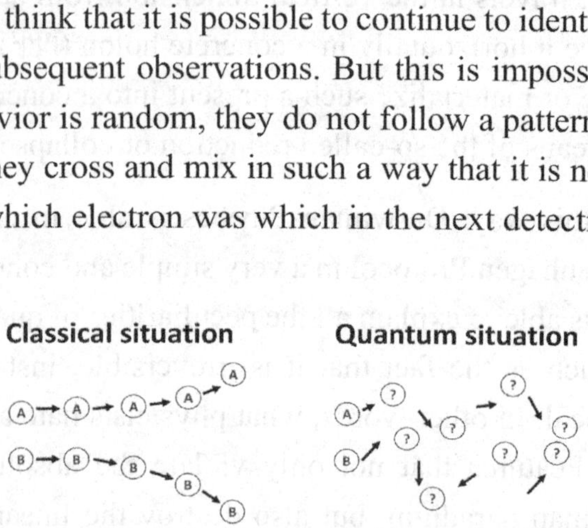

This characteristic is very easy to explain in Darwinian Physics, because it is another proof of its presentism. Of the fact that each instant is unique. It is as if in each instant there were new particles, and so it is; because each instant is an authentically new present, because in each instant Ψ collapses, and it does so in an authentically random way. A collapse that is neither reversible nor predictable. Therefore, it is this real separation between past and future that is the proof of the existence of the "now." Then the real illusion is not the present or the becoming of time, but the entire past and future postulated by Parmenian eternalism.

We have already seen how Darwinian Physics solves the greatest mystery of quantum mechanics, the problem of measurement, but it must also be coherent with all the other peculiarities of quantum mechanics in order to be valid (holistic coherence). To do this, we will start with one of the most notorious features of this fundamental branch of physics, which rightly owes its name to the quantum. This is because many people, including physicists, confuse quantification in quantum mechanics with spatial or material quantification, as the smallest thing that forms things. Like Democritus' idea of atoms. But what is really quantized in quantum mechanics is action. A term from analytical mechanics that refers to something dynamic, representing a process, path, or trajectory of physical systems. Thus it is used in the Feynman path integral, which is based precisely on the principle of minimal action of analytical mechanics. Therefore, the "quantum" or discrete nature of quantum mechanics would be better explained by the presentism of Darwinian Physics, referring to the different "potential nows," an explanation that becomes even more evident if we are more rigorous with the definition of such a principle; to correspond mathematically to the integral of the Lagrangian (kinetic energy - potential energy) with respect to time. It is therefore a function of the entire trajectory, not just of one point (which would be the derivative). Therefore, the real or most probable trajectory or process is the one that corresponds to the minimum action. A balance between not too much kinetic energy, but not too little potential energy. Then, any other action will require more variation, of greater difference between these two types of energy and; therefore, corresponding to a more chaotic process or trajectory. Therefore, it is a perfectly compatible and explainable principle under Darwinian Physics, because like holistic equilibrium, this is the definition of the second type of existence. So that nature seeks the minimum action simply means that it seeks the most stable process, the one that changes as little as possible. And in terms of Darwinian Physics that means to exist more and more, to be more and more ordered, which is the main purpose of the cosmos and its holistic evolution.

This temporal nature of the quantum can be seen most clearly in the case of light, which is not matter per se, but rather energy or interaction. Nevertheless, it is quantized into discrete units that we call photons (this will be discussed in more detail in the next chapter). Thus, thanks to the explanation of time by presentism and nothingness, the essential features of quantum mechanics emerge naturally from the principles of Darwinian Physics. In this way, nothingness would pump present into existence through its fluctuations, which would be mathematically described by Ψ. A probability wave that would propagate towards the cosmos, as if it were expansive waves emitted from an epicenter, crossing the holons and generating its presents, which would be more and more ordered and of longer duration as they progressed towards the most outer or superior layers of existence. As if they were larger and larger gears of a universal clockwork mechanism, forming larger and larger and more stable "now."

Another support for the presentist holoarchic on the part of quantum mechanics, closely related to the above, is nothing less than one of its most famous principles, the aforementioned Heisenberg's uncertainty principle or indeterminacy. A theorem of quantum mechanics that prohibits having absolute certainty about quantum systems, due to the wave-like behavior they manifest through Ψ. And, as has been said in several previous chapters, it means that two related properties (conjugate variables) cannot be known simultaneously with exact precision, being the two parameters most commonly used in quantum systems to illustrate this principle: velocity or momentum (p) and position (x). Thus, the greater the certainty about a particle's velocity (-Δp), the less certainty about its position (+Δx), and vice versa (+Δp → -Δx).

$$\Delta x \cdot \Delta p \geq \frac{\hbar}{2}$$

However, as also mentioned several times in the book, this principle applies to any conjugate or related variable of quantum mechanics, for example: energy (E) and time (t).

$$\Delta E \cdot \Delta t \geq \frac{\hbar}{2}$$

Only in the latter case, the relationship has not yet been demonstrated, precisely because of the lack of a definition of time, as we recently saw with the Pauli's paradox. Something that would be resolved in the new paradigm by integrating time, confirming such a relationship, because thanks to holoarchy, having greater certainty about position (- Δx) means (or is equivalent to) going to smaller and smaller, more precise spatial scales. In terms of Darwinian Physics, we would be talking about going to smaller and smaller holons, and thus with a smaller vertical distance to nothingness. This would explain the loss of certainty with respect to the other parameter, the momentum (+ Δp), which would reach infinite uncertainty, pure randomness, when reaching the maximum precision in position, the smallest possible space, to the spatial point or the Planck unit of space. This is perfectly consistent with Darwinian Physics, since it corresponds to the spatial point of nothingness. The same is true for energy (+ ΔE), since it corresponds to speed (in the sense of kinetic energy). Therefore, going to lower and lower holons means having smaller and smaller (- Δt), shorter, more precise, or faster presents. And therefore again closer to nothingness, by approaching the time point or the Planck time unit. Which again makes sense in Darwinian Physics, since energy is also indeterminate, becoming more and more random as it approaches nothingness.

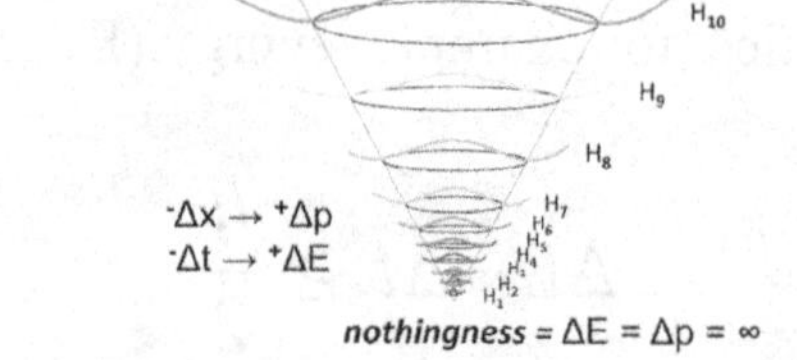

Then, Heisenberg's uncertainty principle would reveal this hierarchy of the cosmos, where as we go up in the vertical dimension, order increases as the measures of energy and momentum become more certain or stable, since we are talking about larger space-time scales (holon-present). On the other hand, as we go down in the vertical dimension, chaos or uncertainty increases in the magnitudes of energy and velocity as we approach nothingness as we are in smaller space-time scales. Therefore, holoarchy perfectly explains this difference in what is measured at large and small scales in nature. It explains why the macroscopic world appears to be classical (deterministic and certain), while the microscopic world appears to be quantum (random and uncertain). Thus, under this new worldview, one can finally understand the behavior of a simple electron isolated in a vacuum, because this situation is equivalent to a very small system, on the scale of the size of the electron itself, since it does not constitute anything larger. This means that one is at a very small vertical distance, inhabiting very subordinate or small holons; and therefore closer to nothingness. Hence its intrinsically quantum or random nature. But if that electron wants to exist higher up, it has to be part of it. To do so, it must form something larger, such as an atom, a molecule, or a rock. This makes it move up in the holoarchy by placing itself in higher holons. This reduces its chaotic or uncertain behavior because if the electron wants to exist, it must respect the existence of others. Maintaining continuity, regularity, coherence with the rest of the members of the system. That is why the measurement collapses the Ψ of the electron (eliminates its quantum superposition), because if the electron wants to continue to exist in the next present and in the next, then it cannot contradict or deny the existence of the photon that was used to detect it, nor of the measuring apparatus that sent that photon, nor of the physicists who planned and carried out the experiment, nor of the laboratory, nor of the city where it is located, and so on... Therefore, Ψ collapses at every instant, because it is the present itself that materializes. One that does so from the bottom up, from nothingness to everything. So, the electron can remain indeterminate if it is isolated, in the void; but if it wants to exist, it has to be part of this cosmos, respecting the existence of the atom of which it is a part, and of the molecule that this atom constitutes, and of the macroscopic thing in which this molecule is located, and so on. Then quantum particles can reveal to us the chaos that still exists in the cosmos by materializing randomly among a range of possibilities at a short vertical distance. However, at a greater vertical distance, they must choose among a few possibilities that do not interrupt the logos, logic, pattern, regularity, or order of existence. For if it did, it would not be part of it, would not be part of this higher present, and would therefore be part of literally nothingness. In this way, the great chaos at a given level or at a short vertical distance, those many worlds of Hugh Everett or those multiple paths of Richard Feynman, end up literally disappearing as the randomness is diluted; as they converge into a single world or path as we move up in the holoarchy. This is how the Logos of Nature unfolds, continuing the story of the cosmos, which has an open ending because it is being written with each new Now. This is the Darwinian physics explanation for a process that physicists call decoherence, completing it by explaining the collapse of Ψ, through presentism.

In conclusion, the interpretation of Darwinian Physics, although it sounds very revolutionary, is actually quite conservative, since it only justifies and explains the official interpretation of the theory, the Copenhagen one. It clarifies its steps and its most controversial postulate, how the random collapse of Ψ occurs. It does so without adding strange and exotic elements such as non-local hidden variables, but by introducing precisely the element that was missing in quantum mechanics, time. A truly elegant solution, which is very economical, especially compared to the infinite parallel eternal universes required by the Parmenian interpretation of many worlds. However, despite its obvious virtues, the quantum interpretation of Darwinian Physics will not be easy for theoretical physicists to accept. Not only because it does not fit their worldview, but also because it violates the most cherished Laplacean dogma of Parmenian physics: the conservation of information. The principle that makes the laws of physics absolute by making them reversible, deterministic, and eternal. Only in quantum mechanics, physicists have changed its name. Now it is called unitarity. The principle that forces quantum mechanics to be deterministic. To see how this classical principle works in quantum mechanics (or how it does not work), we must remember what was said earlier, that causality is not the same as determinism, because there can be several types of determinism, which can be causal or not. The type of determinism that physicists defend is absolute determinism. This is the only type that physicists, due to their lack of philosophical knowledge, believe exists. This type includes the categories of strong (linear) and weak (thermodynamic and chaotic or nonlinear) determinism that we have seen in previous chapters. However, there is also a completely different category: causal determinism. Here, perfect prediction of the future of a system is strictly guaranteed, given perfect knowledge of its present state. But it does not necessarily imply perfect postdiction or knowledge of the past. To understand this, let us suppose two causes: "A" and "B," which converge to produce the effect "C." If we know the initial state of the system, i.e. "A" or "B," we can make a perfect prediction about the future, "C." However, if we are in "C," we could not predict the past, or what exactly caused "C," because we would not know whether it was "A" or "B" that caused it, since there are two possible causes.

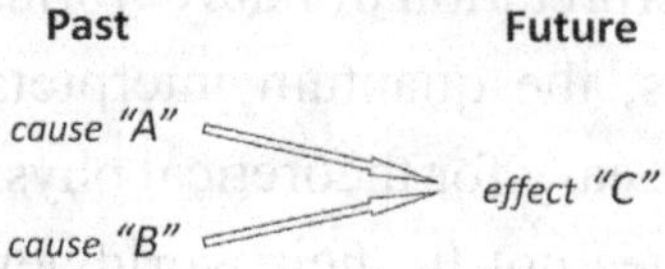

This imprecision or lack of certainty about the past would mean a loss of information, a loss of what exactly happened? If the states do not remain independent of each other and merge into fewer states, we lose information and thus certainty about history. This scenario is exactly what Parmenian quantum mechanics forbids by its unitarity, because a decrease in the number of states or information would mean that the sum of the probabilities of all possible states would be less than "one." Hence the name of the principle. Then the principle of conservation of information is guaranteed by the principle of conservation of probability. This desperate appeal to the mathematics of probability is the last resource that physicists found to justify the absoluteness of the laws of physics, because they produce the same consequence, an absolute determinism. A perfect prediction not only of the future, but also a perfect post prediction of the past, if one has a perfect knowledge of the present, of course. All this thanks to the preservation of the number of states or quantum information. So, at any moment there would be the same amount of information. It does not matter if that instant is in the very distant past or in the very distant future, because if you know the information of one instant, you can know the information of any other instant, no matter when it is. Again, like Laplace's demon in Newtonian mechanics.

In this way, physicists thought they had finally found the definitive argument to support their dogma behind the violation of the principle of conservation of energy and time symmetry. The problem is that, empirically speaking, this principle is clearly and constantly violated every time a physicist makes a measurement of a quantum system. In other words, the collapse of Ψ is blatantly neglected because the measurement postulate is nonlinear. Therefore, it is not compatible with the linearity of the Schrödinger equation; which means that in the measurement if, much of the information of Ψ is destroyed, because the vast majority of quantum states disappear. All the states we do not observe because we measure only one possibility, one state, among the infinite possible states of Ψ. In other words, Ψ has never been seen, it is just an assumption. Therefore, physicists postulate that these other possibilities exist in other parallel worlds to save Ψ, so that Schrödinger's equation remains absolute, like any law. In other words, the unitarity would be fulfilled only before the measurement, in the Pythagorean-Platonic world of mathematics, now called many worlds, but when we observe reality, we get only one single quantum state of Ψ, with 100% certainty, because we observe it, it is there. This is why the collapse of Ψ makes the Schrödinger equation irreversible, because much of the information of Ψ was lost forever when it randomly collapsed at the time of measurement. Similarly, we cannot postulate which dice produced the result 4, because any dice can produce such a result with the same probability (1/6). This is the inevitable consequence of ontological randomness, the nightmare of physicists, because the determinism of absolute laws, or what they now call unitarity, is completely destroyed.

Then, as already advanced at the end of the second part of the book, in Darwinian physics unitarity is not fulfilled, so does that mean that the principle of probabilities is also false? Not at all. As always, it is a matter of interpretation. In mathematics, unitarity will always be satisfied, i.e., the sum of all probabilities will always be one, because that is its world. Therefore, the mistake was to arbitrarily assume that nothingness is not a real possibility, obviously because of your absolutist worldview. However, this is a prejudice of the Parmenian paradigm, because if we consider nothingness as a legitimate possibility, the probabilities after the collapse of Ψ will still add up to one. We would only have to consider the possibilities that disappeared into nothingness, and thus those that we cannot observe in existence.

$$\mathbf{p(nothing) + p(observables) = 1}$$

On the contrary, physicists consider only existential possibilities as valid. The problem is that when measured, the calculation does not add up. This is why the many-worlds interpretation is inescapable in the current paradigm, because to save unitarity one would have to assume that the states or possibilities that have disappeared continue to exist, only in other parallel universes. However, we have focused only on the impossibility of knowing or predating the past, on the reversibility or non-reversibility of physical processes. But we must not forget the most obvious aspect of quantum mechanics, its inability to predict the future accurately. A randomness that is not due to the fact that we do not know enough about the system, as is the case with a coin toss, a gas, or the weather. Because we know everything there is to know about the electron, for example. So even though the technology has improved quite a bit in the last century, with extremely precise measuring devices and highly advanced laboratories, the probability of knowing the spin up or down of the electron has not changed at all, it is still 50%. This means that this probability does not come from our ignorance, from our technological limitations, but is intrinsic to the electron. An ontological randomness. Therefore, if irreversibility or unpredictability with respect to the past implies that information is destroyed, ergo, unpredictability with respect to the future must imply that information is created. This is the true revelation of quantum mechanics and the true explanation of the collapse of Ψ; for if we cannot know from which Ψ a particular quantum collapse or observation came, then we cannot know which particular quantum collapse or outcome will result from a particular Ψ. In fact, it is the latter that is most obvious in quantum mechanics. This is what baffled the giants of physics in the 20th century: the inability to make certain predictions (only probabilities). Therefore, as time moves into the future, the most puzzling thing about quantum mechanics is not its ignorance of the past, but its ignorance of the future. And this is what Darwinian Physics explains perfectly, because it is a strong proof of holistic presentism; because we cannot predict exactly what will happen in the next moment, because it has not yet happened. In other words, we do not have all the necessary information to make an accurate prediction, because that necessary information does not yet exist, because it is being created in each instant, in each new "now," as Ψ collapses and materializes in the present. Therefore, it is the constant creation of new information due to the materialization of new "presents," that is the real cause of indeterminacy at the most fundamental scales of nature. This is what quantum mechanics screamed at us more than a century ago, and what physicists try so hard to ignore. Therefore, the official quantum mechanics, the Copenhagen interpretation, fully supports Darwinian Physics, because its irreversibility and empirical unpredictability are clear evidence of its holistic presentism.

However, in order to understand well this presentist mechanism that collapses Ψ, and how this is incompatible with the unitarity of Parmenian quantum physics, we must first define well what information really means. A concept that lends itself to much confusion and misuse, as happened with the concept of entropy in the previous chapter. To avoid this situation, let us define the concept rigorously, starting with the first theory of information as such, developed by the American mathematician and engineer Claude Shannon (1916-2001). In this theory, presented in 1948, information is defined statistically, by the probability of an event occurring or not. The lower the probability of an event occurring, the more information it contains. This means that if an event is very predictable, very expected, it will contain very little information compared to a very unexpected event. Something that can be measured by the following formula given by Shannon:

$$- \log n\,(p_i)$$

Where "p_i" is the probability of a particular event and "n" is the number of possibilities. For example, a coin flip has "n = 2" (heads or tails) and "p_i = 50%." So, if we have a possibility "a" with 1% probability and another possibility "b" with 99% probability, the information for each would be:

the information of "a" is: -log2 (0.01) = 6.64385....
the information of "b" is: -log2 (0.99) = 0.01449....

Then we have that event "a" has 458 (6.64 / 0.01) times more information than event "b," just because it is a very unlikely event compared to the latter. This is very paradoxical, because according to Mr. Shannon, if we had a random letter generator, thanks to a radioactive isotope, the book that would write such a generator would have much more information than any other readable book. It doesn't matter if we compare it to Darwin's Origin of Species, Newton's Principia Mathematica, or Euclid's Elements. In other words, a book that tells us nothing would always have much more information than any other book that tells us something. So why has Claude Shannon's definition of information been so successful, if it sounds so absurd? Because his definition was never useful in physics, but in the symbolic world. Remember, Claude Shannon was a communications engineer. In fact, the original paper was called Mathematical Theory of Communication. Therefore, his definition has its meaning and utility in a particular human activity, specifically the storage and transmission of electronic information, not in a principle of nature. Therefore, it is a quite useful definition where Shannon applied it, because in order to fulfill its purpose of storing and transporting large amounts of digital information in the most efficient way possible, it is essential to identify or recognize the non-redundant information, the unpredictable, the most uncertain. This would save the telecommunications industry a lot of resources. For example, if you send a video of a painting in a museum, it does not make much sense to send every bit of information because a lot of it is repeated, it does not change. So, it would save a lot of memory and energy if you could detect the redundant information, send it only once, and then simply repeat it several times on the other side of the communication channel. Therefore, in the context of communication, it is much more important to identify the most unlikely data because it is the least dispensable and therefore contains the most information. On the other hand, the most probable information can be stored and forwarded in a process called data compression, which allows the most economical communication with the least possible loss of information.

As we can see, Claude Shannon's definition of information is quite relevant to the field in which he used it and which is still used today, in the world of telecommunications. The problem, however, is that the physicists adopted it practically as it was in their field. This confusion arose because Claude Shannon came up with the idea of using a word from physics, entropy, to define the weighted sum of the different data in a message. Just because the formula bears a resemblance to Boltzmann's entropy.

$$-\Sigma\ p_i\ \text{-logn}\ (p_i)$$

In the example above, the Shannon entropy would be the information of "a" multiplied by its probability, plus the information of "b" multiplied by its probability:

entropy information of "a" and "b":
(6.6438 x 0.01) + (0.0145 x 0.99) = 0.08

A confusion that is compounded by the fact that the relationship between Shannon's information entropy and Boltzmann's thermodynamic entropy is not so far-fetched. As we saw in the previous chapter, the higher the thermodynamic entropy of a physical system, the greater the number of compatible configurations, or microstates, that the macrostate will have. Therefore, thermodynamic entropy can be interpreted as a loss of information, in the sense that it increases the uncertainty about exactly which microstates the system has. For example, if a thermodynamic system has an entropy of 2 (very low), this means that the uncertainty about the exact configuration of the system will be 50% (1/2). On the other hand, if the same system has an entropy of 20, it means that its uncertainty is 5% (1/20). Therefore, the loss of information about the configuration of the system increases with the thermodynamic entropy, because there are more and more possible configurations, of which we cannot know which is the effective one. The problem is that this relationship is reversed in the case of information entropy. As we saw recently, what Shannon postulates is that the greater the uncertainty, the greater the amount of information, whereas in thermodynamics, the greater the entropy or uncertainty, the greater the loss of information. And the explanation for this contradiction is very simple, it is simply the application of the concept of information to different fields. The direct relationship between uncertainty and information makes sense, as we saw above, in the field of computers and telecommunications; but not in the field of physics, because it is absurd to say that something more uncertain or ambiguous has more information.

A big mess that is made worse because, on the one hand, it is said that information is lost as entropy increases in thermodynamic systems, while, on the other hand, in quantum mechanics, physicists say that information is conserved because of unitarity. This confusion is again the product of constantly conflating an epistemic definition of information with an ontological definition. This is because, although the relationship is inverse, both Shannon's information and that of thermodynamics are intrinsically epistemic (what we can know); but again, because of this lack of philosophical skill, physicists ascribe an ontological character to this definition. As a result, if you ask a theoretical physicist what the most fundamental thing is, what nature really consists of, he will most likely answer that it is information. An answer that means "I have no idea," because if you ask him what information is, he will not know how to tell you what it is. However, informationism, the belief that nature at its most elementary level consists of pure information, is very fashionable these days. For example, the eminent American theoretical physicist John A. Wheeler (1911-2008) said, "I think of my life in physics as divided into three periods. In the first, I was wedded to the idea that everything was particles... In my second period, everything was fields... And now my new view is that everything is information. A belief that he immortalized with the phrase "It from bit," which has been very influential and popular in current physics, as evidenced by new theories or hypotheses that have been based on this belief, such as Qbism (see above), but also Wolfram's graph theory, Garrett Lisi's E8, the holographic universe and, of course, the hypothesis of the simulated universe. And we may ask ourselves why, where does this new fashion come from? The answer is simple: cultural. In every era, nature has been described by what was available, by experience, for example: the universe was described as a complicated clockwork mechanism in the 17th and 18th centuries, when complicated windmills and watermills, clocks and other mechanical devices had become popular in the Newtonian era. Then, in the late 19th and early 20th centuries, a particle-interaction worldview became popular thanks to the development of coal and steam engines due to the first industrial revolution, driven by thermodynamics and statistical mechanics, which studied systems composed of many elements, such as gases. Then, thanks to the age of light in the second industrial revolution, where new electronic technologies such as the light bulb, radio and television appeared, a vision of the cosmos based on "waves" and "fields" was promoted, thanks to Maxwell's electromagnetism. And now, thanks to the third and fourth industrial revolutions that have surrounded us with digital technology (PCs, the Internet, smartphones, tablets and artificial intelligence), physicists are beginning to believe that nature is made up of pure information. What a coincidence. So how did we get here, and why has such an absurdity as the idea that we live inside a computer become so popular among physicists? It is all because the same mistake is being made as in the previous chapter. The world is turned upside down, the holarchy of existence is inverted, precisely because of this lack of philosophical ability to understand that one cannot ontologically interpret a concept that is epistemic in nature. In the previous chapter it was mathematics, and now it is information.

So, to clear up this mess, we need to provide a clear and concrete definition of physical information, which is also ontological, to distinguish it from the epistemic or thermodynamic information that has been used (or misused) in physics so far. For this purpose, we will return to the old concept of information, which was defined as a simple property or quality of objects. For example, what is the information of a bicycle? Its size, color, materials, accessories, and so on. In this sense, information is simply the attributes that describe a thing. A definition that once again confirms the conservative nature of Darwinian Physics. Then, under this new (ironically speaking) ontological definition of information, by referring to something real in nature and not to something subjective like our ignorance, we will see the true relationship that would exist between information and thermodynamic entropy.

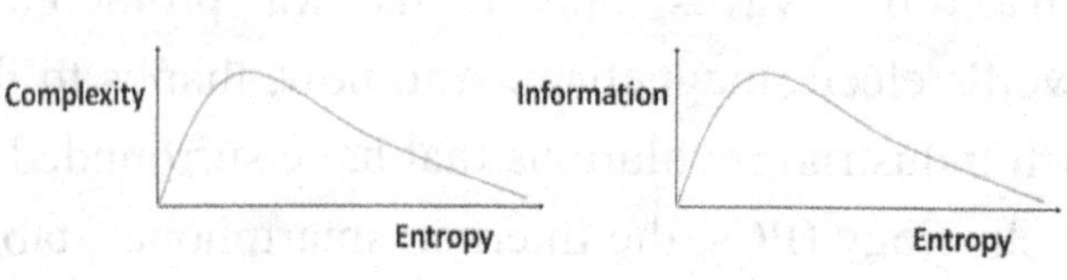

As can be seen in the previous drawing, what would really be related to information would be complexity, not thermodynamic entropy; simply because entropy in current physics is an epistemic concept, so it cannot refer to anything ontological, as we want information to be. What is ontological, however, is complexity, because it refers to concrete things. For example, a dog is more complex than a stone. Moreover, the relationship between complexity and information would be very natural, because a thermodynamic system with very low or high entropy is a simple system, and therefore with very few describable attributes and properties. That is, with very little information. While the more complex states or those with medium entropy are those that would have more information, since they require a greater amount of data to describe them accurately, since they have more attributes.

This more reasonable and ontological definition of information allows us to correct this "it from bit" madness, because obviously things precede information (bit from it). There can be no properties or attributes without a "something" that contains them, just as there can be no smile without a face. Again, something that makes much more sense, since information cannot exist by itself, but must be encoded in something in order to exist. This simple definition of information brings realism back to physics by eliminating idealistic informationism, because information needs a container, a medium, to be encoded. Therefore, it cannot be a fundamental element, on the contrary, it is a strictly derived element, dependent like the smile on the face or physical things on space-time. That is why something has always been needed to store information: stone, papyrus, and now semiconductors; because information cannot exist on its own, it needs something to encode it. Even spoken words are encoded in something, in this case the vibrations in the air. Without air, there is no sound. Therefore, informationism like Qbism is the result of confusing the epistemic with the ontological (again, the container with the content), again due to a lack of philosophical skills.

Then, in order to avoid inverting the hierarchy of existence by falling into idealism (anti-realism), we must define and classify the different types of ontological information, because by defining them as properties or characteristics of things, we will have as many types of information as there are types of things. Therefore, there would be a symbolic ontological information that belongs to the attributes of the elements of the highest principal reality in the holoarchy. This information is not to be confused with epistemic information, because symbolic ontological information does not take into account tastes or preferences (subjective epistemic information), nor our capacity to know or our degree of ignorance (objective epistemic information), as thermodynamic information would. On the other hand, ontological information deals with symbolic reality as it is. In this sense, it is more like semiotics, or the science that studies the signs that make communication between people possible through language. And like any science, it conducts its study as objectively as possible, treating signs in the same way that a zoologist studies a population of parrots, for example. The next type of ontological information corresponds to the most important reality below the previous one, biological ontological information. This refers to the properties and characteristics of living systems. Continuing with the previous example, the physiological and behavioral characteristics of parrots. The ontological information that follows corresponds to the next major reality, physical ontological information, which refers to the properties and characteristics of physical systems: such as stars, planets, rocks, molecules, or atoms. Whereas space-time ontological information would refer to the properties and characteristics of space-time systems, such as the curvature and topology of space-time. Here we could further subdivide each category into as many types of information as there are types of systems. For example, I could further divide physical ontological information into molecular or chemical, atomic, planetary, galactic, and so on. However, as the name of this chapter implies, we will concentrate on the classification of information of quantum systems. In this sense, we must consider the solution of Darwinian Physics to the problem of measurement, so we must again distinguish between two types in order to respect also the properties of quantum systems, which are characterized by having both wavelike and corpuscular attributes. Then, given this duality, we must distinguish between: ontological information quantum physical potential (OIQPP), which would correspond to Ψ as such, and ontological information quantum physical factual (OIQPF), which represents the collapse of Ψ and, therefore, what is observed.

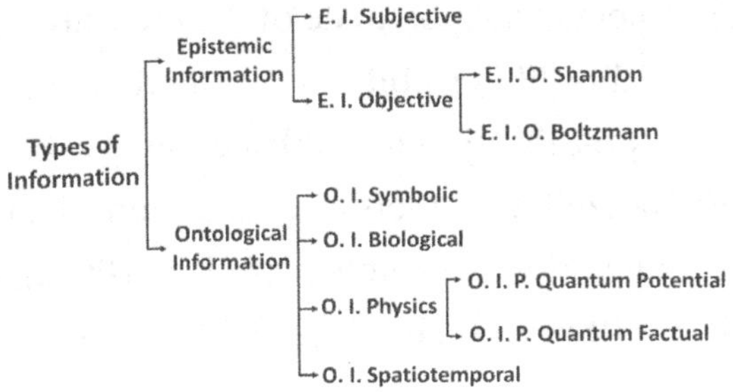

With this distinction between the OIQPP (from now on only potential quantum information) and the OIQPF (from now on only factual quantum information) we can better reinterpret the Copenhagen Protocol to understand it even better. For this we will use the same it from bit principle of John A. Wheeler, but corrected in a realistic way, but also in his writing, because he did not really talk about bit either, but about qbit. This is because Wheeler knew that bit or classical information cannot be fundamental, because what is really fundamental has to be quantum. Therefore, Wheeler meant that things (it) emerge from quantum information (qbit) through collapse, which according to him is caused by consciousness. But since we do not want to be idealistic, but realistic, let us see how quantum information really relates to the measurement postulate. In order to do so, we have to subdivide the potential quantum information again, since we have to apply the procedure that precedes the measurement postulate, the Born rule. In this way, we would have the vertical quantum potential information, which would correspond to the present emerging from nothingness through the vertical dimension, and which would be represented by Ψ or the qbit; and the horizontal quantum potential information, which would correspond to the present already located in a holon (horizontal dimensions), and which would be represented by $|\Psi|^2$ or a pbit (Born rule). Whereas the postulate of the measurement would be the collapse of this $|\Psi|^2$ or pbit into a concrete bit, which would have only one possibility with 100% certainty, being a present already materialized and; therefore, becoming factual quantum information, since it is observed or is (1 or 100%), or is not observed or is not (0 or 0%) such a property in a quantum system.

Interpretation Copenhagen

bit = collapsing Ψ or factual quantum information

↑

$|\Psi|^2$ = pbit: Born's rule or horizontal OIQPP

↑

Ψ = qbit: vertical OIQPP

↑

Nothingness

So, we have the whole process of holistic presentism represented in the Copenhagen interpretation. A potential present (qbit or Ψ) that emerges from nothingness and ascends through the vertical dimension; that arrives at a concrete holon and that, for the same reason, must exclude the vertical dimension since it is already horizontal (pbit or $|\Psi|^2$); to then, finally, make that potential present materialize in a concrete now, through the collapse of Ψ, assigning to a concrete measure a 100% certainty because it is observed (bit or factual quantum information). Thus, any kind of information depends on something, it is a property, a characteristic encoded in a thing, even quantum information, only this time the thing it refers to is time itself. Thus, the whole protocol called the Copenhagen Interpretation describes a process of emergence from nothingness of new potential presents that materialize in nature, generating that becoming or flow of time that we experience.

Then, having finally resolved what information means in quantum mechanics, we can better address what physicists mean by unitarity. As already mentioned in the book, this principle means the conservation of quantum information, understood as all quantum states of Ψ. Therefore, if one wants to interpret information ontologically (realistic or non-idealistic interpretation) and, moreover, to save the absoluteness of the quantum law of collapse through this principle; then one has to interpret randomness in a frequentist way, converting all potential quantum information into factual quantum information. Actual quantum information that we do not observe, so it must exist in other parallel universes. Then no new information emerges, no new chemical properties, no new living beings, no new ideas, because they are all supposedly written in the Ψ of the multiverse, determined from eternity. In contrast, Darwinian Physics is spared the need to believe in infinite eternal parallel universes because it simply recognizes the randomness observed in quantum systems as just another natural property, like any other physical property. And not something that exists only in our heads, as a measure of our ignorance. In this way, we would only have to abandon this dogma of unitarity; and add time, the missing ingredient of quantum mechanics. Because that probability wave (Ψ) that physicists talk so much about in the Copenhagen interpretation is in reality the present that emerges from nothingness. The same one that creates new information when it materializes or collapses. In other words, new actual quantum information is created at every moment out of the potential quantum information sent by the nothingness at every new "now." Thus, simple Darwinian Physics explains what is observed, that quantum properties are spontaneously determined at each instant from an indeterminate or random situation. A nihilistic origin of time, evidenced by the increasing randomness or uncertainty as we go down in vertical distance, approaching the space-time point or nothingness.

In conclusion, in Darwinian Physics, unitarity is obviously not fulfilled, because in reality actual quantum information is constantly being created out of nothing, at every instant, as existence evolves towards "one," towards the absolute ($P(H) = 1$). The problem is that the holistic evolution of the universe is so advanced that it appears to be absolute or unitary. In other words, the degree of order achieved is so high that it gives the physicist the impression that it is eternal, that it has never changed, that it has always been and will always be, because its information has always been conserved. A very high degree of stability and regularity of the universe, which must be so, or else intelligent beings capable of doing physics and science could not have evolved, because there would not have been a sufficient degree of order and complexity to allow it. This is why absolutist mathematical models are so effective, because the universe is sufficiently determined and stable for their equations or laws to work beautifully. However, if we were to look at larger scales of space-time, we would find that the equations would not work, as has already been confirmed by the violation of the conservation of energy due to the expansion of the universe. Then we would be repeating an old mistake in science. One that already happened in biology, because before Darwin it was thought that animals or species did not change, that they remained as they were, simply because no such change could be seen on human time scales. Therefore, it was assumed that they were absolute, created by an absolute being. The same is true today, but in physics. Only it is no longer called God, but "the law of physics." Since it is essentially the same problem, the same solution must be applied. Banish absolutism by Darwinian evolution. However, there is no danger that the phenomenological or instrumental mathematical models of present-day physics will cease to be useful or effective, just as naturalists and zoologists are in no danger that their studies of present-day animals will be rendered obsolete by biological evolution. It is simply that the scales on which Darwinian evolution operates, especially in the cosmos itself, far exceed human scales.

So existence is not absolute because we exist, and therefore there is still complexity, emergence, change, and randomness in the cosmos. However, the Universe is evolving towards a more absolute state, as its degree of entropy or order is constantly increasing. Therefore, this holistic evolution of existence is at the expense of nothingness, its influence or impact on the whole. This means that existence becomes more and more deterministic, more and more concrete. Therefore, the degree of potential quantum information or nothingness would represent the degree of freedom that is still left to existence, those other probable states that are still available to the universe before it is completely enslaved to a single destiny or future with 100% certainty. Therefore, potential quantum information would represent those possible futures, the probable paths that still remain, given the degree of certainty that existence now has. Here the non-deterministic causality of holistic presentism explained in the previous chapter is even better understood, because chance is the product of the non-existence of the past; whereas the future (new present) is caused by the various possibilities (information) that exist in the present. In other words, our destiny is not completely determined; thanks to the fact that there is still that degree of chaos or nothingness in the cosmos, which gives the universe that margin of maneuver or freedom. Thus, neither the absolute nor the nothingness is omnipotent, since it depends on the degree left to it by the other. Therefore, with the great degree of order that we currently have in the universe, nothingness is very much subordinate to existence. This means that the possibilities and, above all, the probabilities of the "futures" yet to come (OIQPP) depend on the "present" we already have (OIQPF). From this follows the principle of causality, how the present influences the future, since nothingness can only act in the now, modifying what already exists. In this way, the previous present conditions the next present, because nothingness works with what is there, cooking from the ingredients it has available, exploiting the potential found in the "now."

Therefore, the more ordered existence is, the fewer degrees of freedom or possibilities are left for nothingness to exploit.

The above may sound very exotic, but again, it turns out to be quite conservative and reasonable. Let's take another example to see this. When you are born, you have a large number of possibilities that obviously depend on your family condition, socio-economic level, health, political system, time, genes, culture, and so on. Although these have great weight, if we take normal conditions of a western country, the possibilities can range from being a criminal, vagabond, serial killer to a tiktoker, engineer, priest, musician, gardener, teacher, scientist, and so on. A huge range of possibilities. A lot of freedom, although each one has different probabilities depending on the circumstances mentioned above. However, as we grow older, some possibilities become more and more likely, while the rest become less and less likely. Until we reach old age and one possibility has become completely dominant. It practically becomes a law because it becomes incredibly stable or difficult to change. Then, as babies, we would be the human analog of the cosmos when it was just emerging from nothingness. Lots of potentiality, lots of possibilities, but nothing concrete, nothing done or realized. Pure potentiality. On the other hand, our old "I" is more existential because it is already a concrete person, with a well-defined identity and personality, with a long history already realized. A great past (causal inertia) that weighs heavily and greatly limits the possibilities that can exist in the future. This is why it is increasingly difficult to change. As the saying goes, "You can't teach an old dog new tricks. A past that, as we already know, is contained in the present. Continuing with the analogy, our past is found in our physical condition (aging), neurological (memories), social ties (family and friends), status, patrimony, and so on. Then in the nothingness there would be maximum freedom, pure randomness. That is why P(H) = 0.5 or exactly 50% (for both states of nothingness) in the beginning. And for the same reason there would be the minimum amount of actual quantum information, i.e. zero, because there is no certainty, no existence, no pattern or regularity. Therefore, in the nothingness, there would be infinite "possible" or "potential" universes, not real parallel universes. Then, when the holistic evolution of nothingness begins, any universe is possible. There is pure potentiality due to ontological randomness. But as holistic evolution progresses, holistic presentism makes some of those possibilities materialize, become concrete, and make existence grow. In this way, the influence of nothingness becomes more and more limited, as it becomes more and more limited by the increase of factual quantum information. Something that happens with every present that rises out of the nothingness in the vertical dimension.

Another way to look at this holistic evolution of the cosmos is to compare it again with biological evolution. In this way, LUCA, the grandmother of all life forms that exist on the planet, would be a biological nothingness; since it had the potential to generate countless living organisms, of which only a tiny part materialized. Nothing more and nothing less than all the living beings that have ever existed on this planet. All the species that have evolved in these four billion years. And this includes not only those that are alive today, but also all those that have become extinct, which is many times the number of animals that exist today. But even though this is an impressively large number, LUCA could potentially have spawned countless other parallel evolutionary trees with an equal or greater number of species. Remember that every sexually born creature has countless potential siblings that will never be born. Well, each of them could have given rise to entirely new evolutionary branches. But as Natural Selection proceeded, one of these virtually infinite possible evolutionary trees was determined. The same is true of the entire cosmos. In fact, again, we can better understand this absurd conclusion of the multiverse reached by physicists, because if chance is eliminated by denying nothingness, the inevitable conclusion is that all these potential universes must become real. But if chance is simply recognized as real, as in the case of genetic mutations in Darwin's evolution, there would be only one universe, just as there is only one tree of life, determined or concretized in each new generation. The biological analog of the new is present in the cosmos.

CHAPTER XIII

The Problem of Light: Special Relativity

The problem in the previous chapter was much more obvious to physicists, but there is a similar problem in the relativity theory of the Jewish-born German physicist Albert Einstein (1879-1955), although no physicist would say so. For if no one understood how the collapse of Ψ occurs in quantum mechanics, then no one knows the reason for the behavior of light. This is the basis of all of Einstein's physics. In this sense, both theories are phenomenological, since quantum mechanics does not explain random collapse, and special relativity does not explain the invariance of the speed of light. It simply assumes this fact for the sake of it and derives its logical consequences from it. One might even say that the problem of this chapter is more serious than that of the previous one, because physicists are not aware that there is a problem in relativity. They have simply forbidden asking about light, postulating that it is a universal law, period. A technicality with which physicists say: "This is so because I have no idea why." The only difference compared to the collapse of Ψ is that the behavior of light is much more compatible with their absolutist principles, being invariant and unchanging rather than spontaneous and random. Therefore, it generates less internal conflict for them by being consistent with their Parmenian paradigm. However, it is fair to ask why light is a universal constant? Because although it may seem very common to us, light is one of the rarest things that exists. For example, its speed is the maximum possible, about 299.792 km/s (from now on we will refer to it simply as "c"). However, light does not accelerate to such a speed. For example, when a photon (the quantum particle of light) is emitted by an electron jumping to a lower energy orbit in the atom, it is already born with this speed. In other words, the photon goes from zero to the maximum possible speed in an instant. As if that were not strange enough, this speed is constant, so no matter how fast you travel, you will always see light traveling at the speed of light. This is very strange because if you are driving a car at 50 mph and another car is passing you at 70 mph, you will see that other car passing you at 20 mph relative to you. Logical. Well, this does not happen with light. Even if you are traveling at 99% of "c," you will not see light overtaking you at 1% of its speed, but always at 100%. And the same thing happens when we travel against it. For example, if we are traveling at 50% of "c" and a light comes in the opposite direction, we will not see the light at 150% of its speed, but again at "c." So the speed of light is neither added nor subtracted. It does not depend on the direction or the speed at which we measure it, light always travels at the same speed. Pretty amazing.

Another very strange characteristic of light is that it does not seem to occupy space. Exactly, as if it were a ghost. This is because the photon is a boson, a classification used by particle physicists to refer to particles that mediate or carry the fundamental forces or interactions in the Standard Model of Particles. In this case, the photon is responsible for the electromagnetic force. Other bosons are the gluon, the particle of the strong nuclear interaction; the Higgs boson, which gives mass to the fundamental particles; and the W and Z bosons of the weak nuclear interaction. The other type of particle is called a fermion, and would be those that undergo or are influenced by these bosons. For example, a typical fermionic particle that suffers the consequences of the photon boson is the electron, since it has an electric charge. Then bosons represent the forces or interactions, while fermions are the particles that make up matter, such as the electrons and quarks that make up atoms. Having clarified this distinction, the bosonic nature of photons allows them to do things that are very unintuitive to us fermionic or material beings. For example, we could indiscriminately put as many photons as we wanted into a small box without it filling up. In fact, the more photons the box has, the easier it would be to add another photon. All because photons do not have to obey Pauli's exclusion principle, which fermions must sacredly respect. A principle that forces the latter to occupy space. Therefore, the electrons cannot all meet at the lowest energy level of the atom, as one would expect. In the same way, if we drop many balls up a hill, they will all fall until they meet at the bottom. However, fermions cannot do this because it is impossible for two fermions to be in the same quantum state or location. They exclude each other. This is fantastic for us, because it allows us to create structure, because if one is at the bottom, the other must be higher, and the other higher, and so on, until all the energy levels or orbitals within the atom are filled. In this way, larger atoms, larger molecules, and larger things can be formed. This is why Pauli's exclusion principle is the origin of all the chemical richness around us, especially in biology; it prevents us, for example, from walking across the floor and falling into the Earth's core due to gravity. Light, however, fortunately does not have to respect this, because thanks to it we can see things clearly, because if electromagnetic waves were to exclude each other and interfere with each other (like water waves, for example), what we would see in front would be interfered with by what can be seen on the sides. Therefore, the vision would be chaotic, not transparent or clear at all. It would be like hearing at a party with lots of noisy people and loud music, where all the sound waves mix and interfere with each other. Remember that sound waves are air waves, molecules, and therefore fermions. Electromagnetic waves, on the other hand, are bosonic, so they can pass through each other like ghosts. In fact, it is this ability to take up no space that makes it possible to make powerful and useful lasers, since many photons can be placed in the same place.

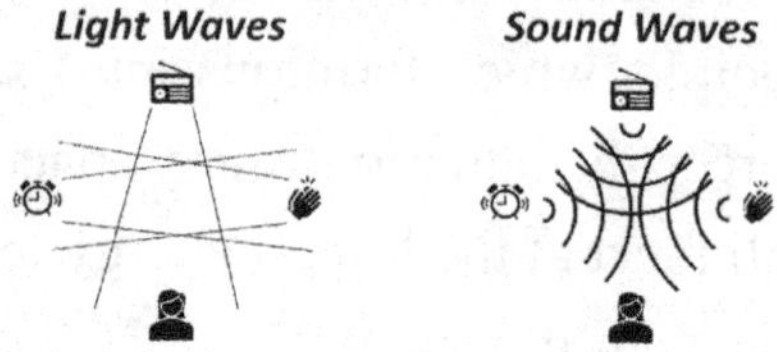

But in spite of these extraordinary properties of instantaneity, invariance, and immateriality of light, it is enough for today's physics to simply describe such behavior mathematically and declare it a fundamental law to satisfy its curiosity. However, since Darwinian Physics seeks to recover the original physics (the physics that seeks not only to describe nature, but also to understand it), we must explain these peculiarities of light. Especially if the solution of the great problems of physics depends on it. Because the theory of relativity is strange and non-intuitive because of light. So, understanding light is the key to understanding the two pillars of modern physics, because the basis of general relativity is special relativity, of course; but special relativity is also included in the Standard Model of Particles, along with quantum mechanics. So light is one of the keys to the holy grail of physics, quantum gravity. No small reason to motivate us to unravel the mysteries of light.

But before we can fully explore this mystery, we must first explain how physical reality came to be in the context of Darwinian Physics, because there would be no light if this reality did not first exist, since practically all the phenomena we know in science depend on electromagnetic interaction. From particles to atoms to molecules and everything we observe, including life. In this way, we can perfectly say that the physical reality is the world of light, so we must understand how this reality was first born in order to understand the reason for the behavior of light. So, to begin with, the most characteristic of our reality as compared to our mother reality (the space-time holon) is the emergence of multiplicity. This is because before physical reality, there was only unity, the "one," the holons; but there were no things or entities horizontally separated in space and time. There were many holons, yes, but they were separated vertically, stacked on top of each other in the vertical dimension. In their reality, however, there were no things of which it could be said that they were here but not there; or that they were now but not before or after. This great distinction appeared when a great degree of complexity was reached thanks to the VHE. In this way, physical reality emerged from space-time reality, analogous to how biological reality evolved from physical reality (from chemistry). And although it may seem very radical and heterodox, in reality the emergence of physical reality from space-time occurred in a very similar way to the origin of the universe as described in current physics. Recall that the space-time holon would correspond to the reality of quantum fields in the vacuum, from which fundamental particles such as electrons and photons emerge as local excitations of those quantum fields. So, the official story is that at the beginning of the Big Bang, the universe was so small and hot that not even the nuclei of atoms could exist, so there was a plasma of quarks and gluons. When the entire observable universe expanded to the size of our solar system and cooled down a bit, only 10^{-5} seconds after the Big Bang, the first nucleons (protons and neutrons) could form. However, it took much longer, about 380,000 years, for the universe to cool enough (about 3000°K) for the first atoms, mostly hydrogen and helium, to form. This is because during all that time, the electrons still had too much energy to be captured by a proton to form an atom. So, there was a state called plasma, like that found inside the sun. In fact, this is why we cannot see the interior of the sun, because light cannot travel freely inside it because it is constantly interacting with charged particles. But when the universe expanded enough to drop its temperature below 3000°K, it became transparent as the first atoms were formed. And since these are neutral (positively charged protons and negatively charged electrons), light or photons could travel freely through them without having to interact with the now electrically neutral matter. This first light released into the universe is what physicists call the CMB. It is the oldest light, dating back to 380,000 years after the Big

Bang, according to current estimates. That is why it is so cold today, because when that first light was released, it was 3000°K; whereas today, after 13.5 billion years, it is only about 2.7°K above absolute zero (-270.45°C). This determines the temperature of the entire universe, because that light is everywhere.

This is the official story. To reinterpret it in terms of Darwinian evolution, one would only have to interpret the expansion, antiquity, and cooling of the universe as the natural ordering process resulting from holistic evolution. As we already know, the cosmos seeks to expand and increase its antiquity by increasing its level of order due to causal inertia. An order that translates into a decrease in temperature or an increase in entropy. Both are equivalent to an increase in order, since both mean changing less, reaching a more stable or permanent state. This is why things around us do not heat up by themselves, but cool down, because they are trying to lose energy. The cosmological equivalent of this is the expansion of the universe, because as the volume of a gas increases, its temperature decreases. The universe works the same way, so it cools or becomes more orderly as it expands. An expansion that in Darwinian Physics happens through VHE or the emergence of new holons. Then, by the very purpose of the cosmos to be more and more existential, more and more holons emerge, which also increase the degree of complexity due to the superposition and accumulation of their holistic equilibrium. And one of the consequences of this increase in complexity was the evolution of multiplicity (physical things) from unity (holons). To understand this extraordinary phase transition of the cosmos, we need to imagine holons as a pixelated network. This is a reasonable assumption since they originated from nothingness, a point in space-time. It is therefore reasonable to believe that they are formed from it, since they are essentially a projection of nothingness.

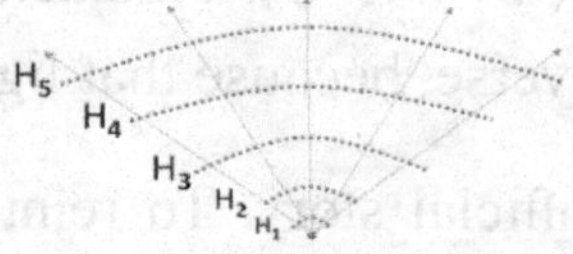

Thus, these pixels that would form the holons would behave similarly to the original pixel, randomly oscillating between two states. In fact, this would explain the reason why the higher or larger holons are also more ordered; at the same time, it justifies the need for existence to expand, since a larger holon contains more pixels. And if these fluctuate randomly, it makes the holon more globally homogeneous. This is the same principle of Boltzmann's thermodynamics, the principle of large numbers, which we have seen several times before. To see it in this case, let us take the typical example of the coin, which is equivalent to a pixel. So, if we have a single coin, the probability of hitting the result will be 50%; however, the more we increase the number of coins, the more likely the expected value will be. For example, if we flip 1000 coins, the probability that the result will be 500 heads and 500 tails will be very close to 100%; because if the result is, say, 493 heads and 507 tails, then we will have hit the number of heads with 98.6% probability. A probability that gets even closer to 100% the more coins we flip. The same happens with holons, which gain certainty, stability, and homogeneity the larger they are, because they would contain more pixels. In other words, a macrostate (holon) with more microstates (pixels) or entropy will be more probable or stable. So, it makes sense to create larger and larger holons to order the cosmos, because having more pixels allows it to remain more stable.

However, as already mentioned, this ordering process has a side effect: the increase of the degree of complexity. An increase that, at a certain point, allows for the preservation not of the global state of the holon as a whole, but of a microstate itself. A particular configuration of the holon. In other words, the holon was not preserved as such, as it had always been, but its pixels. An event that physicists interpret as the preservation of some excitations (pixels) of the quantum fields (space-time holons), which became stable particles such as electrons and quarks. Therefore, before physical reality, there were only holons, empty quantum fields, without stable particles. A global existence, because the excitations of the quantum fields were not conserved, they disappeared fleetingly, fluctuating frenetically in a random way. Something that was called in physics: virtual particles. Excitations of empty quantum fields that are so ephemeral that they cannot be directly detected. Hence their name, since their existence is practically virtual. A phenomenon confirmed with the highest precision in physics, 11 decimal places, under the theory of quantum electrodynamics or QED. An intrinsic energy of empty space that Darwinian Physics can explain again by explaining its justification, the Heisenberg uncertainty principle. Then this frenetic and spontaneous creation and annihilation of virtual particles and antiparticles would be these two states of the pixels of the spacetime holons. Since the pixels of spacetime holons are the product of nothingness, being its projections, if nothingness randomly oscillates between two states, so do the pixels. This explains the quantum fluctuation of empty quantum fields. In other words, Darwinian Physics explains the reason for the existence of two manifestations for physical things: matter and antimatter, which are the expressions of the states of nothingness in physical reality. A great mystery of modern physics, which would be easily solved in Darwinian Physics, since it is a property inherited from the dual nature of nothingness. Again, a more reasonable and satisfactory interpretation than the one currently given, which defines antimatter as matter that would travel backward in time. In fact, this is how it is interpreted in Feynman diagrams. An indispensable tool for quantum field theory calculations. In Darwinian Physics, on the other hand, antimatter would simply be the second state that nothingness needs in order not to be. The state that allows it to fluctuate randomly. So, matter and antimatter annihilate each other and return to nothingness.

So, by knowing what the space-time holons were like and why they behaved the way they did, we can now discover how physical things could emerge from them, thanks to the VHE providing that disruptive transition, what physicists call a phase change (like the transformation of liquid water into ice when it cools). Something that normally happens in every evolution, just as it happened with the emergence of life from complex organic molecules, the emergence of the Renaissance since the Middle Ages, or the transition from nomadic to sedentary life for Homo sapiens. Only this phase transition produced one of the greatest disruptive changes in the cosmos, the birth of multiplicity from unity, thanks to the fact that one of the two pixel states of the space-time holons was preserved. Specifically, the state of matter instead of antimatter. In terms of current physics, it is interpreted as the completion of the process of creation and annihilation of particles, through a slight but important baryonic asymmetry. Recall that there is an equivalence between energy and mass, due to Einstein's equation ($E = mc^2$). Then, when the universe was much smaller and more energetic, such energy (photons with very small wavelengths) could be spontaneously converted into particle and antiparticle pairs by exciting quantum fields. Two very similar objects, but with opposite charges, e.g. the negatively charged electron and the positively charged positron. The problem is that when these two meet, they annihilate each other, turning back into energy, photons. So, at the beginning of the universe, when it was much smaller and hotter, there was a constant conversion of very energetic photons into particle and antiparticle pairs, which annihilated, producing high-energy photons, which produced particle and antiparticle pairs, and so on. Until the expansion of the universe removed enough energy from the photons to prevent them from forming the masses required for these stable particles ($m = E/c^2$). That's why we don't see the light around us turning into matter, because it has too little energy to form the mass of an electron and a positron, for example. The problem is that, according to the sacred laws of physics, this spontaneous generation of particles and antiparticles had to be perfectly symmetric. This means that for every particle of matter created, there must have been an equivalent particle of antimatter. Then all matter must have been annihilated, leaving only light. This is the prediction of today's physics. But fortunately for us, for every 10,000,000,000,000 particles of antimatter, one extra particle of matter was created. It is because of this insignificant difference that all these billions of galaxies exist. In other words, according to the absolute laws of physics, everything we see should not exist.

On the other hand, on the side of Darwinian Physics, this milestone in the universe is far from being a great anomaly, since it was an expected event under the VHE. In other words, it is the holistic equilibrium that is responsible for this violation of baryonic symmetry, because each new holon is a new symmetry, pattern or order; but one that must irrevocably break with the regularity or symmetry of the previous holon, because otherwise it would be the same holon or a lower reality. Therefore, the new realities (higher holons) are fundamentally asymmetrical in relation to the parent reality (lower holon). Something that again we can find its biological equivalent, for that reality was also the fruit of an important asymmetry. One that scientists have also been unable to explain. A mystery that stems from the fact that the organic molecules fundamental to life have a property called chirality. This means that such molecules have a partner that is their mirror image. A version that would correspond to their reflection in a mirror, like our hands and feet. Well, the organic molecules essential to life on this planet have the same property. There are two versions that are very similar, like gloves; but like gloves, they cannot overlap. That's why you can't put your left shoe on your right foot and vice versa (at least not very comfortably). The extraordinary thing is that nature creates these two versions in equal numbers, because they are symmetrical. This means that there is no preference for one over the other, since both versions cost the same to produce. Nevertheless, life prefers only one type, although in principle life could also exist with the other mirror image. For example, all amino acids that form proteins are "left-handed" (L-form, i.e. levorotatory) and not "right-handed." This is called homochirality. A phenomenon so unusual that it is even used as a biomarker, a sign of life, because it is unique to biological chemistry. For this reason, the most accepted explanation for this great asymmetry in living reality is that it is the product of evolution. This means that at the beginning of life on the planet, there must have been many primordial biological organisms with both chiralities, based on left-handed and right-handed organic molecules. But since they reproduced by making copies of themselves (asexual reproduction), neither could feed on organisms with the opposite chirality, since they could not assimilate those nutrients (because they did not fit). Thus, the competition for survival favored one chirality over the other, and the only survivors were the descendants of LUCA. This is why all life on the planet is based on left-handed amino acids.

So homochirality or asymmetry in general would be strong evidence for Darwinian evolution, as we just saw. And surprisingly, it is also something that is very common in physical reality. For example, we have already seen that there is a clear baryonic asymmetry, since fortunately we only observe matter in the universe. But there is also a clear parity asymmetry, just like biological homochirality, because the weak nuclear interaction only recognizes the left-hand component of particles. In other words, this nuclear force is left-handed, even though fundamental interactions, such as laws, should be perfectly symmetric. In addition, physicists have been waiting for decades to discover supersymmetry, an equality or invariance between bosonic and fermionic particles. However, despite physicists' great confidence, no sign of it has yet been found. And if you think that these asymmetries are modern things, that physics is not what it used to be, we can certainly find asymmetries in Maxwell's own equations of magnetism. This is because the electric field is clearly different from the magnetic field, because we can find single electric charges, like electrons and protons; but single magnetic charges have never been observed, like a magnet with only one pole. All known magnets have two poles. However, because theoretical physicists are so fond of symmetries, they have already predicted magnetic monopoles. And, like supersymmetry, they take them for granted, even though there is no evidence for their existence.

In conclusion, all these asymmetries are great mysteries of the foundations of physics, whereas in the case of Darwinian Physics they are the natural product of the VHE, which forces the cosmos to choose, to generate a new regularity, and thus to create a new reality or holon. In the beginning it was one of the two states of nothingness to create time and space. Later it had to choose between matter and antimatter to create physical reality. Then it had to choose between the left-handed or right-handed component in the weak interaction, and in living reality, between the left-handed or right-handed chirality of organic molecules. Thus, existence itself and its realities are the product of asymmetries, the conservation of one of the two states; for if both states continued to exist simultaneously, there would literally be nothing. In the same way, if matter and antimatter existed, there would be no physical reality, because they would have been annihilated. A consequence that even applies to biological reality, since if life existed with both chiralities, organisms would not be able to feed each other because they would not be able to assimilate such molecules. They would be inedible and even toxic to the organism of the opposite chirality. Therefore, one of the two possibilities had to be chosen to allow life on Earth.

Then matter evolved, and with it local or particular existence. The most characteristic of physical reality. A very radical disruptive change, because until then there was only a fundamentally temporal and global existence, in the sense that it mainly continued to be. Hence the holon. Therefore, except in the case of holarchy or vertical distance, it made no sense to ask where you were in each holon because everything was that holon, that reality. But now, with local existence, it makes sense to ask where you are horizontally, because you can be there, but not there in that holon. So, existence evolved into a different kind of being because it went from being essentially temporal to also being spatial. Therefore, it also had to evolve into change, into spatial change, into what we call motion. For if one could not change being when time or the present arrived, by preserving the state of the pixel, then this change, this present, must manifest itself in another way, as a change of place. However, the emergence of spatial change did not completely replace temporal change, but made it evolve as well, becoming a new characteristic of being. A good example of this coexistence of these two types of changes is the electron, because it not only moves by moving from one place to another, but also changes its being by randomly fluctuating between a spin ½ (up) and -½ (down). A characteristic or property shared by all other particles. Only the value of the spin varies.

Clearly, the emergence of physical reality or multiplicity was a major milestone in the evolution of the cosmos. A great disruptive change that brought many important innovations, not only spatially, with the emergence of local existence and its movement, but also temporally, as time also had to adapt to these new circumstances. And this adaptation of time is what we call light. In other words, light would be the manifestation of time in physical reality. This will sound very crazy at first, but it makes a lot of sense for several reasons. First, modern physics already recognizes an intimate relationship between energy and time through Emmy Noether's theorem. This is not surprising, since energy is simply the ability to change, and in physical reality this means to move. This is quite clear in the case of kinetic energy, but it also applies to potential energy. Something we will see later when we analyze mass. So, Noether himself demonstrates this connection between time, energy and motion; because motion is fundamentally change, and change is time. So, what causes motion in physical reality is ultimately light. This is true even when you push something, because what allows you to push objects is the electromagnetic repulsion between the electrons of the atoms that make up that object and your hand. A force that is transmitted by the photons (light) that exchange these electrons. Thus, any new movement, change in direction or speed can ultimately be explained by the interaction of photons. This relationship between motion, time and light is also supported by thermodynamics itself, specifically temperature. This is because temperature can be defined as the average kinetic energy of the components of a macroscopic thing. The higher the average motion, the higher the temperature of such a thing. Something that can be known by the light or color emitted by objects. The higher the temperature, the more the spectrum of light is shifted toward blue (shorter wavelength), while the lower the temperature, the more it is shifted toward red (longer wavelength). This is why the coldest stars are red, while the hottest stars are blue (the sun has a medium temperature, so it is white). So, light is what indicates how much the components of a thing are moving. Moreover, there is a strange similarity between light and temperature, because just as something with mass cannot reach the speed of light, only approach it infinitely, neither can an object reach absolute zero temperature (0°K), only approach it asymptotically. This is the third law of thermodynamics, and the explanation given by Darwinian Physics is simple: that absolute 0° cannot be reached because time cannot be stopped, since that would mean that there would be no change or movement, as if the subatomic particles were completely frozen. Which would mean that time has stopped. Something that is simply impossible in Darwinian Physics, because how can you eliminate nothingness with more nothingness? So you can't stop it from fluctuating randomly by pumping presents. The same explanation is given for Heisenberg's uncertainty principle ($\Delta x \cdot \Delta p > \hbar$), which forbids the absolute determinacy of a particle's velocity ($\Delta p = 0$), which would be zero if it were not moving. And as mentioned in the previous chapter, it is nothingness that explains this principle of quantum physics, so it also explains the third law of thermodynamics.

A relationship between motion, light, and time that must be presentist in order to be consistent with quantum mechanics and its inherent randomness. This requirement can be clearly seen in quantum systems, such as the motion of an electron jumping from one atomic orbit to another. If the electron is in a minimum energy orbital in an atom, it will remain there unless a photon (light) of the specified frequency arrives, allowing it to jump to a higher energy orbital and excite the electron. So far, this is to be expected. It was the light that caused the jump by giving the electron the necessary energy to create this displacement. It's like seeing a ball that was on the floor, but now it's on the table. Somebody had to pick it up. But if the electron is already excited in a higher energy orbital in the atom, it will emit a photon equal to the energy it had to lose to jump to a lower energy orbital. This is a very natural tendency, since everything tends to a lower energy state. So, things cool down when they are hotter than their surroundings, not the other way around. But the most interesting thing about this case is that the emission of the photon by the excited electron is spontaneous. That is, it cannot be predicted because nothing causes it, literally. To return to the ball example, it is as if the ball returned to the ground without anyone or anything pushing it off the table. So, the electron's jump is random, because if something existential caused it, it would be part of a regularity, an order, and therefore predictable. But since nothing caused the electron's jump to a smaller orbit, it is random. This not only confirms the necessity of light to intervene in the movement or spatial change, but also, when nothing intervenes, it happens randomly, which again reveals the nihilistic origin of time, this time together with light, through the spontaneous emission of a photon by the electron to place itself in a lower energy orbit.

This connection between presentist time, motion, and light would explain many things in modern physics, such as the abstract nature of energy, because in modern physics energy is not a thing in itself, but rather a number, a mathematical quantity, a property of things. It is not that there is loose energy out there, as many shamans and New Age fraudsters preach, because energy depends on things. In other words, there is no energy without things, because it is simply the capacity for change that such things have. Energy is not radiation either, because radiation also consists of particles, i.e. things. For example, alpha radiation is two protons and two neutrons emitted by a heavy atomic nucleus, beta radiation is a neutron that becomes a proton and emits an electron and a neutrino, and gamma radiation is very high energy photons. So, the explanation that energy seems to be something intangible, inherent in things, is that it is a product of time itself. That which things need to move, to change. Making the small emphasis that radiation also occurs spontaneously, that is, again caused by nothingness.

So if light is the manifestation of time in physical reality, and Darwinian Physics explains what time is, then Darwinian Physics should be able to explain all the peculiarities of light, and it does. For example, it easily solves the first mystery of light, the reason why it does not need a medium to propagate, the famous luminiferous ether, even though it is an electromagnetic wave. This is because every traditional wave needs a medium to travel, and is only a disturbance of that medium, like a wave in the sea, sound in the air, or a wave made by people in a stadium when the audience stands and sits in coordination. Therefore, it remains a mystery how light can propagate in a vacuum without requiring a physical medium, this ether; this despite the fact that at present such a mystery is covered by calling it a quantum field, because no one knows what the latter is, what it refers to. On the other hand, Darwinian Physics can provide a satisfactory answer to this mystery, because light propagates through the vacuum because it is time itself. In other words, it is the various presents emanating from nothingness that use the holons themselves, or space, to undulate. For if time is change, then it makes sense that time itself is an oscillation or disturbance (change) of existence itself. And if light is a time wave, it automatically solves the second great mystery of light, the one that puzzled and fascinated Einstein, the reason why no massive thing can reach the speed of light, let alone exceed it. The explanation is very simple, because nothing can move faster than time itself. Logical, isn't it? If light is the manifestation of time in physical reality, it means that it moves at the speed of the vibrations of nothingness. Therefore, if motion is produced by the fundamental presents that come from nothingness, then it makes sense that the speed of light is the maximum possible speed, since there are no smaller or faster changes. Therefore, if something wanted to go faster than light, it would have to get more time, which makes no sense, since nothingness is the only origin of change, since it is pure change. While existence itself seeks the opposite, not to change.

Moreover, since light is the manifestation of time in physical reality, it makes sense that it always travels at the same speed. This is true for all existing types of light, whether we are talking about photons thousands of kilometers in size (radio signals) with very little energy, or photons with an incredibly small wavelength, the size of an atomic nucleus (X-rays), with a lot of energy. All types of light travel at the same speed, the speed of time (the vibration of nothingness). Only that time can manifest itself in very different ways depending on the scale or vertical distance, because each type of photon corresponds to a different "now" of a different physical holon. This is very strange to current physics because there are very different types of light or photons such as: x-rays, ultraviolet, visible, infrared, microwaves, and so on. And yet they all travel at exactly the same speed "c." This is not necessarily the case for physicists. In fact, there are many theories that predict differences in speed between these types of photons, such as the famous quantum loop theory. However, no evidence has been found for this, although measurements have been made using very distant supernovae or quasars, billions of light-years away. And yet, with all this time traveling from so far away, everything indicates that radio light and X-rays, for example, arrive at telescopes and antennas at the same time. The same has been demonstrated with gravitational waves. This great mystery, which in current physics is simply justified by a principle or law of nature, is perfectly explained by Darwinian Physics, since all light travels at the same speed because they are different manifestations of the same thing, time. In other words, the speed is fixed by presentism, so they are already moving at the speed of time. The only thing that changes is their energy, which is very well represented in the first equation of quantum mechanics formulated by Max Planck ($E = \hbar/\lambda$), where the energy of light (E) depends on a fixed value ($\hbar$), called Planck's constant, which would represent the fundamental present or the oscillations of nothingness (hence it is constant); and on the wavelength of the photon (λ), which is inversely proportional to the energy. This means that the larger the size of the photon, the less energy it will bring, which is perfectly consistent with the holarchy of Darwinian Physics, because the more superior the holons are, the larger they are, which makes sense for them to be present with longer wavelengths. And since they are farther away (greater vertical distance) from nothingness, the origin of change and time, it makes sense that they would also have lower energy. In the same way, the farther we are from a campfire, the less we feel its heat. And conversely, the lower the holon is, the smaller it will be, and the shorter or briefer its present (lower λ), and therefore the more energy the photon will have, being closer to nothingness. Thus, a photon with a very large wavelength, for example of meters, such as radio signals, carries very little energy, having a large vertical distance; compared to a photon with a very small wavelength, for example a gamma ray of 10^{-12} m (0.000000000001 m), which would be very close to nothing. The light we can see, on the other hand, would have an intermediate energy, corresponding to a range of 400 to 700 nanometers. A very small spectrum of all existing light. Then the entire spectrum of light would represent nothing more and nothing less than the very holarchy of Darwinian Physics.

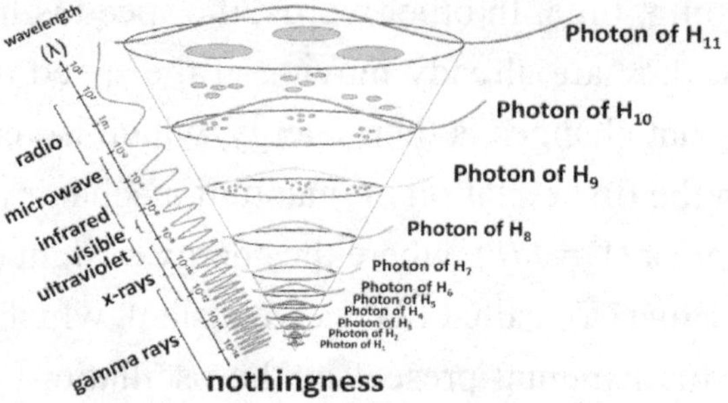

As we can see in the image above, each type of photon corresponds to the present of a holon. Therefore, the size of the photon, its wavelength (λ), represents the size of that holon. That is why there are photons the size of the planet, the sun, the galaxy, even the size of the universe itself, because there are obviously holons of that size. And the energy of such photons is inversely proportional to their length, because, as already explained in the book, the higher presents are formed from the fundamental presents, from the vibrations of nothingness. Then, a photon belonging to a much higher holon (a large λ) generates a change (vibration) only when it accumulates all the fundamental presents it needs. For example, if a photon is one meter long, it will need 10^{35} fundamental presents to generate its change, its tick-tock, because they are the oscillations of nothingness itself (Planck length: $1.6 \cdot 10^{-35}$ m) that fit into this higher present. Then, such a need to accumulate so many fundamental presents to cause its change, its tick-tock, is what we interpret as the low energy of this photon, since it would be 10^{35} times weaker. On the other hand, a gamma ray photon needs much less fundamental present (only 10^{23} if we consider a wavelength of 10^{-12} m, as in the previous example) to cause its tick-tock. Therefore, we perceive it as very energetic, since it would be 10^{12} or 1,000,000,000,000 times more powerful than one meter of radio light. Therefore, light works in a similar way to a mechanical clock, where the hour wheel is equivalent to 60 tick-tocks of the minute wheel to produce its tick-tock. And yet, all types of photons travel at the same speed, the speed of time, because their effect or consequence manifests itself in the present, the only moment that exists. Therefore, only their effect varies, the intensity of the "now," which is weaker the greater the vertical distance of the holon. A description of light that also explains its discretization, the reason why it comes in packets of energies or photons, corresponding to the "nows" of each holon. Higher presents that always need a determined quantity of fundamental presents to be, explaining their quantization.

A presentist explanation of light that must be consistent with the solution that won Einstein the Nobel Prize: the photoelectric effect. A phenomenon that occurs in some metals that emit an electric current when illuminated by light. This was not what puzzled physicists at the end of the 19th century, because it could be understood that the light gave extra energy to the electrons in the metal atoms, allowing them to jump out of their orbitals and generate the electric current. The extraordinary thing is in the how, because by increasing the intensity or brightness of the light, that is, by having more energy, no more electric current was generated. In fact, what creates the photoelectric effect is the type of light. If the light was ultraviolet, that is, of high frequency or short length, electric current was very easily emitted, even if it was of very low intensity or brightness (low total energy). On the other hand, if it was visible light, of lower frequency or longer wavelength, even though it was a very intense light, very mowing (much total energy), there was no trace of electricity. This was very paradoxical, because if the light was a typical wave, the energy of the light would depend on its brightness or intensity, which is equivalent to the amplitude of the wave. This is very well understood with ocean waves. If the waves are very small, a few centimeters high (low amplitude), we say that the sea or the waves are very calm, right? But if the waves are 10 meters high (high amplitude), we will say that we have a tsunami. In this example, the frequency does not matter much, because the small waves could be more frequent, reach the coast faster or more often, but by remaining small, we would not say that the swell is more dangerous. On the other hand, if the swell is 10 meters, even if the frequency is very low, just one wave of that size would leave a clear devastation on the coast. Now, increasing the intensity of the light would increase the size of the sea wave, as if it were a large tsunami of light, and yet the electrons in the metal are hardly aware of its presence. On the other hand, a very weak light, of very low amplitude or brightness, but of very high frequency or very small length, such as ultraviolet light, ejects electrons from the atoms of the metal very easily.

This contradicted everything that was known about waves at the time. A great mystery that was partly solved by postulating the wave-particle duality of light (again, pragmatism in physics). However, this phenomenon would really be understood if we integrate the presentist holoarchy, since the low frequency light does not affect the electrons of the metal atoms, even though they are many (high intensity), because they are simply at a different vertical distance. It would be like asking ourselves: why do airplanes not move us, do not hit us, if they weigh dozens of tons and we weigh only a few kilos? And the answer is very simple, because they pass several kilometers above us. The photoelectric effect would be similar. Light would come in discrete packets, photons, because they are present; but such packets or photons have their effect on their particular holons, which are located at different vertical distances in the holoarchy. Therefore, no matter how many photons there are, if they are in one holon, they will not affect things that are in more distant holons. In this sense, the intensity of light would be a horizontal amplitude, so the more intense the light, the more photons or energy will be in that holon, but again, these will not affect things that are vertically below. Going back to the previous example, it would be like increasing the number of airplanes flying overhead, which clearly will not affect those walking miles away on the ground. Applying this to the photoelectric effect, the electrons would simply be far below vertically because they make up the atoms. So, they need a light or presence of that vertical height, that is, of high frequency or low wavelength. On the other hand, whatever happens higher up, by a light of low frequency or high wavelength, no matter how bright or intense it is, the electron will not be aware of it because it is at a greater vertical distance, being in a higher holon.

But we can continue, because this presentist explanation of light applies not only to a problem already solved by Einstein more than a century ago, but also to the rest of its peculiarities. For example, its ghostly nature. The reason is that the photon, like a boson, does not seem to occupy space, as fermions do. The reason would be simple: the photon is not a particle of matter, of space, but rather a particle of time, a "now." Hence the confusion of interpreting the photon as a particle of matter rather than a particle of time. That is why many photons can be collected horizontally at the same point, because it is rather different " instants " that are collected at the same place. That is why photons do not have this limitation that pixels (fermionic particles) have, because being temporal, practically all photons of the same frequency can be grouped together in the same space. This is how lasers are made, by concentrating many photons in the same place. On the other hand, fermions, such as electrons and quarks, are a very different case, since they represent the holons, specifically their pixels, rather than their presence. That is why they occupy space, because being bound or directly related to a holon, they are localized, fixed in the holarchy. Hence the Pauli exclusion principle, the reason why no two electrons can exist in the same place or in the same quantum state. A postulate that is assumed axiomatically, just because, but that would have a very simple explanation in Darwinian Physics, because there is only one holon. Then the other electrons, in order to be, would have to manifest themselves or be located in other holons, which would correspond to the orbitals of the atom. In other words, it is precisely the limitation imposed by holarchy that forces the excitations of the physical holons (fermions) to create structure, matter in the universe, through the vertical accumulation of fermionic excitations in layers that allow the construction of ever larger and more complex physical things as they ascend in the vertical dimension. A coupling that does not begin with the electrons, but with the quarks, since these are also structured in layers or orbits within the protons and neutrons in the nucleus of the atom.

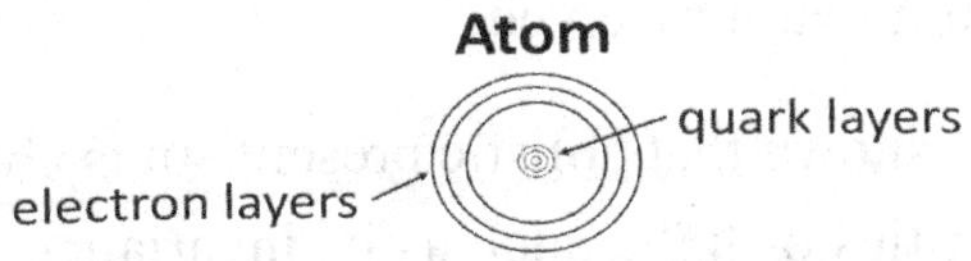

Once we are clear about what we mean by bosons and fermions, we can better understand what physicists mean by the interaction between them. This is because the official interpretation of bosonic particles is that they are messenger particles that carry the fundamental forces. Let's take the case of the photon, which would be the messenger particle of electromagnetism, the one that tells electrons, for example, how to repel each other. So, an electron emits a virtual photon that is received by the other electron and warns it to move away. This is what is interpreted in modern physics, a game of villar with many kinds of balls. In contrast, what is interpreted in Darwinian Physics is that the bosons would represent the bosons themselves that exist corresponding to each physical holon. So, for example, when a photon is absorbed by an electron, it would simply correspond to that which is present when it reaches that electron. This means that it must change, move. A very abrupt change at such discretized scales, because from one moment to the next, it literally has to move from a lower energy orbit to a higher one, for example. What physicists call a quantum jump. The instantaneous transfer of an electron from one atomic orbital to another. So, it makes sense that there is no continuous travel from one orbital to another, because nature is fundamentally quantized, not only in space (holons), but also in time (the present). That is why the electron is relocated or materialized in another place, in another instant, because it is its new "now" that corresponds to another holon. A present that arrived in the form of a photon that it received.

Having shown that holistic presentism is able to explain all the peculiarities of light, such as its invariance, immateriality, and instantaneity, we can now turn to the theory of special relativity. We will begin by defining one of the most complex and elusive concepts in physics, mass. We will understand it as the internal structure of a physical object. And since all physical things, including what is inside other physical things, move (our cells, molecules, atoms, particles, and so on), this obviously means that they also use time. Therefore, mass would represent the internal time of such physical objects. To see how this works, let's imagine a fundamental massive particle, such as a single electron, as a photon bubble, where the bounces of that photon inside it equal its mass, its internal motion, or tick-tock. To see that this analogy works to represent mass, let's make the comparison to an inflated balloon, where the air or molecules bouncing around inside would be the equivalent of the photon bouncing around inside the electron. So just as the air exerts a pressure inside the balloon, the bouncing of the photon inside the electron exerts an internal force. And since the bounces are random, due to the nature of nothingness, the internal pressure manifests itself in all directions (because there is no privileged direction), so the pulls neutralize each other. In other words, if the photon pushes in one direction inside the bubble, it will push in the opposite direction the next time. The end result is a net force of zero outside the bubble. So, if the massive particle is moving in one direction, it will not be disturbed by the internal motion of the bouncing photon. It would appear that nothing is happening inside the massive particle. Therefore, it maintains its speed, its motion (inertia). This simple analogy of mass as a bouncing photon explains all of its properties, such as resistance to acceleration or deceleration, or simply to changes in velocity. To understand this, let's go back to the analogy of the balloon, if we push it, the internal molecules that go in the direction of the push, probably half of them, will generate less pressure because that part of the balloon is moving away from them because of the acceleration. And conversely, the internal molecules going in the opposite direction of the push, the other half, will generate more pressure because that side of the wall is getting closer to them. The end result is a difference in internal pressure going against the direction of the thrust. In other words, a force is created that resists acceleration.

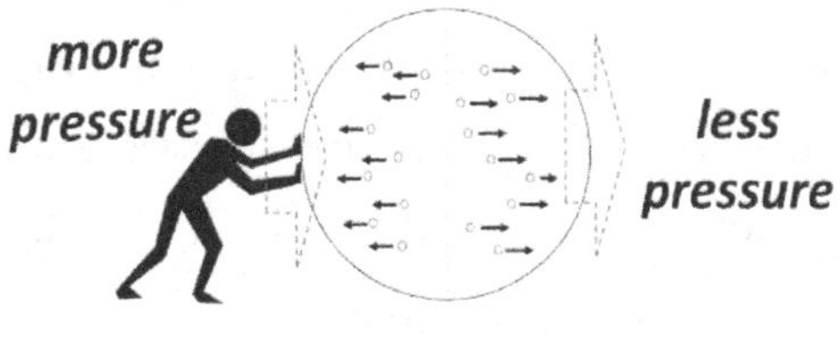

So the more molecules the balloon has, the more it will bounce, and the more difficult it will be to accelerate it, because a greater internal pressure difference will be created. The same thing happens when you try to slow it down, creating the same internal pressure difference, but this time on the opposite side of the trajectory. A resistance to deceleration. A resistance that quickly disappears when the balloon stops pushing or braking, because this pressure difference between the internal photons dissipates quickly as they are distributed among them, interacting or colliding with each other. Therefore, due to the homogenization of the energy between the molecules after the push, the speed, the inertia, is conserved again, but this time a new speed is conserved, a higher one, due to this injection of energy from the previous push. A new speed that would cost even more to accelerate again, because due to the injection of extra energy there is now more internal pressure, which would generate even more pressure difference when pushed again. This explains why it costs even more to accelerate the same already accelerated object, because the internal pressure (mass) will be higher. So, the "inertial mass" increases with speed.

Now that we understand what is essentially the mass of an object, we can better understand what makes light so strange. To do this, we will use Einstein's clock, which consists of a photon bouncing between two mirrors. This is very similar to the analogy of the electron as a bubble in which a photon is bouncing around. So, each round trip of the photon marks a tick and a tock, an instant. Now compare this to something more familiar, the bouncing of a tennis ball off the floor. So, each bounce from hand to floor will also be a tick-tock. Now let's imagine that the bouncing of the ball is happening inside a moving train, as Einstein himself imagined it. A person watching from outside the train would see the ball traveling farther because it is zigzagging in the direction the train is moving. And he will also see the speed of the ball increase as the speed of the train is added. Then two people, one inside and one outside the train, will coincide in the tick-tock, in the duration of the "now" of the ball hand system. However, it will be at the cost of seeing different speeds and distances. The person inside the train will see the ball move slower and travel a shorter distance than the person outside the train. A difference that allows both to measure the same tick-tock or "now."

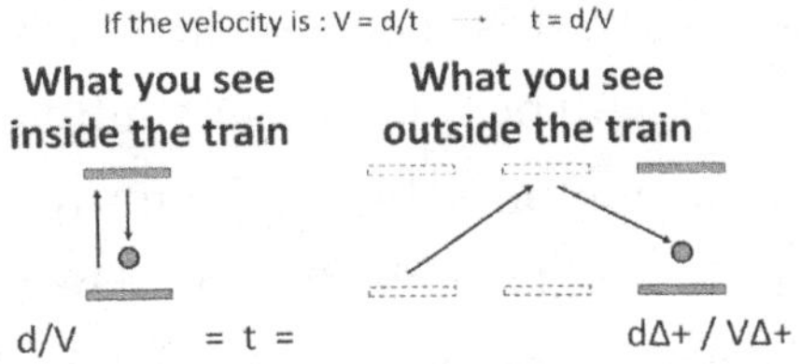

Makes sense, doesn't it? If you are inside a train and you are walking in the direction the train is moving, your speed will appear faster to someone outside the train, because the speed of the train is added to your speed. This is Galilean relativity, which says that everyone measures different (relative) speeds according to their frame of reference, at the cost of keeping time (tick-tock) and space (distance) constant. What seems so logical to us does not apply to Einstein's clock. Recall that the speed of light is constant for all observers, regardless of velocity or direction. So, if the definition of speed is the distance traveled in a given time (v = d/t), and the speed of light is fixed, then something else must vary, and that must be time "t" and space "d." If we return to the train example, but this time we replace the ball and chain system with the photon bouncing between the two mirrors, we would find that both people, both inside and outside the train, must coincide with the speed of the photon, that is, "c." This has surprising consequences, because if both people (the one outside and the one inside the train) must coincide in the speed of the photon, then they will have to differ in time, in the tick-tock that the photon will make. This is because the person outside the train will see the tick-tock slower or longer, because the photon will have to travel a longer distance at the same speed, zigzagging due to the movement of the train. On the other hand, the person inside the train will see a shorter or faster tick-tock, since they do not see the horizontal displacement that the photon makes, because it moves with the train. The end result is that both people measure different tick-tocks or "nows" for the same photon between the mirrors. The tick-tock observed by the person outside the train will be slower the faster the train moves, since the photon has to travel a longer distance at the same speed "c". While, the person inside the train will always see the same tick-tock. This difference in tick-tock is what is called time dilation in special relativity. The fact that two observers see different time rates when moving at different speeds.

The extraordinary thing about all of this is that this amazing phenomenon is not unique to Einstein's clock, but also applies to the tennis ball, the person, and the train itself, because we are all essentially Einstein clocks, if we define mass as the product of photon bounces. For example, the electron as a bubble of photons is a small Einstein clock. Therefore, such photon bounces inside the electron are equivalent to the tick-tock of the electron, to its internal time. So, the faster the electron goes relative to one, the slower we would see the electron tick-tock. Its time would slow down. This may seem difficult to comprehend or prove because the electron is a stable particle, so it seems to live forever. It does not age. However, this time dilation is much more obvious for an unstable particle that decays in a short time, such as the muon. A twin brother of the electron, but heavier (200 times more massive). In fact, time dilation made it possible to discover this particle because the muon decays very quickly, in 2.2 microseconds. Therefore, when muons are produced in cosmic ray collisions at the top of the atmosphere, at about 15 km altitude, they should not have enough time to reach the surface of the Earth. It does not matter if they are traveling at 99% of the speed of light, because with those 2.2 microseconds of life, they would only have enough time to travel 660 meters. Yet these particles have been discovered here on the surface, how is that possible, do they travel faster than the speed of light? No. They just live longer. Thanks to time dilation, traveling at such a high speed, the tick-tock of the muon is much slower for us. Slow enough to make it all the way down here. That's why, thanks to special relativity, one muon passes roughly through your hand every second.

This conception of massive particles as bubbles of photons explains in a very simple way the reason why they cannot travel at the speed of light, let alone exceed it. To see the reason, let us turn the electron into a transparent bubble. We would see the photon zigzagging inside, just like Einstein's clock. Since the speed of the photon is always the same, this zigzagging takes away its motion to move in that direction; compared to another photon that uses all its motion to move in a straight line.

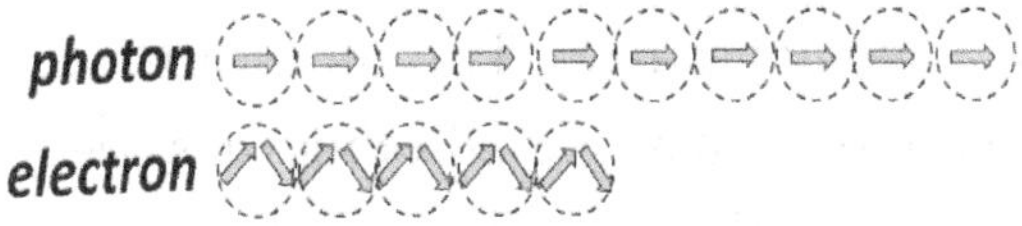

That is why light has the maximum speed, because it devotes all of its motion, which is given to it by holistic presentism, to moving through space. On the other hand, in the case of particles with mass, part of this movement of the photon is used to build up this mass, this internal bounce, this tick-tock. Therefore, the most massive particles are also the slowest. However, this principle applies not only to massive fundamental particles like the electron and the muon, but to all massive physical objects. Yes, you too, because all physical things are essentially bubbles of photons. Only this time they are bigger and more complex bubbles because the mass is bigger and more complex. In fact, we can think of the process of creating new physical holons as the confinement of photons in ever larger and more complex bubbles. Thus, fundamental particles such as electrons and quarks would be created by confining a photon, while the creation of an atom is due to the confinement of those electrons and quarks (protons and neutrons). And to create a molecule, those atoms had to be confined, and to create a cell, those molecules had to be confined, and to create a multicellular living being, many cells had to be confined, and so on. This means that all massive things, from electrons to people to planets to stars to galaxies, are essentially beings of light, or bubbles of bubbles of bubbles of bubbles of photons. Each bubble is the tick-tock of an increasingly complex internal mechanism. So, our tick-tock is formed by the tick-tock of our cells, which are formed by the tick-tock of their proteins, which are formed by the tick-tock of their atoms, which are formed by the tick-tock of their fundamental particles. This means that we are essentially an Einstein clock, just like all other massive things. Therefore, all physical reality should suffer the consequences of the invariant speed of light, and it does. Obviously, we did not realize this before, because the speed of light is very fast compared to the speeds we are used to. For this reason, relativistic effects such as time dilation are simply negligible. For example, a car traveling at 100 km/h is traveling at 0.000000009% of the speed of light. So, time dilation is ridiculously small. Only if we were traveling at a significantly large percentage of the speed of light could we appreciate its effects. However, as subtle as relativity is on a day-to-day basis, there is something very familiar to us that simply wouldn't work if we didn't take these relativistic effects into account. And that thing is the Global Positioning System, or GPS, because of the high speed of the satellites that orbit us to triangulate our position, their clocks run slower compared to the clocks that are on the surface of the planet. So, if this difference between the clocks in orbit and those on the Earth is not taken into account, it would simply be impossible for the GPS to work because our position would have delays that would accumulate giving errors in the position of the margin of 11.3 kilometers per day.

In fact, this analogy of physical objects as photon bubbles is so powerful that this is how mass is created in fundamental particles in the first place, because this is how the Higgs mechanism works. This is because the Higgs field works by generating a back and forth or zigzag between the two components of the fundamental particles, traditionally called the "left" and "right" components. Therefore, this oscillation between these two components is the equivalent of Einstein's clock, because the more coupling or interaction the fundamental particles have with the Higgs field, the more back and forth it generates between these two components, which generates more delay in their displacement in space (they move slower as they zigzag more), which we perceive as more mass in the fundamental particles. This means that without the Higgs field, all particles, including electrons and quarks, would travel at the speed of light because they would essentially be photons. So, quarks could not form protons, and protons could not capture an electron to form an atom, because everything would be moving at maximum speed all the time, without slowing down, because they have no mass. So, there would be no structure in the universe. Just a soup of particles moving at the speed of light. This was the prediction of the Standard Model of Particles in the early 1960s, because it is the product of the gauge symmetry of quantum fields. Only by spontaneously breaking this symmetry could the fundamental particles be given mass. This is exactly what the Higgs mechanism does, making particles like the electron tick-tock.

This description of mass as a bubble of photons also explains very well the famous equivalence between mass and energy ($E = mc^2$), because if energy is essentially light, photons, and mass is confined photons bouncing around inside massive objects, then this conversion of mass into energy is very easy to understand. It simply means releasing the internal motion of massive objects, converting that potential energy (mass) into kinetic energy. While the conversion of energy to mass would be the reverse process, trapping external motion to convert it to internal motion in an object, or converting kinetic energy to potential energy. To this must be added the explanation of one of the most extraordinary properties of light. For it is one thing that light is the maximum speed, the speed that nothing can exceed; but another thing, something even more surprising, is that this speed is also constant. Remember what was said at the beginning of the chapter, that no matter in which direction, no matter how fast you travel, you will always see light traveling at the same speed: "c." An incredible phenomenon that physicists prefer to ignore, convincing themselves that they understand it simply by giving it a name: Lorentz invariance. A symmetry of nature, a law. On the other hand, Darwinian Physics, respecting the principle of sufficient reason, can explain this great mystery, which is very easy to understand if we use this analogy of the bubbles of photons, because all other changes or higher presents are built from this fundamental present or "c" of nothingness. Remember that to measure any speed we need a clock (v = d/t). The problem is that all existing clocks use light to operate because all massive objects are essentially Einstein clocks (photon bubbles). So, it makes perfect sense that all physical things, including our most accurate measuring instruments, measure exactly the same speed "c" for light, because they use light to function, to tick-tock. Therefore, it is this dependence on light that explains why we cannot measure any speed other than light, because light is not only out there, but also inside all things, being the origin of mass. The dependence on light is due to the fact that light is the manifestation of time itself in physical reality, so it makes no sense to want to measure another speed for time using the same time. So, we will always measure the same speed for light, no matter what we do or how fast we go, because we also use the same time to function. All of this is because light basically allows for all motion in physical reality. In fact, to see light traveling slower, we would have to process the information from our senses faster, which is impossible because we cannot count shorter or faster changes or tick-tocks, since light is the shortest or fastest present, being the product of oscillations of nothingness. And vice versa, we cannot count longer or slower changes, because all changes, all dynamics, come from nothingness, which is pure change. In conclusion, the explanation that the speed of light is invariant is simply because time is limited by the present. So, if it were possible to observe different speeds for light, it would be due to having more or less time available, like the supertime of Chapter XI, which makes no sense.

This invariance of the speed of light is so important that the entire theory of special relativity can be summed up in a single sentence: "Everything moves at the speed of light in space and time." Exactly, you also move at the speed of light, just like a photon. The only difference is that in addition to moving through space, you also move through time. This was Einstein's real contribution, the unification of space and time, a product of the fact that the speed of light, or distance divided by time ($c = d/t$), is a constant in Maxwell's electromagnetism. This fact was formalized by Hermann Minkowski in 1907 and is called Minkowski space-time. Here the Pythagorean theorem is used to measure the distance (hypotenuse) between two events or points in space-time. Nothing new so far, as the ancient Babylonians and Egyptians used this formula to measure crop fields and to build temples and pyramids. Minkowski's novelty, however, was to add a small modification. In this way, the internal time "T" (the time an object experiences) is equal to the variation of time "t" (time for the rest) minus the variation of distance "x" in space.

$$\mathbf{T^2 = \Delta t^2 - \Delta x^2}$$

This sign "-" explains everything there is to know about space-time and the phenomena of Einstein's special theory of relativity. For example, if you stand still in relation to another person, your experience or passage of time will be the same as that of the other person. In other words, you are simply moving forward into the future because all the available motion is internally destined to move, for example, every molecule in your cells. This can be seen very well in the above formula, because if you do not move, your variation in space (Δx) is zero, so your internal time (T) would be equal to the variation of time (Δt).

$$\textbf{If: } \Delta x = 0 \quad \rightarrow \quad T = \Delta t$$

However, when you move relative to that other person, part of that motion at the speed of light is now destined to move through space. As a result, the motion through time must slow down, aging slower (time dilation). Again, this is clearly seen in Minkowski space-time, because the more we move through space, the more it subtracts from internal time.

$$\textbf{If: } \Delta x > 0 \quad \rightarrow \quad T < \Delta t$$

In other words, the more we move in space, the less we move in time. All because of this sign (-) introduced by Minkowski. This means that in space the shortest distance between two points is a straight line, but in Minkowski space-time a straight line (constant speed) gives the longest internal time between two events. A non-straight line always implies a longer distance, but also less internal time. In current physics it is not known why this is so, it is just a fundamental assumption, not a conclusion that follows from deeper principles. Something that is obviously different in Darwinian Physics, since this fact is perfectly explained, and it is again due to having an explanation for time. So, this subtraction is due to the fact that time is limited by the presentism. Therefore, time must be divided between internal motion or the passage of time (T) and external motion or displacement through space (Δx); because there is simply no more time or change available. This can be seen again in the Minkowski equation by simply rearranging the terms.

$$\mathbf{T^2 = \Delta t^2 - \Delta x^2 \quad \rightarrow \quad T^2 + \Delta x^2 = \Delta t^2}$$

If we add the two sides of the equation by Δx^2, we can clearly see that the (presentist) time is fixed (Δt) and that it must be divided between the internal time (T) plus the displacement in space (Δx). Thus, this fact discovered by Minkowski in Einstein's theory of special relativity can be explained and derived from a more fundamental principle: that there is only what we observe and experience: the now. It is as simple as that. Therefore, in Einstein's theory, everything moves at the speed of light, because all motion, all change, whether through space or through time, is always constant: "c."

In conclusion, the presentism of Darwinian Physics explains Einstein's theory. One of the most important branches of modern physics, because it is the basis of General Relativity, of course; but it is also included in the Standard Model of Particles. All thanks to Darwinian Physics, which recognizes and explains time through holistic presentism. This will be incredibly ironic to physicists, because this theory has traditionally been used as an argument against the A-theory of time, or presentism, because its time dilation destroys simultaneity, since the "now" depends on one's speed and even position (which will be discussed in the next chapter). For example, for one person moving relative to another, "one now" may be equivalent to "two nows" for the other. Thus, two events cannot be said to be truly simultaneous, since the "nows" of each event may have different durations. However, this fact does not refute all presentism, only the most basic and simplest type, absolute presentism, or Newton's concept of time, which postulates a universal clock for all things. In other words, one second here is the same second on Mars as it is in the Andromeda galaxy. But what Darwinian Physics defends is holistic presentism, which is a kind of relative presentism. Exactly what one would expect from non-absolutist physics. Remember that in Darwinian Physics you have a temporal holarchy that is the product of different holons, each of which has its own oscillation or present. Therefore, there are as many "nows" as there are holons in Darwinian Physics. One would just have to conceive of the present in a more complex way, coming from a complex existence.

Another criticism of presentism using special relativity is that it proves the existence of the "future," since it allows us to travel faster towards it, again due to time dilation. For example, a person who undertakes a one-year space voyage, traveling at 99.4% of the speed of light, will be equivalent to 9 years for all the people who stayed on Earth. Therefore, when the astronaut returns, he will have the impression that he has traveled into the future, since everyone (including the planet) is 8 years older than he is. This would be a confirmation of eternalism, showing that it is possible to travel to the future, because the future is real, it is there. But this interpretation is completely wrong, because everything that has happened has happened in the present. The only thing that has happened is that the present of the astronaut has slowed down 9 times more than the present of those who stayed on Earth. Nothing more. In fact, this situation is completely equivalent to putting the same astronaut into hibernation or cryogenic suspension on Earth and waking him up 8 years later. Although the perception for him would be that he had immediately traveled to the future, as happened to the character Fry in the cartoon Futurama, no one would say that he had actually traveled through time, gone through a wormhole, or anything like that. That person's present just slowed down, or in the cartoon example, froze, compared to everyone else's. So, you're not really traveling to the future. Just the present, the Tictacs are out of phase.

In conclusion, Darwinian Physics is not only perfectly compatible with special relativity, but explains it by revealing its reason or foundation, light. It was only necessary to conceive the present in a more realistic, more complex way, with a temporal holoarchy. In this sense, in order to better visualize this structure, let us return to the analogy of macroscopic physical objects as a series of bubbles of photon bubbles, where each one bounces inside the other, marking different tictacs. So, we start with a photon, a massless particle, which bounces inside the first bubbles, which would correspond to the fundamental particles with mass, such as the electron, marking its tick-tock. And these in turn would bounce or oscillate inside another bubble called the atom, producing its tick-tock. And the atom would also bounce inside another bubble, tick-tock off the molecule, which would bounce inside another bubble, tick-tock off the cell, which would bounce inside another bubble, tick-tock off, say, you. And while you have to think of this bouncing as a metaphor, all particles, atoms, and molecules are actually vibrating, constantly vibrating. For example, the official definition of the second is: 9,192,631,770 vibrations of a caesium-133 atom. Thus, if we consider a human present as a second, it would correspond to more than 9 billion presents of an atom. Therefore, the temporal holarchy would be the hierarchy of these presents or internal times of the physical holons, which are constituted in such a way that the presents of the higher holons are formed from the presents of the lower holons. And since the lowest or fundamental present is constituted by the oscillations of nothingness, all higher presents are ultimately constituted by nothingness. The same that gives the photon its constant and maximum speed. This is why Darwinian Physics remains a presentist theory, because ontologically speaking there is still only one present; only that, due to the VHE and the multiplicity of physical reality, different higher presents began to be constituted as new holons emerged. Higher presents that are different from each other, that are relative to each other, only because they are constituted by the same fundamental present, the smallest possible now, which is the product of nothingness itself.

The robustness of holistic presentism lies not only in making sense of special relativity, but also in resolving the unknowns that have lingered about this theory, either because it has not been possible to answer them, or because their answer has not been satisfactory enough. One of the most famous of these unknowns is the twin paradox, which refers to relative motion and time dilation. To understand this, let's use the classic example of an astronaut twin who embarks on a spaceship to Alpha Centauri, some 4.37 light years away. Meanwhile, the other twin waits on the planet. The interstellar twin will be constantly traveling back and forth at 90% of the speed of light (0.9c). Due to the time dilation suffered by the twin astronaut, the entire trip (4.37 light years out and 4.37 light years back) took only about 4.2 years, as his time passed much more slowly; while for the twin who stayed on the planet, the trip took as normal, almost 9 years. So far so good, except for the small detail that one of the axioms of special relativity is that "No point of view or point of reference has priority over another." This means that from the perspective of the astronaut twin, it was his brother, along with the entire planet, that was moving toward and away from him at 90% of the speed of light. Therefore, from the astronaut twin's perspective, it was his terrestrial twin who suffered the time dilation and who aged almost 5 years less. And therein lies the rub. How can it be that both twins are almost 5 years younger relative to each other? It obviously makes no sense.

The current physical explanation is that this paradox is locally true. In other words, both are actually simultaneously older in terms of their local reality. However, if one were to compare them to confirm this absurdity, there would be a leap in time, a leap into the future. For from the perspective of the Earth twin, if he were to look at his astronaut brother's watch, he would see that it is running slower and slower, both when he goes to Alpha Centauri and when he returns to Earth. The same is true from the perspective of the astronaut twin, since from his point of view it is his brother who is moving away at 90% of the speed of light, along with the entire planet. However, according to current physics, when the astronaut twin arrives at Alpha Centauri and returns to Earth, the terrestrial twin's time jumps. In other words, he would be much, much older all of a sudden. And even though he would see his terrestrial twin age more slowly when he returned to Earth, the time jump would be so large that his terrestrial twin would be almost 5 years older when they met anyway. If you don't get it, don't worry, neither do I. The reason the "leap into the future" is so unsatisfactory is that it is an "epistemic" rather than an "ontological" explanation. Again, the problem of lack of philosophical literacy. This is because it does not explain what is happening at every moment, to every twin. Because it is ontologically impossible for both twins to rejuvenate each other at the same time. The official explanation is limited to explaining what would happen if and only if they met again. In other words, it is not a realistic explanation because it depends on what the twins know about each other. That is why it has been said that it is an epistemic solution, because according to it both twins would be ontologically younger, which is absurd. But when the twins see each other again and can compare and measure each other, the situation becomes coherent. But as long as one does not measure, does not compare, according to the official explanation, the absurdity is real. Both would be older in relation to the other.

The solution offered by Darwinian Physics to this paradox is very different, since it consists in changing this official epistemic perspective to an ontological one. Thus, the answer to the question of who suffers time dilation is very simple: it is the one who accelerates. Who stops traveling at a constant speed after having acquired more energy to increase his speed. This is because in order to travel at the same speed as before, nothing needs to be done. This is what physicists call inertia, something that is only postulated, taken for granted, a dogma in today's physics, but which is explained quite well in Darwinian Physics, because everything continues to maintain its speed if nothing is done, because it continues to receive the same amount of time, the present. It is as simple as that. On the other hand, if you want to accelerate, you have to redistribute that time, giving more changes to the external motion, to its displacement, to the detriment of its internal change or time. This is what time dilation means. Something that happens because of an external intervention, or as physicists say, a force. So far so good. The real dilemma is to determine who is really accelerating, since all points of view are valid. Therein lies the problem, because although the twin who gets into the rocket is clearly the one who is accelerating, the perception of the astronaut twin who sees his terrestrial twin accelerating away from him along with the entire planet is also valid. This is precisely the epistemic perspective, since any individual or subjective frame of reference is valid. However, it is possible to objectively determine who is really accelerating by postulating a single, special frame of reference, and that would be the viewpoint of the entire universe. In other words, if we consider not only the subjective perception of the twin on Earth and the twin in the rocket, but also the subjective perception of the Moon, the Sun, Jupiter, Saturn, the star Alpha Centauri, Betelgeuse, the Andromeda galaxy, and so on. When we put all the subjective perceptions together in a coherent way, we have objectivity; because from the perspective of the twin astronaut, it is his brother who is accelerating away from him along with the Earth, but it is not the same as he sees the Moon and the rest of the planets and stars. Therefore, the relative motion with the entire cosmos can be determined. In physics, this is called the Co-Moving Frame, or the reference frame of motion with respect to all things in the universe.

This cosmic frame of reference is supported by special relativity itself, because in this theory not everything is relative, because we must not forget that here there is an even more fundamental axiom: "the speed of light is the same for all frames of reference." The time of the photon is not relative, it does not rejuvenate or age according to our speed. Therefore, there would indeed be a preferential point of view, but not among physical things, but with respect to time itself. In other words, to special relativity we would have to add a universal frame of reference that is above all physical things. The present itself or nothingness. In this way, motion would remain relative among physical things, because the universal frame of reference would not be in the horizontal dimension, but in the vertical dimension, with respect to nothingness (the nucleus of the cosmos). Therefore, if we take all the objects of the universe as a reference (which in Darwinian Physics would be equivalent to comparing them to the nothingness, the origin of time), it is perfectly clear who is the twin that accelerates and therefore who has to suffer time dilation. So, although locally each observer cannot know who is moving relative to whom, the whole cosmos, the nothingness itself, knows, because we are moving relative to time itself. That is why the speed of light is the same for everyone, regardless of our direction and speed. So, the dilation that the Earth twin sees of his astronaut brother is the real one; while the time dilation that the astronaut twin sees of his Earth brother would be an illusion, the product of the photons he observes from his brother and the planet having to travel farther and farther to reach him as he moves away from him and the planet. Whereas the jump in time that the astronaut twin would see upon returning to Earth is the product of the abrupt change in direction as he now collects the photons coming from Earth, the same photons he previously fled from as he moved away from the planet. Therefore, the Earth twin's time dilation is an effect of the photons it emits, which the astronaut twin sees; not of the internal photons tick-tock inside it. Therefore, the terrestrial twin's time does not actually pass more slowly; because the time dilation that the astronaut twin sees in the terrestrial twin is a consequence of the travel of the photons that the latter emits, not of the photons tick-tock inside it.

Then, thanks to holistic presentism, there would be a privileged frame of reference, but it would be nothing physical, but nothingness itself. The same that justifies the other postulate of special relativity, the invariance of the speed of light. So, this ontological interpretation gives more realism to Einstein's phenomenological theory, because not all relative motion is valid, because it can be misleading. For example, if we were to see a very distant star traverse the entire firmament in a single night, from our perspective it would appear to be moving through the galaxy at several times the speed of light, when in fact this motion is not real because it is the product of the Earth's rotation on its axis. Therefore, relative motion cannot always be relied upon because it can be completely illusory. This abuse of relativity would be corrected by a universal frame of reference, time itself, nothingness. The same thing that explains the absoluteness of the speed of light. Something that is already used in modern physics, since it is equivalent to space-time itself, which we saw in Chapter XI, a Cartesian coordinate system in which one can clearly see who is accelerating and who is not, at all times.

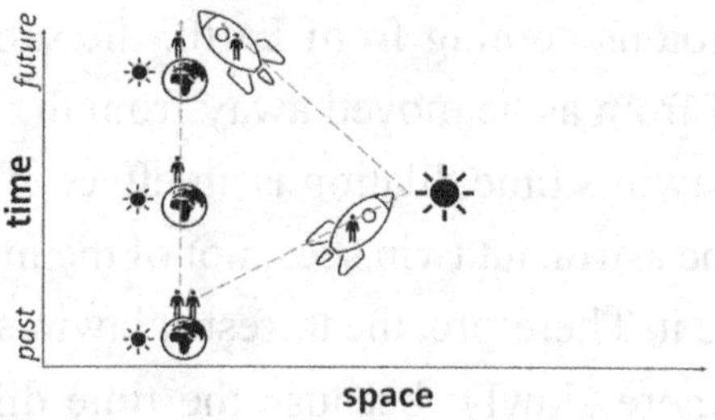

As we already know, the horizontal axis represents the displacement through space (the three spatial dimensions summarized in one, left-right), while the vertical axis represents the movement in time, being down the past and up the future. As can be seen in the figure, the twin that remains on Earth does not accelerate or change its velocity through space because it is together with the planet, which is fixed along with the Sun and the rest of the stars in the galaxy (all moving in the same way, inertia). Therefore, its time displacement remains the same. It just moves forward into the future as usual. In contrast, its astronaut twin literally uses some of its time to move faster and faster through space. Hence the diagonal line to the right as he travels to the star Alpha Centauri, and then to the left as he returns to Earth. Thus, in the space-time diagram, it is always clear who is aging relative to the other, knowing who is moving faster through space and who is not. The same perspective that time or nothingness would have.

Another outstanding mystery in special relativity that would be solved in Darwinian Physics would be the speed of light itself, in the sense of knowing its exact speed. Exactly as you hear it. Even the vast majority of physicists do not know this secret, and the fact is that the famous speed of light of 299,792,458 m/s is just a convention. To understand what this is all about, we have to remember how speed is measured, which is obtained by measuring the distance something travels divided by the time it takes. So, when we measure the speed of light, we have to put one detector at the beginning of a given distance, say one meter, and another detector at the end of that meter to measure the time it takes light to travel it. For the path time measurement to be accurate, both detectors must be perfectly synchronized. The problem is that if we want maximum accuracy in the synchronization of both clocks, we have to use light. Exactly what we are trying to measure, so it is no help. Another possibility would be to synchronize both clocks when they are together, and then separate them at the desired distance, say the meter. The problem is that by moving one of the clocks to separate them, such a clock undergoes time dilation. And we already know what that means, the delay or slowing down of the clock as it moves, which causes the synchronization between them to be lost. We can keep trying other creative solutions, but in the end the only possible solution is to use a single detector at the beginning and at the end, by placing a reflecting mirror at the other end of the required distance. This way, the same detector measures the light there and back, as the light bounces off the mirror and back to the same detector.

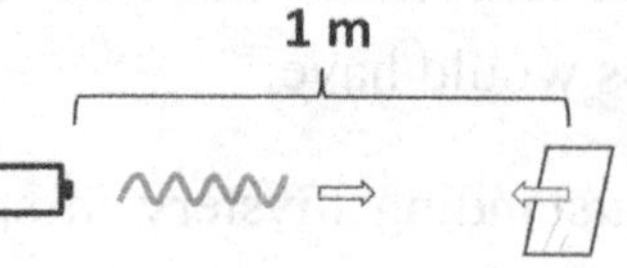

However, this only possible solution measures two trips. Therefore, it is conventionally assumed that light took the same time to travel there and back, because it is impossible to measure the time of only one trip. All because special relativity has killed simultaneity and thus perfect synchronization between clocks. So, the problem with all this is that we will never know the speed of light in a single direction. Always back and forth. A problem very similar to the twin paradox, since in that case the spacecraft also had to go back and forth to compare the twins and determine which one had aged relative to the other. Therefore, since it cannot be known in only one direction, it is established that the outward velocity is identical to the return velocity, but only by convention. In fact, this is what Einstein establishes in his famous 1905 publication, which is why it is known as Einstein's synchronization convention. In this publication, he makes it very clear: "It is neither an assumption nor a hypothesis about the physical nature of light, but a stipulation that I can voluntarily make in order to arrive at a definition of simultaneity." In this way, Einstein himself makes it clear that there is no ontological explanation for the speed of light, because light could very well travel at 0.5c (half its official speed) on the outward journey (which takes twice as long), but instantaneously or at infinite speed on the return journey. And this convention would be perfectly valid, because it yields exactly the same experimental results. In fact, all possible combinations of speeds between 0.5c outward and ∞ return would work because we would not be able to tell the difference. The phenomenological or epistemological nature of special relativity is once again evident. Something that would cease to be so under Darwinian Physics, since, with an explanation of time, it can ontologically justify this official convention of the speed of light. A solution very similar to the one used to solve the twin paradox. And that is that the speed of light must be the same there and back, because both are the product of the same thing, time. In other words, it would make no sense for the speed of light to vary according to whether it is coming or going, because it makes no sense for time to change according to direction in space. Therefore, light cannot travel at any speed other than "c" because it is fixed to be the speed of the present. Then the speed of light can be established, no longer by convention, but by necessity; for light, being the manifestation of time in physical reality, can only move at the rate of the oscillations of nothingness. In short, it is the same "now" on the outward journey as it is on the return journey.

Despite this great compatibility between holistic presentism and special relativity, we cannot forget to explain the other relativistic phenomenon that is the counterpart of time dilation, the Lorentz contraction. To understand it, let us use the situation of the discovery of the muon (the heavy brother of the electron). As we said, this fundamental particle decays very quickly, in 2.2 microseconds. Therefore, when muons are produced in cosmic ray collisions at the top of the atmosphere, at an altitude of about 15 km, they should not have enough time to reach the Earth's surface. But since they are traveling at 99% of the speed of light, they experience a significant time dilation, so that they live for us not 2.2 microseconds, but more than 50 microseconds. Enough time to travel the 15 km to the surface at that speed. But this is from our point of view, because for the muon it does not live longer, but exactly the same 2.2 microseconds as usual. And even though time has not slowed down for the muon itself, it still manages to reach the surface. So how can the muon explain that it traveled 15 km if it neither lived longer nor could it travel faster than light? The only way is that the distance was reduced. In other words, it no longer had to travel the 15 km to the surface, but only 660 meters, which allowed it to reach the surface in those 2.2 microseconds.

This is relativity at its best, because the same situation or event is perceived completely differently depending on who you ask. From our point of view, the muon suffered time dilation, because we do not see that the layer of atmosphere has shrunk by less than 5% of its thickness, fortunately; rather, we observe that the tick-tock of the muon slows down, allowing it to live longer to reach the surface. However, the same situation, but from the point of view of the muon, the time did not vary, it did not feel that it lived longer; rather it perceived that the distance was the one that was shortened, allowing it to reach the surface. Recall that if the speed of light is constant, then time and distance must vary (v = d/t). Therefore, when time varies, we call it time dilation (what we observe in the muon and what the terrestrial twin observes to the astronaut twin), but when space varies, we call it Lorentz contraction (what the muon and the astronaut twin observe). This second phenomenon of special relativity is explained by Darwinian Physics with the same solution, holoarchy. Recall that not only does each holon have different nows to explain time dilation, but the holons themselves have different sizes to explain Lorentz contraction. Therefore, both relativistic phenomena would be the fruit of this intimate relationship between present and holon; the same that justifies this intimate relationship between space and time in Einstein's theory. So, if an object like the muon is moving faster, it is because it is accessing a shorter present. This means that it is descending in the holoarchy by locating itself in lower holons that have these faster presents, at the expense of a slower tick-tock (internal presents of the muon) due to presentism. However, these same lower holons are also smaller, which produces the Lorentz contraction when a shortening of length in the direction of motion is perceived, because the space itself, the holon in which it is located, is of smaller size. In other words, both time dilation and Lorentz contraction are due to a reduction in the vertical dimension, forced by the constraints imposed by the presentism of Darwinian Physics. Then, to return to the case of the muon, by moving faster it not only accesses a shorter present, suffering time dilation from our point of view, since this higher velocity must be at the expense of a lower internal velocity, due to the fact that there are no more changes or time available, due to presentism. But on the other hand, it shrinks its distance by accessing smaller holons. It is as if the muon takes a shortcut through the vertical dimension, because when it transits to lower holons it accesses shorter spaces, being closer to nothingness (the space-time point). But from our point of view, the distances have not shrunk because we remain in the same holon, maintaining our usual speed (non-relativistic or very low speed compared to "c"). However, since the muon was created at a height of 15 km, with a lot of energy, it travels at a very high speed (0.99c); so it descends considerably in the vertical dimension to capture those faster present, but it also accesses smaller holons, shrinking its distance, generating the Lorentz contraction.

Again, the two phenomena of special relativity would confirm Darwinian Physics, since they would be a consequence of holistic presentism. In fact, through them it can be seen even more clearly that light is the manifestation of time in physical reality, being the product of nothingness itself. This is because light, the photon, as the only one that travels at "c," undergoes both phenomena of special relativity, but to the extreme. Therefore, its perception is the same as that of nothingness itself, because every photon, whether it is a gamma ray or a microwave, suffers the maximum possible time dilation when traveling at "c." This means that its perception of time is so slow that its tick-tock has slowed down so much that it has frozen completely. So, if a photon is traveling freely through space (which is the vast majority of photons in the CMB), without interacting with anything, it will perceive the entire history of the universe, all those billions of billions of billions of years, in just one instant, at one point in time. And in turn, by also undergoing the maximum possible Lorentz contraction, the distance or length of this entire vast universe, these billions of billions of billions of light years of distance, will be shortened so much that the diameter of the entire universe will be one spatial point. In other words, there is no time or space for photons, because everything happens in a space-time point or nothingness. That is why it has been said that light is the manifestation of time, because it is the same perception that nothingness, the origin of time, would have. This confirms that light is simply the present, because only we, massive physical things, perceive the flow of time and space lengths by using these presents, by using photons to be. Different presents, at different vertical distances, produce different types of light, such as: infrared, visible light or gamma rays. But from the point of view of light (time) there remains only the space-time point, the nothingness..

Then, holistic presentism explains special relativity, allowing it to cease to be a phenomenological theory by explaining its foundation, the invariant speed of light. Thanks to this, we can better understand how holistic presentism works, for which we will use ourselves as an example, since it is easier to realize that cellular presents cannot correspond to the past of our present (its higher present), because they are included in it. The cellular presents are within those 200 milliseconds that we had considered as our "now." Therefore, a true human past would be the previous 200 milliseconds. And this is precisely what holistic presentism denies exists. There is only the "now" in each reality, in each holon; only they are "nows" of different sizes, because the same holons to which they belong have different scales. So, for the "now" of a given reality, the "lower nows" that make it up are in a sense atemporal, because they belong to another time, the present of a parallel reality, because behavior or change in a reality does not operate on those lower temporal scales. We simply do not behave at the speed of cells, it is impossible. Neither do they behave at the pace of molecules, nor at the pace of atoms, nor do galaxies behave at the pace of humans, and so on. This can be better understood by returning to the analogy of the mechanical clock. For example, the past of the hour is the previous hour, not 30 minutes. If the gears move to mark only one time at a time, as in the old mechanical clocks, what happens within the hour is completely irrelevant to the gear that marks the time, because its perception of time is one hour at a time. Whatever happens in between simply isn't "there" yet, because it doesn't move until the entire cycle is complete. Your perception would be similar to time-lapse photography, where several static photographs give the impression of movement. Since the intervals in which the photograph is taken are the present of each holon. Therefore, what happened between each photograph is simply not in that time lapse. The same thing happens in the cosmos. Thus the past of the minute is the previous minute, not 54 seconds; or the past of the second is the previous second, not 7 deci-seconds. Therefore, the quantum jumps of the electrons require exactly the amount of energy necessary to go from one orbital to another in the atom; because the exact present that corresponds to that holon, to that orbital, is required. So, if 95% of that energy is given, the electron rejects it completely. It does not do 0.95 of the jump. It either does or it doesn't. And if you give it more energy than it needs, say 120%, it accepts only what it needs and rejects the rest, the 20%. This behavior is easily explained in Darwinian Physics, because the amount of energy required to make such a quantum jump, this quantification, corresponds to the present of that holon. Something that is amplified by having to receive a photon (present) with the exact wavelength to make that holon or orbital vibrate. Going back to the clock analogy, the quantum jump would be like building a tick-tock; for example, if the present of an orbital in the analogy is one minute, 54 seconds would not be enough because it would still be only a few seconds. On the other hand, if we have 83 seconds, we can only make one minute, because the rest would still be several seconds. In this case, 23 seconds. In conclusion, this apparent continuity of time that we perceive is due to the different smaller and smaller presents that are located in the vertical dimension; so there has always been the present, only that there are different sizes, due to the holoarchy.

In this way, holistic presentism would be able to explain not only the instantaneity of quantum jumps, the reason why they are discrete and not continuous, but also the spontaneity in how they occur. A concept that is very uncomfortable in current theoretical physics, almost taboo, since it implies a change in the state of a system that has not been provoked externally. The best example of a spontaneous phenomenon is radioactivity. A phenomenon in which, for example, the nucleus of a very massive atom, such as uranium, emits an alpha particle, which is a helium-4 nucleus (2 protons and 2 neutrons); or a massive particle decays into lighter particles, such as a muon decaying into an electron, a muon neutrino and an electron antineutrino. In both cases, the event is spontaneous because it is caused by literally nothing external. Therefore, it cannot be predicted because it is not caused by any law. And that is why it is a random phenomenon. The big difference now is that in the new paradigm it makes sense to be caused by nothingness. In fact, it would confirm it again, since radioactivity, not being caused by an external agent, cannot be causal. At least not in the traditional sense. This lack of causality makes it impossible to predict, so its behavior is stochastic. This is perfectly consistent with Darwinian Physics, since randomness is a product of the lack of causality, the absence of a past. In other words, radioactivity is caused by nothingness, so it cannot be predicted because there is no law or regularity that produces it. This is why spontaneous phenomena are very common in quantum systems, because they are generated directly by the "now" that comes from nothingness. Therefore, they are truly random phenomena.

Then, to really understand spontaneity, we have to rethink the meaning of causality, because in reality there would be more than one type. The traditional causality, which refers to the fact that every present effect has its cause or origin in the past, which we will call from now on horizontal causality; while the second type would be a new causality for physics, which we will call vertical causality. This second type of causality is better known in the field of philosophy and is called hierarchical causality. In fact, it is a very old causality, having been used by the ancient Greek philosophers, although today it is best known through Thomas Aquinas' five ways of proving the existence of God. However, it is often misunderstood when it is confused with horizontal causality because people do not know that there is another type. This is because the first way is that each moment is supported by a lesser moment. For example, to walk we need the contraction of our muscles, which needs cellular activity, which needs the production of proteins, which needs the chemistry of carbon atoms, which needs the quantum effects of electrons, and so on. This is why in philosophy it is called hierarchical causality, because every higher reality or system depends on a lower reality or system for its action. So, it is a presentist causality. Something indispensable and natural in Darwinian Physics, where neither the past nor the future exists. And they do not exist because they are simply not necessary, because what sustains or causes our "now" is the smaller "now." Returning to the previous example, our "now" is supported by the "now" of our cells, and these by the "now" of the molecules, and these by the "now" of the atoms, and so on. On the other hand, in today's physics, its only causality refers to the fact that in order to walk in the "now" we must first have left the house, and first have dressed for it, and first have gotten out of bed, and so on. This is where the conclusion of eternalism in modern physics becomes clearer, because it is absolutely necessary. If there is no vertical causal chain to sustain the "now," then we need a previous moment to sustain it. And to sustain that previous instant, we need another previous instant, and another previous instant, and so on, until all the infinite instants of the past. And lest our present also fall forward, it must be supported by the next instant, and the next, and so on into the infinite future. In other words, even if there is no evidence for the existence of the past or the future, they are indispensable to the physics of the present because they are what sustain our "now." A "now" that is indistinguishable from the other instants of the block universe, so there would be no "now" in reality either. On the other hand, vertical causality is very different, since it consists of a causal chain that affects or influences the "now" itself, not the past. Therefore, it does not need all the instants of the past and the future, since they are redundant, since it has a temporal holoarchy that supports the present itself with other inferior presents.

So, understanding the difference between these two types of causality, we can now understand the first way of Thomas Aquinas; for what he argues (something he learned from Aristotle) is that all movement is due to a motor, which in turn is moved by another motor, and so on. Like the chain formed by man, muscle, cells, molecules, atoms, particles, and so on; in the previous example. In order to avoid a regression ad infinitum, i.e. that each motor would need another more fundamental motor indefinitely, it would be necessary to postulate a primordial motor. A start for vertical causality. For if the vertical causal chain were infinite, motion would never reach us, because there would be an infinite chain of motors, so that motion would be impossible, as Parmenides would have it. Thus, Thomas Aquinas is not referring to the beginning of horizontal causality or the big bang, but to the beginning of vertical causality, the beginning of "now" itself. This is a very different problem of origins that not only confuses physicists, but also confuses the famous English evolutionary biologist Richard Dawkins, whom I admire and respect very much, but who in this case was wrong to criticize Thomas Aquinas in his book The God Delusion. Because even Thomas Aquinas accepted the possibility that the universe was infinite in the past, i.e. horizontally eternal; because it has nothing to do with the problem of vertical origin (the first way).

So, having clarified this distinction between the two types of causality, to solve the problem of vertical origin, Thomas Aquinas invokes the famous immovable motor, or the God of Aristotle. A motor that does not move, because if it were to move, it would need another motor to produce such movement. So, the first engine must be motionless. Something coherent for them, since the absolute, the eternal, must be the immutable; as we already know, due to the definition of God and the laws of physics. However, this solution suffers from the serious problem of justifying how something immovable can cause motion. In fact, it is the same problem with the absolute laws of physics. That is why many reputable physicists preach that time, motion, does not exist; because they do not know how to justify the obvious dynamics or becoming in nature through its immutable laws. You know, the famous Parmenian solution. Or they are believing physicists, invoking God to justify becoming from the laws, which would be his will or his word. On the other hand, the situation of Darwinian Physics is completely different, because it does not have this absurdity, because it does not have to justify how the immobile can produce the mobile, because it comes directly from the mobile. In other words, the vertical first cause is nothingness itself, a primordial moving engine. The origin of time. In this way, change is explained in the only possible way, from change. Therefore, all the dynamics we experience and measure come from maximum change, from pure chaos, nothingness. This perfectly explains vertical causality as the product of the very flow of presents or changes bombarded from below (in the vertical dimension) by the fluctuations of nothingness. In other words, nothingness would be the fastest and smallest gear that sets in motion the rest of the increasingly larger and slower gears (higher holons) (higher presents) in this complex clockwork mechanism called the cosmos. Thus, in summary, horizontal causality corresponds to the sequence of presents of the same holon or reality. To the properties or attributes of the physical object that seek to be preserved or remain between the exchanges of presents, or to the memory (causal inertia) that is preserved between the "nows." In this way, this causality has a maximum speed: "c," since it corresponds to the sequence of the shortest or fastest presents, the fundamental presents, or the fluctuation of nothingness itself. This means that it corresponds to the speed of light, which physicists also call the speed of causality, since it is the maximum speed at which information can be transmitted. That is, to transmit a regularity or pattern. Vertical causality, on the other hand, is the sequence of those present in the holarchy who belong to the different realities or holons in the vertical dimension. Therefore, it is what is happening in the "now," instantaneously, but in an interholonic or vertical manner.

Thus, thanks to the integration of this second type of causality in physics, we can perfectly explain spontaneous phenomena that are not caused by any law, any regularity, because they are truly random. And, therefore, phenomena without explanation in modern physics, although they are quite common on the smaller or lower space-time scales (low vertical distance), as happens in the case of radioactivity, particle decay, the collapse of Ψ, or quantum jumps. A spontaneity that manifests itself randomly because it is caused in the very present (nothingness) and not by previous presents (causal inertia). That is why vertical causality is fundamentally random, because it has independence with respect to the presents that have already been (past). On the other hand, horizontal causality is much more stable, especially at a greater vertical distance, because it is intimately related to the previous presents, to the factual information or memory (regularity) that one seeks to preserve. This is where its predictive power comes from, because the information of the previous present (past) seeks to be repeated in the next present (future). So, we can use this factual information in the "now" to predict what will happen later, in a new "now." On the other hand, at small vertical distances, vertical causality becomes more relevant and dominates unpredictable behavior because it represents what is changing, what varies from moment to moment, what is directly caused by nothingness. Unlike horizontal causality, which essentially comes from existence itself, from the regularity and information that seeks to be preserved in each new present. Thus, by empirically proving the existence of randomness through quantum mechanics, one also proves the existence of the "now" and thus of vertical causality and nothingness. This was the important ingredient that physics needed to really fulfill the principle of sufficient reason, because, as we have said, eternalism needs an infinite sequence of moments to support and justify the "now," because the moments are simply aggregated side by side, supported by themselves, similar to the grains of sand that support each other in a hill. An argument that is neither economical nor satisfactory, since it commits the logical fallacy of regression ad infinitum. Hence the eternal problem of the horizontal origin, or what caused the Big Bang. On the other hand, holistic presentism, with its vertical causality, renders the infinite past and future (as well as the infinite parallel universes) completely unnecessary and useless; because it actually justifies the "now" by revealing a "necessary connection" between the lower presents in a hierarchical manner, which support and justify each other. In this way, what sustains and enables the movement of a higher reality is the movement (the motor) of the lower reality, and so on. Until the whole hierarchical chain of motors is justified and given meaning by nothingness itself, which is pure change. Thus, the principle of sufficient reason is truly fulfilled, since every fact or event has a reason for being, a justification for the entire cosmos, thanks to nothingness.

Before concluding this chapter, we will take advantage of this new causality to better explain two physical phenomena discussed in previous chapters: quantum entanglement and strong emergence; because only by integrating vertical causality can such phenomena be fully understood. Let us begin with strong emergence, because spontaneity can fully justify how change can exist as an authentic novelty in Darwinian Physics. This is because vertical causality provides the only ingredient capable of making strong emergence possible, chance; because it is the only thing that can make something truly non-deterministic emerge by truly breaking this chain of horizontal causality or law. A chain that for physicists is unbreakable, absolute. That is why it was not possible to imagine the emergence of new things, such as the emergence of space-time, of matter, of new complex molecules, of life, of consciousness, or of new ideas. Everything had to be fixed forever. In Darwinian Physics, however, the origin of these natural phenomena can be addressed, because it recognizes the emergence of authentic new things, because they can emerge from nothingness. Therefore, they are true novelties because they do not emerge in their entirety from the past, from pre-existing information; since in the "now" new (factual) information is also created, as we saw in the previous chapter. This avoids the absurdity that the fact that you are reading this book right now has been set in stone since the Big Bang, as current physics claims.

In this sense, there is room in Darwinian Physics for true agency for complex beings. True free will. For since the "now" arises fundamentally out of nothing, there is true freedom. A true independence from the past product of vertical causality. Even if it is a very limited and diminishing independence due to holistic evolution. However, despite the advanced degree of order that existence currently has, it is a freedom that still exists, especially at a smaller vertical distance, because existence is not yet absolute. A small rupture of horizontal causality that is impossible in modern physics because the initial conditions, like the laws, are absolute. And yet it happens, because it is caused by time itself. And although this indeterminacy is extremely small compared to the abyss of the cosmos at its current level of evolution, it can still have very significant consequences, thanks to the butterfly effect of chaos theory, which can amplify this small freedom that still exists at a small vertical distance and make it relevant in the higher realities we inhabit. A process that can take place precisely in our brain, the most complex object known so far, but we will leave a holistic explanation of free will for the second book. What is relevant here, however, is that the introduction of vertical causality makes it possible to explain not only typically spontaneous phenomena such as radiation, but also more complex phenomena such as strong emergence and free will, thanks to the final overthrow of the absolutist determinism that has reigned for millennia, making these natural phenomena simply unimaginable.

On the other hand, in the case of quantum entanglement, which we saw in the previous chapter, vertical causality allows us to better understand the spooky action at a distance that so frightened Einstein. A stronger correlation than the classical one, which occurs no matter how far apart the entangled particles are. It doesn't matter if one is here and the other is 2.5 million light-years away in the Andromeda galaxy. It is as if they were communicating at infinite speed. This very rare phenomenon of current quantum physics is very well understood in Darwinian Physics because it is the product of vertical causality or the "now." It is therefore instantaneous and random. Thus, the two entangled particles are part of the same quantum system, the same Ψ. In fact, this is how the calculation is done in actual quantum physics itself. Then, if one were to separate these two entangled electrons after bringing them together (this is important because entanglement between two particles does not occur just because, but because they have interacted); one would also be extending Ψ for the whole distance in which such particles are separated. Therefore, when one of the two particles is measured, it is actually collapsing both particles, since it is the same Ψ or quantum system. And since what collapses Ψ in Darwinian Physics is time itself, this determination of the properties of the two entangled particles is instantaneous because it is happening in the same present. And it does not matter how far apart the two entangled particles are, because it is a horizontal distance. Instead, the present emerges from below, in the vertical dimension, and manifests in the whole holon, potentially. Thus, horizontal distance has no relevance to time itself, since time itself is omnipresent (there are, of course, no places without time). So, it is not that the measurement of one entangled particle causes the horizontal determination of the other electron, but that a present materializes where both electrons are connected, they share properties. A phenomenon that violates locality, understood in an absolute way; but that respects horizontal locality, understood as horizontal causality, because the latter is characterized by the transmission of (factual) information, by the continuation with a certain regularity or pattern. However, quantum entanglement, as a vertically caused phenomenon, does not transmit (factual) information because it is a product of the now, not of the past (memory or causal inertia).

Therefore, what exists in entanglement is a random correlation because it is vertically caused by nothingness.

In short, vertical causality is what happens in a single "now." That is why vertical phenomena are instantaneous and random, such as radioactivity, entanglement, or the tunnel effect. Horizontal causality, on the other hand, must be present in order to cause, to produce an effect. Therefore, horizontal phenomena are characteristic of existence, which is recognized by current physics, because they use factual information. Therefore, they can be predicted by following and transmitting a pattern, a regularity in the sequence of presents. Therefore, they cannot be instantaneous by using successive presents to cause, the shortest possible presents being the fluctuations of nothingness itself. Hence there is a maximum speed for horizontal causality, "c." If it is still difficult to assimilate this difference between these two causalities, let us, as usual, give an analogy. Imagine that these two types of causality are like two light bulbs connected by the same electrical wire. When they are plugged in, both bulbs light up at the same time, no matter how far apart they are, because they are connected by the same wire. This current flow turning on both bulbs would represent vertical causality, and the length of the wire would be the vertical distance. While the interaction between the two bulbs would represent horizontal causality, which depends on the distance between them because they move at a finite speed compared to the electric current, which seems instantaneous.

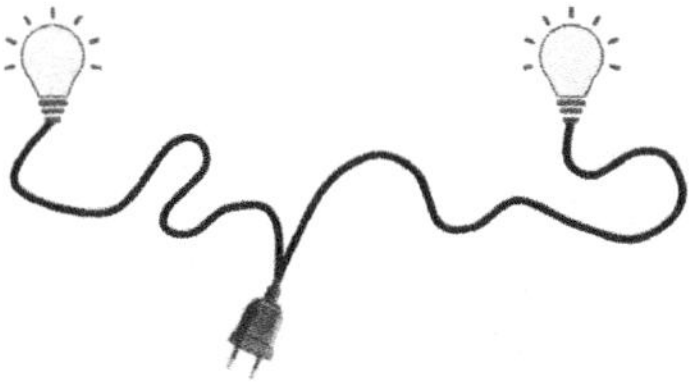

So, if we flip a coin and randomly turn the power on and off, there will be a perfect correlation between the two light bulbs because they will both be on or off at the same time, which would be entanglement. Something that we know is not because one affects the other, but because both depend on the same energy source. This is the mistake of modern physics, again forgetting the projector that gives life to the "now," time.

CHAPTER XIV

THE PROBLEM OF GRAVITY: GENERAL RELATIVITY

We have finally arrived at what physicists themselves consider the most grave problem in physics: gravity. An irony that is not only due to its name, but also to the fact that it was the first mathematical description of a fundamental force or interaction that gave birth to modern physics, thanks to the impressive book Mathematical Principles of Natural Philosophy, published in 1687 and written by the English physicist Sir Isaac Newton (1642-1727). Probably the greatest physicist of all time, he not only created the first mathematical theory of a fundamental interaction with which he was able to unify terrestrial and celestial mechanics, but he also invented the mathematical tool necessary for its application, the calculus. And yet, although it is the first, it is by no means the simplest. In fact, it is the most difficult in physics. This is because it has not been possible to integrate it with the rest of the fundamental interactions (electromagnetism, weak and strong nuclear forces) in the Standard Model of Particles. So, to understand why it has been so difficult to make gravity compatible with the rest of the interactions, we first need to understand the most accurate description of this interaction to date. One that we owe to the only genius who could unseat Newton. We are talking, of course, about Albert Einstein, who achieved this in 1915 thanks to his theory of general relativity. An extension of his previous theory (special relativity), which completely abandoned the Newtonian concept of gravity as a "force" acting instantaneously between all bodies with mass, and replaced it as a product of the geometry of space-time itself.

To understand what this means, let's start where Einstein himself started, with his happiest idea, in his own words. One known as the principle of strong equivalence, which he had when he imagined, looking out the window, a person falling into an elevator. This person would float in the elevator as he fell, because all things fall at the same speed. This was discovered by Galileo Galilei. Therefore, the person would not feel his weight. However, let us imagine that the elevator was floating in space, far away from any celestial body. The person would still have the same experience. He would be floating. Therefore, if the person could not see what was happening outside, he would not be able to know whether he was falling toward the earth or floating in space. They would be two indistinguishable situations. We must add that to orbit is also to fall, just as the Earth orbits the Sun, or the International Space Station orbits the Earth. Therefore, although they are spinning in a circle, they do not suffer from centrifugal force like a ball that we spin around our head with a rope, for example.

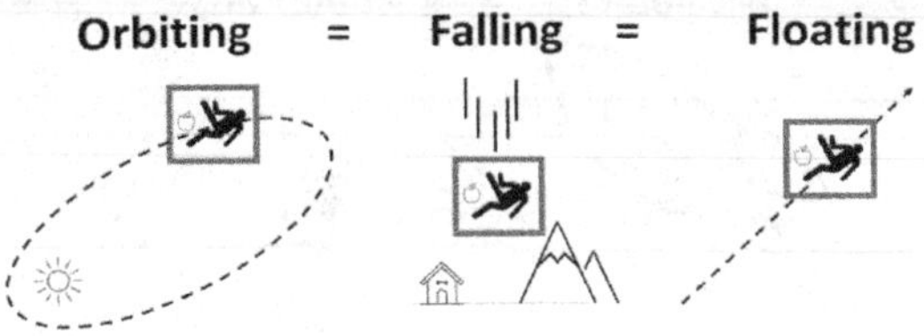

This is very strange because they are very different situations. In one, you are falling and accelerating at 9.807 m/s^2, and in the other, you are floating in space at a constant speed; and yet, for Einstein, they are completely equivalent situations. This is why future astronauts going to the International Space Station can practice their stay in an airplane going up and down in free fall, because the sensation is exactly the same. And this is because the International Space Station is also falling, only it is falling at such a speed that it can fall permanently without having to crash into the planet, because at that speed it can make a complete turn around the Earth.

An equivalence that becomes even more extraordinary when we consider that standing still on the Earth is exactly the same as being accelerated by a rocket at 9.807 m/s^2. In other words, using the example of the person in the elevator, he would not know if the elevator was standing (still) on the surface of the Earth or if it was being pulled by a rocket accelerating at 9.807 m/s^2 in space, because in both situations he would feel the same force that keeps him glued to the floor of the elevator. This means that the force or pressure you feel against the seat of your car when accelerating rapidly is exactly the same as the pressure or weight your feet feel on the surface of the planet. In other words, artificial gravity is created every time you accelerate, whether in a car, a rocket, or any other vehicle.

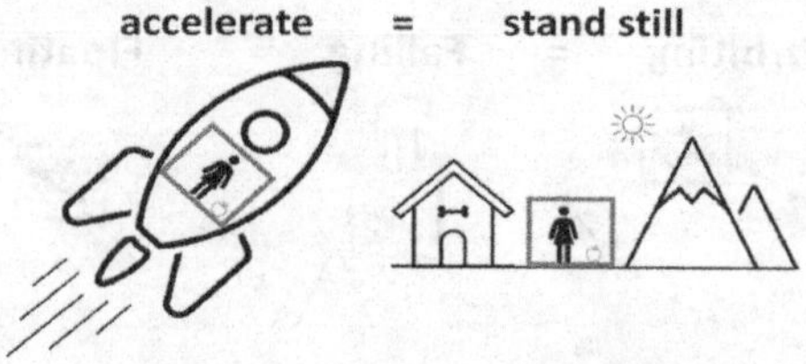

This is the principle of equivalence, that the effects of gravity are equivalent to those experienced in an accelerated frame of reference, because every object experiences the same acceleration in a gravitational field (when falling), regardless of its mass. This is not the case with other forces, such as electromagnetism, where the consequences are very different if one electric charge is greater than another. Instead, gravity pulls everything in the same direction and with the same acceleration. So how did Einstein solve this paradox: how can falling be equivalent to traveling at a constant speed, and how can acceleration be equivalent to standing still on Earth? The answer Einstein found was the geometry of space-time. To understand what this means, let's take it one step at a time. First, in a very basic sense, space is what allows us to locate ourselves, to be somewhere concrete. While time allows us to change, to move through that space. Then, according to the form that space-time has, it will determine the way we will move. They are like roads for cars. Depending on the shape of the former, whether it is straight or curved, whether it goes up or down, will be the trajectory that the cars that pass through it will take. Similarly, the geometry of space-time determines how physical things will move through it. This is nothing new, it has always been this way, except that for Newton, space could only have one possible shape, Euclidean geometry, also known as plane geometry. The same one we are taught in school. A geometry that was believed to be the only possible one for more than 2000 years, where the famous Pythagorean theorem, $\pi = 3.14159...$, the interior angles of a triangle always add up to 180°, and parallel lines never cross.

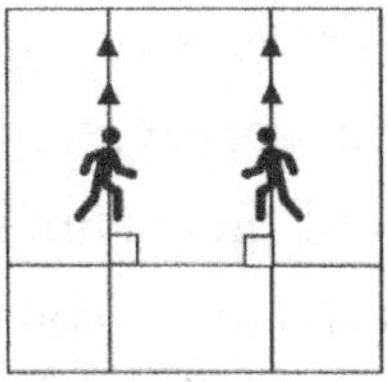

So far we have special relativity or Minkowski space-time. A flat, static and infinite one. But to describe gravity, a phenomenon that does not distinguish between falling, floating or orbiting, it is not possible to do so by means of a force, as Newton did, because floating is a force-free phenomenon. So, if a force does not intervene in space-time, then space-time itself must change, specifically its geometry. And so it did. If Minkowski's space-time was flat or Euclidean, then Einstein's new space-time would have to be non-Euclidean, i.e. curved or dynamical. This new geometry was developed long before Einstein knew of its existence by the great German mathematician Georg Friedrich Bernhard Riemann (1826-1866). Here there can be positive curvature, also called closed geometry, where the straight lines are curved, $\pi < 3.14...$, the interior angles of a triangle add up to more than 180°, and the parallel lines end up crossing each other.

As you can see above, it is the geometry that, for example, the surface of our planet has, whose π is almost 2 (less than 3.14...), since our planet is almost a sphere. Here, the shortest path between two points is no longer a straight line, but a curved line called a geodesic. That is why when you look at the flight path of an airplane on your seat screen during an intercontinental flight, you see a curved line. This is not because the pilots deliberately extend the trajectory to burn more fuel and accelerate global warming, but because it is the shortest trajectory on a curved surface (much to the chagrin of terraplaners). So, as shown in the figure above, if two people start walking north in a straight line from the equator, they will end up meeting at the pole, forming a triangle whose interior angles add up to 270° (90°+90°+90°). Obviously larger than the 180° of the interior angles of a traditional triangle drawn on a sheet of paper. Now, before Einstein, this new geometry was just considered an incredible feat of mathematics. An impressive curiosity with no real-world application (except on the surface of the earth). This changed when Einstein discovered it (at the suggestion of a friend) and applied it to the shape of the universe itself. This was extraordinary, because just like the situation above where two people walked parallel to each other from the equator and met at the North Pole, if they were terraplaners and did not believe that the Earth was round, they would end up interpreting what happened as a mysterious force attracting them, because from their perspective they always walked in a straight line. This was exactly the reasoning of Newton and his contemporaries when they interpreted gravity as an attractive force. All because they ignored the fact that space-time can actually be curved. So, the planets actually move in a straight line around the sun, only they do so in curved space-time, which forces them to orbit in elliptical orbits. But they continue to move inertially, so we do not feel a centrifugal force that would shoot us out into space.

So this was Einstein's great genius, to apply the curved geometry developed by Riemann in the 19th century to the structure of the universe itself. In doing so, he discovered that Newton's gravity is not a force, because nothing physical pushes or pulls, but is the consequence of traveling through warped space and time; finally explaining why objects of different mass or weight always fall at the same speed in a vacuum (which Galileo Galilei had already discovered in 1589), since it is due to the fact that objects continue to move inertially, in a straight line and at a constant speed, i.e. free of external forces, only this time they do so in a curved space-time that is the same for all those who are there. This is why a feather and a hammer fell at the same time on the moon during the Apollo 15 mission of astronaut David Scott. The same thing happens here on Earth, but in a vacuum to eliminate air friction during the fall. However, we can easily replicate this experiment by dropping a crumpled piece of paper in the shape of a ball (more aerodynamic) next to a hammer or something heavier. In short, we are always talking about inertia when we talk about gravity (free of forces or disturbances). Only when space-time is flat or Euclidean, the motion is straight and at constant speed. But when space-time is Riemannian, motion is curved and accelerated.

So far so good, but one important question remains: what determines the degree of curvature of space-time? In Newton's gravity, it was mass. The greater the mass, the greater the gravitational force. However, since Einstein, we know that mass is simply another kind of energy (potential energy), as was made clear in the previous chapter ($E = mc^2 \rightarrow m = E/c^2$).

So, it is energy that gravitates. Then, all kinds of energy bend space-time, and this relationship is embodied in a very precise way in the equation of general relativity, which can be summarized as follows: "Things (energy) tell space-time how to bend (geometry), while space-time tells things how to move."

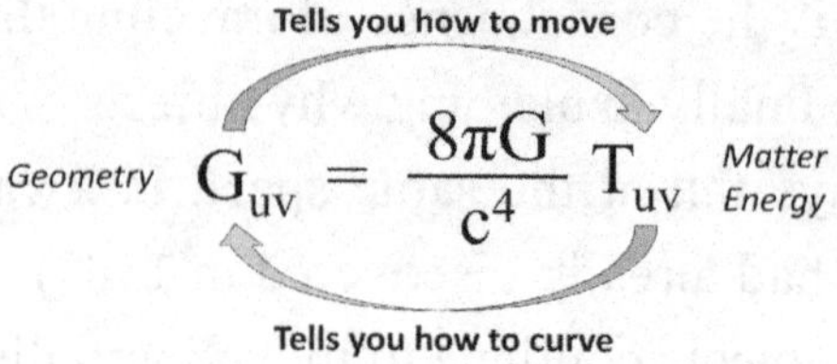

So under Einstein's theory of gravity, even light gravitates, because although it has no mass, it apparently contains a lot of kinetic energy, or momentum. This is why Einstein was able to test his theory in a solar eclipse, since the light from the stars behind the sun bends as it passes around the sun.

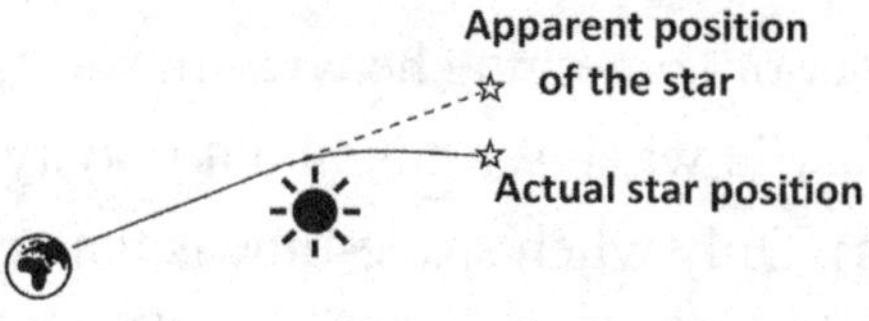

This is gravity according to current physics. A phenomenon that has ceased to be a force, as it was in Newtonian physics, and has become the inertia of a curved space-time. But although Einstein made it quite clear in his equations, physicists have not wanted to accept that gravity is different from the other fundamental interactions. A resistance that has grown over time, as the vast majority of theoretical physicists of the last century did not specialize in gravity, due to the great success of the Standard Model of Particles in the 20th century and the difficult mathematics of general relativity, which is not linear like quantum mechanics. As a result, Einstein's theory of gravity has been largely abandoned. This has led theoretical physicists to misunderstand the meaning of general relativity and, even worse, to view it biased from the perspective of particle physics, applying the techniques they learned in the latter directly to general relativity, forcing the unification of the four fundamental interactions by quantizing space-time in the same way they quantized the rest of the fields of the fundamental interactions. This strategy turned out to be disastrous because it is based on an idea that is already meaningless. The idea that space-time cannot be a fundamental element of physics, that it must be an emergent phenomenon, the product of something even more fundamental, for example strings. This is the same problem we discussed in Chapter X, which can be summarized as confusing the container with the content.

This has obviously led to many problems. For example, the quantum fields on which the interactions of the Standard Model are based, such as the electron field, are necessarily located in space-time. A space-time that must also be flat or Euclidean. So where would the quantum field of spacetime be if it were spacetime itself? In fact, this dilemma has its own name in physics, the problem of microcausality, which means that if there is a disturbance in one region of the quantum field, another region cannot immediately feel it. You always have to respect the speed of light, which is the maximum speed in the universe. Then you apply this condition to the quantum field and modify it to conform to it. In this way, you determine which regions of the quantum field arc causally related and which are not. But how do you impose this condition on the curvature of space-time, because what tells you which regions are causally connected is the curvature of space-time itself, but that is exactly what you want to get from the field. That is, you want to modify a thing based on the very thing you are modifying. It is paradoxical.

But the problems are just beginning. Physicists believe that Einstein's gravity may be equivalent to an interaction mediated by a hypothetical bosonic particle with spin 2, called a graviton. Similar to how two electrons repel each other by sending a photon at each other, or how two quarks attract each other by exchanging a gluon. In other words, physicists believe that gravity, which is inertia, is equivalent to a traditional gauge interaction like the electromagnetic force. Again, this does not make sense for many reasons, since transforming spacetime into a quantum field implies having all the properties of a quantum system, such as quantization. This may be intuitively possible, since in the same way that electromagnetic waves (light) are quantized into photons, gravitational waves could also be quantized into gravitons. So far so good, but quantization is not the only characteristic of quantum systems, there is also superposition. What about the superposition of spacetime, i.e. the existence of an infinite number of spacetimes, each with a different curvature? But how can space itself be in many spaces at the same time, and time in many times at the same time? What does that mean, again infinite parallel universes? A big mess. And to all this we must add the fact that current quantum mechanics lacks a fundamental element that is indispensable in general relativity, the variable "time." Let us recall that current quantum mechanics is a timeless theory that cannot define this variable due to the problem of measurement, as was well explained in Chapter XII. Therefore, it is paradoxical to want to quantify space-time when we do not even have a definition of what time is. In fact, they preach that time does not exist. For all these reasons, it is not difficult to understand why theoretical physicists have not been able to quantify space-time, since they are based on a strategy that is already invalid, since they confuse the container with the content.

The fiasco of the theory of everything has been so great that it has become the greatest failure of physics. This is ironic, because both theories are essentially about the same thing, the vacuum. What curves around the sun, for example, is the empty space-time through which the planets travel. The same empty space-time in which particle physicists place their quantum fields. Remember that quantum fields extend throughout space. That is, they are everywhere. Therefore, every cubic millimeter of space contains all quantum fields. The problem is that even empty quantum fields contain energy, due to Heisenberg's uncertainty principle, which produces a frenetic creation and annihilation of virtual particles. Though unobservable individually, the effect of all these particles can be measured. One consequence is the so-called zero-point energy of quantum fields, which obviously cannot be eliminated, because no more vacuum can be obtained from vacuum. This means that all the vast space between planets, stars and galaxies contains energy. Even the empty space inside the atoms, which is 99.97%. If we add to this fact what Einstein says, that all kinds of energy gravitate, we have that the energy of the vacuum of quantum fields must have a considerable consequence in the geometry of the universe. Exactly what was discovered in 1998, to the great surprise of physicists, when they observed that the universe was expanding at an accelerated rate. The famous dark energy. So far a happy surprise, since quantum field mechanics predicted that empty space must have energy. An element that would also be easy to integrate into general relativity, since only one parameter would have to be included in the equations, the famous cosmological constant (Λ). However, the tragedy came when physicists calculated how much this energy would be and compared it with the energy observed in the expansion of the universe, and discovered that they differed by 10^{120}. In other words, the theoretical prediction is a 1 followed by 120 zeros greater than the observations. This is the largest discrepancy in the history of science. This is why physicists say it is one of the biggest problems in physics, because the above number reflects quite well how far theoretical physicists are from achieving quantum gravity. And worst of all, they don't even know how to approach the problem. Which is wrong.

So, again because of extreme specialization, theoretical physicists have focused on the right-hand side of Einstein's equation, on the world of energy and matter represented by the Standard Model of Particles. The pride of theoretical physicists. While the left side was left behind. Which made them forget that there are two different realities (physical holon and space-time holon). So, they have tried to apply what works on the right side to the left side with very little success; because they believe that space-time is just another quantum field (something physical). They deny the fact that they are two very different things. On the one hand we have a smooth, continuous and certain geometric world, which Einstein compared to marble; on the other hand we have the discrete and uncertain energetic world, which Einstein compared to poor quality wood. And as different as they are, quantum mechanics is implicitly included in Einstein's equation, on the right-hand side; but how or what makes energy change the geometry of the cosmos is completely unknown. The only hypothesis that comes to the mind of physicists is that matter particles exchange gravitons. A strategy that, as we already know, did not work.

So, to avoid the same mistake of confusing the container with the content, we must first answer what is space-time, the fundamental element of general relativity. And as might be expected, the space-time fabric would correspond to the mother or lower holon of physical reality. The holon that precedes us, that supports our reality. This is strange, because in Darwinian Physics a lower holon is usually smaller than its upper holon, producing that structure of matryoshka dolls or Russian nesting dolls that has so often been used to characterize holoarchy. However, Einstein's space-time is what surrounds us, what is around us. It is the container that holds all the stars, planets, and galaxies. Therefore, the space-time holon turns out to be much larger than our holon, the physical reality. This apparent contradiction is due to what makes our reality special, which is multiplicity. The appearance of the physical holon changed everything, because before us, existence consisted exclusively of holons. The entity that extends through space (horizontally), as physicists define fields. However, since physical reality is made up of local existences (the pixels of these holons), it makes sense that physical reality is the first upper holon that cannot be larger than its lower reality. Thus, if physical reality had continued to evolve holistically like the other realities before it, it would have been a new outer layer added to the cosmos. Its new surface. Fortunately, however, the new reality reached such a level of complexity that it allowed the emergence of local existences, as described in the previous chapter. However, this did not prevent or change the way our mother reality evolved. In the same way, bacteria and archaea continued to evolve along with fish, dinosaurs and mammals. So, the space-time holons continued to expand as always, adding layer after layer, while the new physical holons continued to evolve, but this time within themselves.

Then, due to the emergence of multiplicity, we have empty spaces or space-time holons not only inside molecules and atoms, but also around us, as in the outer space between planets, stars and galaxies. This makes the way physical reality evolves completely different than before; because it is no longer possible to grow (+Δ space) and age (+Δ time) in the conventional way. This caused another strategy to emerge in physical reality, which was the union between local existences to create larger and larger and more stable structures. In this way, the electron and the proton naturally come together, they attract each other because they separately have more energy (less order) than their conjunction, the hydrogen atom, which is more stable (more order). And the atoms, in turn, naturally attract each other to form molecules because they have more energy (less order) when separated than when they form a structure (more order). The same thing happens with quarks, which group together to form composite particles called hadrons, such as the proton and neutron, which have less energy than the quarks separately. This ordering process (loss of energy) of physical reality began to receive additional help when it reached a certain size or vertical height. This is because the electromagnetic interaction allows these atoms and molecules to conglomerate up to a certain point, forming a gas or dust. From there, however, something extra was needed to continue this ordering process and form larger balls of gas, like the first stars. Guess what? That help was gravity.

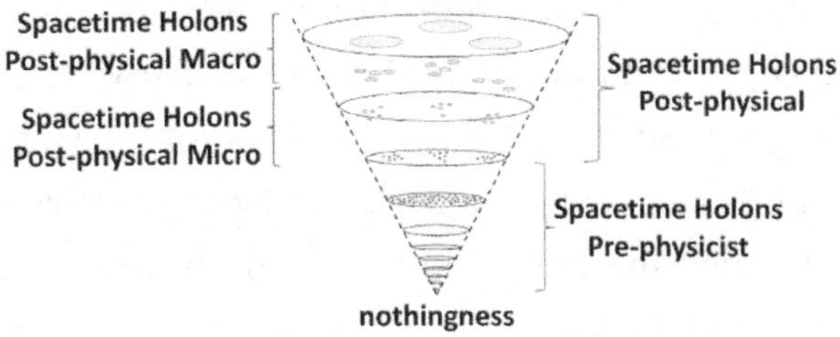

As shown in the figure above, there would then be two main types of spacetime realities. The post-physical spacetime holon, which would be the spacetime that surrounds us (macro), including the empty space between planets and stars; but also, the space inside us (micro), between molecules, between atoms, and even inside atoms. A spacetime holon that performs the important task of locating physical things and allowing them to move. In other words, Einstein's space-time (macro) and the space-time of particle physics (micro). On the other hand, the novelty for current physics would be the pre-physical space-time holon, corresponding to the space-time reality that emerged before physical reality, before multiplicity, and thus a smaller reality than physical reality. One would expect that there would be nothing in current physics that could refer to this type of space-time, to being pre-physical; but there is. In fact, physicists have given it a name: the Great Desert. A place on energy scales ranging from the heaviest particles observed at the LHC (10^{-20} m) to the Planck scales (10^{-35} m). A big space, even if it seems to be the opposite, since it corresponds to a range of scales that separates bacteria (10^{-5} m) from planets (10^{10} m). A large, seemingly empty space that worries physicists a lot, as they still predict many new fields such as supersymmetry and dark matter particles. String theorists even expect infinitely many new particles. The big problem with this strategy is that these new particles have to be even more massive, so they must be coupled to the Higgs field to obtain their mass. However, this coupling has to generate corrections to the mass of the Higgs boson. And since physicists expect a series of new particles up to the Planck scale, then the Higgs boson should have that mass too. However, we already know that the Higgs boson does not have that mass, since it was discovered already with a much smaller mass. Therefore, the discovery of the Higgs mass is great evidence of the nonexistence of these new particles beyond the energy scale of the LHC. However, physicists are impressively very far from this conclusion, the simplest and most obvious of all. This is why the community only talks about the naturalness problem, which assumes a priori the existence of all these new, increasingly heavier particles, so what is wrong is the mass of the Higgs boson. This means that there must be another parallel contribution that counteracts such coupling. In other words, the mass of the Higgs boson is much heavier, but there is another large contribution that corrects its mass downwards, leaving that tiny mass that has been observed at the LHC. The problem with this is not only assuming the existence of this countless number of new particles, nor conjecturing the existence of another enormous inverse contribution; but also that both gigantic contributions have to be eliminated almost exactly to leave a tiny mass for the Higgs boson. In other words, it is another enormous fine tuning problem. Instead, Darwinian physics again proves to be more empirical, because it simply assumes what has been observed, the mass of 125.11 gigaelectronvolts for the Higgs boson. This indicates that there are no heavier particles, because there are simply no more physical fields beyond that scale, since from there it corresponds to pre-physical spacetime holons. In the same way that if we go back further than the Triassic we will not find more dinosaur fossils, because they had not yet evolved.

So, the pre-physical spacetime holon corresponds to that great desert, to that reality that emerges directly from nothingness, before physical reality evolved. While, the spacetime holon post-physical micro corresponds to Minkowski spacetime, which is practically flat or Euclidean and therefore with very little gravity. This is why particle physicists do not need to take gravity into account in their calculations and predictions, because in the Standard Model of Particles the curvature of spacetime is negligible. Therefore, gravity, as a curvature of space-time, would be a phenomenon that emerges in the higher space-time realities, in the space-time holons of the post-physical macro, or in our scale of reality. The help that allows dust bound by electromagnetic forces to continue to clump together to form planets, stars and galaxies. So, the key question is: why does the spacetime holon postphysical begin to bend at these higher scales? The explanation provided by general relativity is energy concentration. When a certain energy density is reached in a certain region of spacetime, spacetime begins to curve. It is precisely at these higher or larger scales that a critical density is reached. However, in Einstein's theory, the how is answered incredibly well, with great precision, but the why is again abandoned. Again, Darwinian Physics is in a privileged position because it includes, as always, the variable that physics lacks: time. What then generates this curvature in Einstein's space-time is precisely the constraint imposed on it by the present. Let us remember that the cause of the curvature of space-time is energy, and the origin of all kinds of energy is obviously time. Therefore, it is logical to think that space-time has a limit to the concentration or support of energy, just as time has a limit. Only this limitation of the present manifests itself in a different way, because in the case of special relativity the limitation applies to motion, whether internal (time dilation) or external (Lorentz contraction), whereas in the case of general relativity the restriction applies to energy.

Normally in physics it is thought that energy has no limit, that it can be infinite. This is because although there is a limit to speed, a massive particle can accelerate as much as you want without reaching the speed of light. As each approach becomes more energetic. For example, the incredible energies achieved at the LHC are due to the acceleration of protons to 99.9999991% of c. However, particles from space (cosmic rays) have been detected with a velocity of 99.99999999999999999999951% of c. That is a lot of energy, the equivalent of a single proton having more mass than a bowling ball. And with each additional 9 decimal places, the energy scales exponentially. So mathematically there would be no limit to adding more and more 9s to the speed, so the energy could tend to infinity as it gets closer and closer to the speed of light. As always, Darwinian Physics does not like infinity because it believes that there are conditions of physicality. And here, of course, it is no exception. In fact, this energy limit can be clearly observed in nature due to the existence of black holes. That is why you cannot indiscriminately put as much energy as you want into a zone of space-time, because it will saturate and collapse into a black hole. Therefore, space-time has a limit to the tolerance of energy, and this is due to the fact that it is limited by time. Only this time, that time limit manifests itself vertically, because the speed limit is more directly caused by the duration of the fundamental present. It is therefore a more horizontal constraint, being the product of the sequence of these presences. On the other hand, the limitation of energy is more vertical, since it tends to infinity, that is, to nothingness. Which, as we know, is something finite, the unit of space-time. In other words, the limit of energy is in nothingness, from which time originates, and therefore the ultimate origin of energy. It is therefore not surprising that at the center of black holes there is a singularity, a point in space-time (we will return to this later). Then it should not be surprising that physicists consider gravity as a kind of negative energy, because the curvature of spacetime compensates for the concentration of positive energy that physical things have. Only in Darwinian Physics would this negative energy make sense, since it would be interpreted as the price the macro-spacetime holon has to pay for containing something that also consumes energy, i.e., time. And since the latter is limited by presentism, since there is no more time, the concentration of energy that produces the massive physical things has to be compensated.

A presentist explanation of gravity, supported by the most peculiar feature of general relativity. One that, strangely enough, is almost entirely omitted from the disclosure, and that is that much of the reason we fall toward the Earth is not due to the curvature of space, but to the curvature of time, because the planet deforms extremely little of the space around it. It is so little that the curvature is extremely difficult to measure. And yet we are falling at 9.8 m/s^2; because we are doing so mostly because of the curvature of time. A curvature that is much easier to measure, in fact, you just have to put two atomic clocks on top of each other to observe the difference in the speed of time, because one is closer to the gravitational field at the center of the Earth than the other. A curvature in time that has been observed even with millimeter differences in height and that physicists call gravitational time dilation. A name very similar to the time dilation of special relativity, because they are actually the same phenomenon. The only difference is what is causing the increase in speed. In special relativity, what pushes is something else physical, like your hand or rocket fuel. On the other hand, in general relativity, what pushes is space-time itself (inertia) due to its curvature. Therefore, in both cases, time dilation is produced by a difference in speed. And just as in the case of time dilation in special relativity, the faster time passes, the slower it passes, since the increase in the use of the fundamental particles to move must be at the expense of a lower internal use of the fundamental particles. A compensation that is again due to the limitation of holistic presentism, since the only thing that exists are those fundamental presents. Only this time, this time dilation is obtained simply by being at different heights.

This can be quite surprising because it is a time dilation that occurs even though you do not perceive that you are moving. In other words, you do not have to be literally falling to experience gravitational time dilation, because it depends on your position in space. For example, for a person standing at sea level, his internal time will pass more slowly than for another person standing at the top of a mountain. This means that the person at sea level ages more slowly than another person standing in the mountains. How is this possible? It is because they are actually moving at different speeds, even though they do not appear to be moving, because this difference in speed is represented by a difference in weight. The person at sea level weighs a little more than the person on top of a mountain at 7000 meters, exactly 0.22% more.

The same thing happens between your head and your feet, only the difference is even more insignificant. In fact, for every second that passes for your head, it takes 1.000000000000000415 seconds for your feet. So don't worry, the age difference won't be very noticeable. But even if it is a very small difference, it exists, it is real, because it is measurable with atomic clocks. A difference that is due to the fact that weight is a resistance to acceleration, just as the pressure you feel on the back of your car seat is a product of the resistance to acceleration of your car. So, for example, when we freefall in a parachute jump, we are accelerating purely. Before the parachute opens, of course. So, the weight disappears as it transforms into that acceleration. But when we are not falling and we are standing on the ground, all that acceleration is transformed into weight, into the resistance of the surface of the planet to that acceleration.

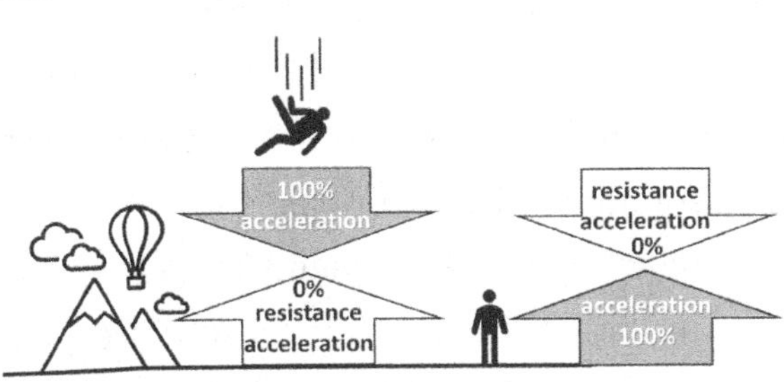

Therefore, the pressure on the seat of your car when you accelerate it and the weight on the Earth are the same. Only that one is the product of the acceleration of a physical object, in this case a car, while in the other case it is space-time itself that is pushing. Therefore, when you are at sea level, you are traveling at a higher speed because you weigh more than when you are in the mountains. And as with any increase in speed, internal time has to pass more slowly to compensate, due to presentism, since there are no more fundamental presents or extra time available.

So holistic presentism explains gravity, but a very important question remains: how does it bend? Again, Darwinian Physics is privileged, because such a bend is a literal collapse of existence itself. In other words, it is a descent in the vertical dimension. As noted above, since the cause of gravity is energy, this is a more vertical phenomenon where there is holarchy. A relationship between holons and present, which would now represent the space-time of general relativity. Then, as in the post-physical space-time holons there is a high degree of order because they are at a great vertical distance from nothingness (the largest scales), this concentration of energy or time due to very massive physical objects disturbs its container by increasing the degree of chaos of those spacetime holons so ordered in that zone. Therefore, the restriction of time forces such space-time holons to bend, sinking such massive physical objects to place them at a vertical distance according to their degree of chaos. That is, at a smaller vertical distance or closer to nothingness (toward smaller scales). A descent in the holarchy that allows the surrounding physical objects to access lower holons and presents, which perfectly explains this change in their trajectories and in the acceleration of these objects as they fall, since they access smaller spacetime holons with shorter and therefore faster presents.

In conclusion, both the time dilations and the curvature of space-time itself (decrease of the vertical dimension) are due to the same cause, the restriction imposed by holistic presentism. Thanks to this, we have obtained a theory of quantum gravity, because let us remember that Darwinian Physics is the first authentic quantum theory. All of this is due to the recognition of a fundamental element that current physics has always tried to ignore: the emergence of time. In this way, we can see even more clearly the mistake that theoretical physicists have made with quantum gravity, because gravity is really an emergent phenomenon, like everything else in Darwinian Physics. But what emerges on the largest scales of the cosmos is not space-time itself, but its curvature, its sinking into smaller space-time. Therefore, Einstein's space-time does not need us because it does not emerge from physical reality. We need time and space to be, not the other way around. This is why we can observe gravitational phenomena that are completely independent of physical reality, such as the expansion of the universe, black holes (extremely curved empty space-time), and most importantly, gravitational waves, which are ripples of space-time that propagate freely through the universe without interacting with matter. Thus, there is much evidence that space-time is not only real, an ontological object, but that it is something independent and more fundamental than anything physical. This again confirms the conservative nature of Darwinian Physics.

Then, once it is clear how general relativity works under the new paradigm, we can do cosmology, which is the application of this theory to the whole cosmos to discover the shape or geometry of the whole universe. To do this, it is necessary to make an assumption. Since we cannot see the entire universe, the simplest assumption is that all matter and energy content is evenly distributed throughout the universe. This assumption is known as the cosmological principle, which simply says that no place in the universe is privileged over another because the universe is equal in all directions. This is why it is also known as the Copernican Principle, after the Polish astronomer Nicolaus Copernicus (1473-1543), who took us out of the center of the universe by replacing Ptolemy's geocentric model with a heliocentric one. Now history is repeating itself, but on a cosmological scale, because neither our sun is the center of our galaxy (this position belongs to the supermassive black hole Sagittarius A, with a mass of four million suns), nor our galaxy is the center of the universe, since all galaxies are moving away from each other. A principle that, although a philosophical position, is a scientific fact, at least in the observable universe (90 billion light years in diameter), confirmed with great precision in the CMB. Thanks to this principle, we can reduce the infinite possibilities of the shape of the universe (square, triangle, star, dog, and so on) to only three options: a sphere, a saddle, and a flat universe.

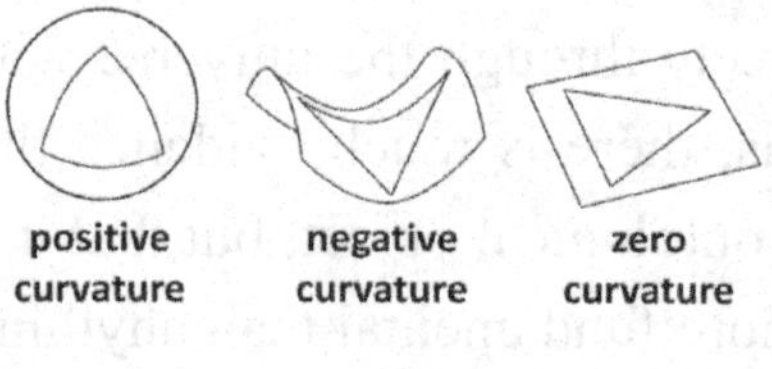

Which of these three depends on the type and amount of energy contained in the universe. Since Einstein's space-time has a life of its own and can expand despite being empty, cosmologists call the critical density the amount of energy the universe must have to compensate for the behavior of space-time in order to obtain zero curvature. Thus, if the amount of energy is equal to the critical density, then the universe would have a flat geometry; it would extend in all directions without end, just as parallels lines always remain parallel in a Euclidean geometry. That is, it is infinite. On the other hand, if the energy density is less than the critical one, the curvature is negative, producing a hyperbolic or open geometry, where the parallels diverge and separate more and more forever. That is, it is also infinite. The only finite geometry is that of positive curvature, where the energy density is greater than critical. Here the parallels lines converge, cross each other, and are a closed geometry. It is therefore the most realistic geometry of the universe, avoiding all those absurdities and paradoxes regarding infinity that we saw in Chapter V.

But it is not only the finiteness that makes the spherical geometry the most likely geometry for the universe, but also the many problems with the other two options. For example, planar geometry requires incredible fine-tuning, requiring the exact amount of matter and energy to match the critical density; because any deviation from this would either close or open the universe. So, it is not a very stable or likely scenario. Hyperbolic geometry, on the other hand, exists only in the Platonic world of mathematics; because although a graphical representation has just been presented as an example, the classic saddle seat, it is impossible to faithfully insert a two-dimensional space of constant negative curvature into a three-dimensional Euclidean space. Therefore, the saddle is only an analogy. Moreover, it has many exotic properties that make one seriously doubt its feasibility, for example, at a point where a parallel passes, an infinite number of parallels can also pass, or if you throw a stone very far away, it can come back and hit you in the face. But the most categorical indication that it is not a realistic geometry is that our universe would require a negative cosmological constant (the energy of the vacuum itself) ($\Lambda < 0$). However, we have long known that the cosmological constant we measure is positive ($\Lambda > 0$), regardless of its exact value (cosmological constant tension). This is why the 2011 Nobel Prize in Physics was awarded to the discovery that the universe is expanding at an accelerated rate. Therefore, even though it is completely excluded, this geometry is still considered only because it is a simpler geometry without a singularity. In this way, physicists can do cosmology, quote, unquote, without having to deal with the problem of the origin or singularity of the Big Bang. This is why it is so popular, for example, in the conjecture: AdS = CFT. As we saw in Chapter X, where the AdS stands for anti de Sitter, and corresponds to a hyperbolic geometry with a negative cosmological constant; which allows them to play at doing gravity in an imaginary universe that has no relation to ours, because they simply cannot do realistic gravity.

On the other hand, Darwinian Physics could produce a realistic quantum gravity applied to our universe, which would have a positive cosmological constant and would also be finite, or with closed geometry. This is because the radial dimension of the sphere used as an analogy would have a real, physical referent representing the temporal dimension. Recall that the first pilot model of Darwinian Physics can predict the dimensionality of our universe. In this sense, the cosmos would be a sphere, but a very special sphere, because this radial or internal dimension would not be another dimension equivalent to the two dimensions that form the surface of the sphere; it would be a fractal dimension, which we have called the vertical dimension. The sum of all holons. This means that at the center of this cosmic sphere there would literally be nothingness. This makes a lot of sense for general relativity, since this vertical dimension would represent the temporal dimension, since it is where the present emerges from the nothingness. Also, remember that this dimension is where holons emerge, so to say that the entire universe began at a singularity, a point we call the Big Bang, sounds very much like the emergence of ever-larger holons from the spacetime point of nothingness. Exactly what would explain the fact that our universe has been expanding since the Big Bang. This can be understood very well by a very popular analogy that has been used in popularization to explain the expansion of the universe, that of an inflating balloon. In this way, our universe would be the surface of the balloon, and you can see how each point on the surface, where the galaxies would be, moves away from each other as the balloon fills with air.

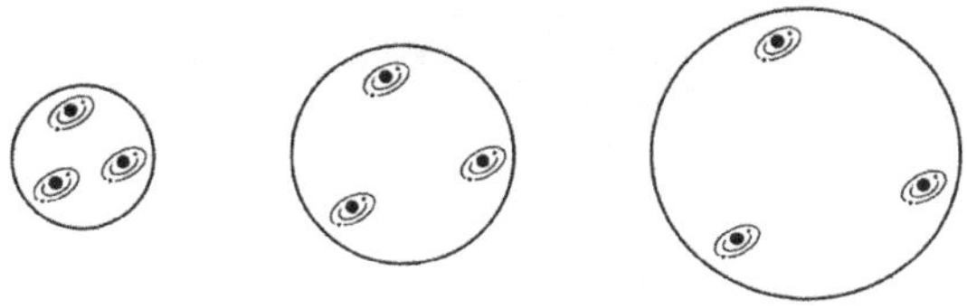

Well, this, which is only an analogy for Parmenian physics, would literally be what happens to the universe according to Darwinian Physics, since while in current physics the interior of the globe has no physical meaning (it is only part of the analogy), in this new paradigm the radial or vertical dimension is essential, since it leads us to nothingness. The center of the globe is exactly what we call the universe. Therefore, in Darwinian Physics, this analogy of the balloon is very representative, since the nothingness would be the lung that pumps time into the cosmos with each breath, materialized in holons that expand and order the entire universe.

Then, in this globe universe, the radial or vertical dimension would function as the temporal dimension, in the context of holistic presentism; since it points inward, it marks the direction of the past, when the universe was a smaller globe. A finite past, because it ends at the center, in the singularity of the Big Bang or the nothingness. Whereas when it points outward from the radial dimension, it marks the direction to the future. Exactly where the universe is expanding. So, we would be on the surface of the globe, in our holon, representing our present. While the inside of the globe would correspond to the past and the outside would represent the future. Exactly where the universe is expanding due to dark energy.

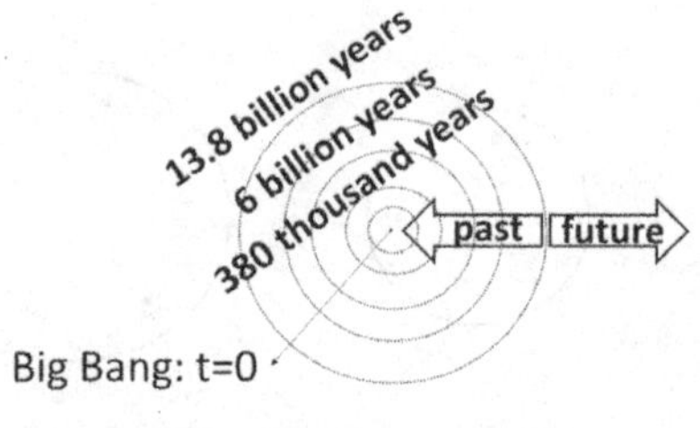

This is very consistent with what is actually observed in astronomy, because in every direction we look in the universe, we are pointing at the center of the sphere, the big bang. The reason for this can be better understood with a sphere of lower dimensionality. Imagine that our universe is the two-dimensional surface of a three-dimensional sphere. At every point in that universe, we would be pointing toward the center of that sphere. In fact, this is exactly what happens with the Earth, because if we were to dig a hole at any point on the surface of the planet, whether it be in Chile or Mongolia, we would end up at the Earth's core. The same principle applies with one dimension less, in a one-dimensional universe, which would be the edge of a circle.

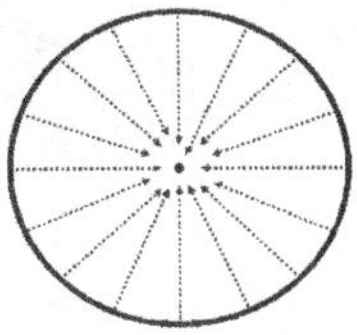

That all points in space point to the same center is a general property of all spheres, including our cosmic globe. This means that if we were looking at any point in our universe, we would be looking at the center of the cosmos, at the singularity from which the whole universe originated, the Big Bang. So, the farther we look in any direction, the farther we go into the center of the cosmos, the closer we get to the beginning of the universe. And this is exactly what happens when we point our telescopes at any point in the sky, because the speed of light is not infinite. So, when we see the moon, for example, we see it as it was a little more than a second ago, because that is the time it takes for the light coming from the moon to reach our eyes. We see the Sun as it was 8 minutes ago, Mars as it was 20 minutes ago, Alpha Centauri as it was 4 years ago, the Andromeda galaxy as it was 2.5 million years ago, and so on. In short, the farther we see, the farther back in time we go. That is why we can see the CMB, something that happened 13.5 billion years ago. As far back as we can observe. So, in all directions that we point to the big bang, we just have to look as far back as we can. Something that is completely coherent in general relativity, because in Einstein's theory, all the geodesics, the timelines of the universe converge or end at the Big Bang, because that is where they came from. So, in general relativity, a singularity is inevitable.

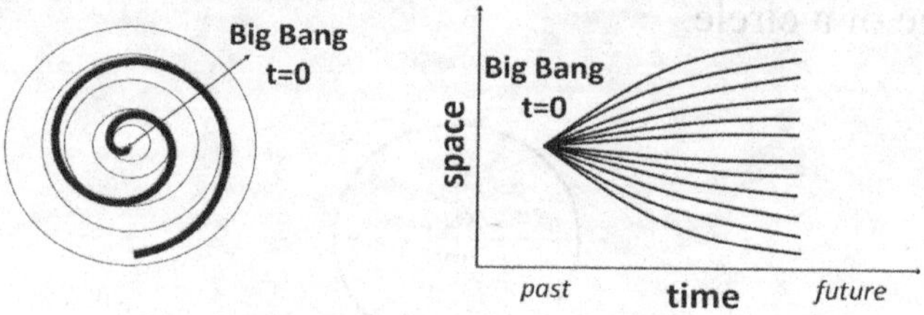

The lines in the two figures above would be the geodesics, or paths of light, and general relativity says that they all converge to a singularity in the past, which we have called the Big Bang. In the first figure, the spiral line represents a geodesic that falls irrevocably toward the singularity in the center. It represents whatever direction we point in the sky, which always ends at the Big Bang. While the figure on the right shows the same, but for all geodesics that our universe would have, all directions and paths, which also end at the same point in the past, when t = 0.

This convergence of geodesics towards the past is very easy to understand if our universe is a sphere, since it would simply be the growth or expansion of that sphere from its center. So far, this is not so different from current cosmology, although the latter preaches a flat and infinite universe; however, the most radical thing in the new paradigm will be how this radial dimension is interpreted. This is because the dimension of time will no longer have an eternalist interpretation, being equivalent to another spatial dimension (the fourth dimension); it will have to be interpreted in a presentist way, so that the radial dimension, which marks the expansion of our universe as a sphere, would actually be a fractal dimension. That is why this radial dimension is mathematically represented in general relativity by the scale factor, which increases with time, which is perfectly consistent with holistic evolution. In fact, this geometry is much better understood in Darwinian Physics because it is equivalent to the formation of a holarchy in this vertical dimension; which explains why this radial time dimension is so different from the horizontal spatial dimensions. In the same way that we can dig a hole at any point on the two-dimensional surface of the Earth to descend to the core of the Earth, we can descend vertically at any point in our universe to the core of our cosmos, which is nothingness. The space-time point or singularity that pumps time into the cosmos. So at any point in space we can descend to the nothingness because time arrives everywhere. This means that the nothingness is at the center of the sun, at the center of Jupiter, at the center of the earth, but also at the center of your hand. Everywhere in the universe. Just go to smaller and smaller scales everywhere. This is the great discovery or revelation of Darwinian Physics, that the nothingness and the big bang singularity are exactly the same thing. Not only because they are both points in space-time, but because time is born or originates in both. The only difference is that in the Parmenian interpretation of general relativity, the Big Bang is in the distant past, about 13.8 billion years ago; whereas in Darwinian Physics, the Big Bang is right here, in the now, because, remember, it is a presentist theory. So, there is no past, there is no distant future, there is only the present.

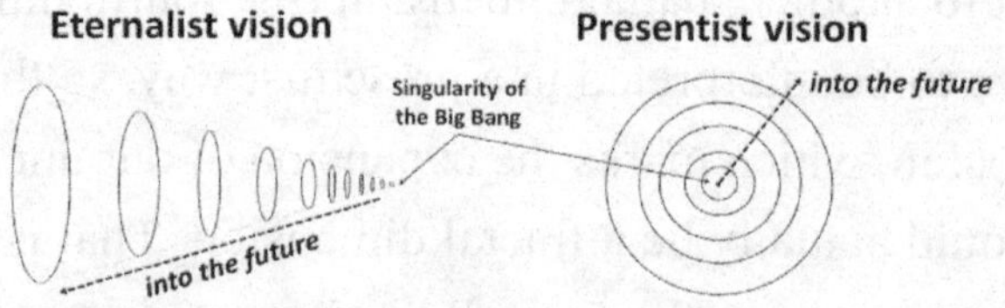

This means that, contrary to Parmenian physics, according to Darwinian Physics, there was not a single Big Bang in the universe, but countless Big Bangnes. But do not be alarmed. We are not talking about a multiverse, because these Big Bangnes are happening in this universe, because every moment is actually a Big Bang. Remember that according to relativity, the center of the cosmos is the big bang, the same place where nothingness would be in Darwinian Physics. Therefore, every new present created by nothingness is actually a big bang, which is where time originates. Something that particle physicists have been proving for decades without knowing it, because particle accelerators like the LHC would be true drills of the cosmos, digging directly towards its center, towards nothingness or the Big Bang. This is because the smaller the scale you want to explore (discovering smaller and smaller particles), the higher the energies you need to generate. That is why colliders are getting bigger and bigger. In fact, this is how particle physicists interpret the results of particle colliders as a way to explore the first moments of the universe after the Big Bang, since they would reproduce the same conditions of very high temperature that existed in the early universe, when it was much smaller and hotter than today. The only difference with the Darwinian interpretation of physics is that the comparison is again taken literally, rather than just as an analogy. In other words, it is not that you are simulating the conditions after the big bang in particle accelerators, but you are really exploring the moments after the big bang, because the big bang did not happen only once 13.8 billion years ago in a Platonic-Pythagorean past, but it happens at every instant, because it creates the new now, it creates the flow of time. So, the particle accelerators are actually observing the conditions after the big bang, because the big bang is actually in the present. It's just that during the book we called it the nothingness or the space-time point. So, to call particle accelerators drills is quite accurate, because as they go down the vertical dimension to smaller and smaller scales, it is equivalent to penetrating and accessing deeper and deeper and older sediments or geological layers, getting closer and closer to the origin of time, the singularity of the Big Bang, or as we say here, to nothingness.

Then, according to the new paradigm, every present would really be a Big Bang; for if the latter is the origin of time, then before it there must have been nothingness, "no time," a singularity. Thus, the passage of time is the fruit of this series of Big Bangnes emerging from nothingness. This is why it has been said that it is a completely different worldview from the current one, because it reinterprets general relativity in a completely different way. Remember that determinism or the block universe actually represents a single moment (as explained in Chapter XI). A timeless statue. So, it makes perfect sense to describe it in terms of a single big bang, because that is where everything came into being. All information, all energy, even the end of the universe. But if the universe evolves and grows gradually, in each instant, organically, then it makes much more sense to conceive of each instant, each new present, as an authentic Big Bang in itself. In this way, the emergence ex nihilo could be understood as a process of crystallization, where each new big bang (present) allows the creation and conservation of regularities or patterns, both spatial (things) and temporal (laws). Allowing this cosmic globe to grow or expand layer by layer, holon by holon, again as conceived in general relativity under a closed geometry. Thus, our universe would be composed of a series of concentric holons, similar to an onion, like smaller and smaller spheres that reach into nothingness, to the singularity of the Big Bang.

The key to the transition to the new paradigm, then, is a more correct interpretation of the temporal dimension, which can no longer be a horizontal dimension equivalent to the spatial, as it is currently interpreted under eternalism. For example, if the radial dimension were another traditional spatial dimension, it would be very difficult to conceive of a finite or closed universe, because the curvature of our universe is too small. This means that if our universe were really the surface of a sphere, the radial dimension would have to be monstrously large. For example, if we discover in the future that the curvature of the universe is 0.25%, then the diameter of the entire universe would be 250 times larger than the diameter of the observable universe, which we already know to be 93 billion light years. This means that the diameter of the entire universe would be more than 23 trillion light-years away, so the center of the cosmos would be more than 11.5 trillion light-years away (the radius). A lot of space, which in today's Parmenian physics would serve no purpose except to record the entire past, every moment. The problem is that it should also contain all the future, so not only the 23 trillion light years inside would be real, but also the unimaginable trillion of trillion of trillion of light years outside the sphere into the infinite future. A block universe without limits.

Another reason that this radial dimension would be very problematic if it were horizontal is that the center of the cosmos would have to be nothingness, so does that mean that nothingness is also 11.5 trillion light-years away? Fortunately, no. Here we must consider the dimension of time (radial or vertical) as a completely different dimension from the horizontal spatial dimensions. Remember that nothingness is the smallest thing, that which is at the tip of your finger or nose, so it makes no sense to measure the radial or vertical dimension in light years or meters. The right thing to do here is to measure it by the scale factor, which in particle physics is equivalent to the energy scale. That is why you need cosmic drills, particle accelerators, to explore smaller and smaller scales like the size of an atom or the size of a proton. And this is exactly what reveals how far away nothingness is, because you might think that nothingness is very close, after all it is the smallest thing, a space-time point, but nothingness is incredibly far away. Only this vertical distance is measured in energy, which has a non-linear but exponential behavior, as we saw with the energy that particles with mass acquire as they get closer and closer to the speed of light. So, again using the LHC as an example, the world's largest particle accelerator, with a diameter of 27 kilometers, can only study proton-sized scales, for which it needs to generate energies of 10^4 GeV (gigaelectronvolts). But where there is nothingness, energies a thousand trillion times higher would be needed. That is, about 10^{19} GeV (Planck scale). This means that we would have to build a particle accelerator of the diameter of the entire galaxy. An impossible feat not only technically, because even if it were possible to build it, it would collapse into a black hole when reaching those energies. Therefore, it is now quite clear how far away nothingness is. So, the vertical dimension works perfectly as the radial dimension of our cosmic globe, which is a very special dimension that at the same time confirms that time is very different from space, although they are closely related.

Having understood what the geometry of the universe would be like, we can now better understand what general relativity means in Darwinian Physics. The key is to distinguish this radial dimension from the horizontal spatial dimensions by interpreting it in a presentist and fractal way. For example, this vertical dimension would explain very well why the curvature of time is so much greater than the curvature of space in general relativity, because gravity would be mainly the product of sinking or falling in the vertical dimension. It would also explain why all quantum fields are at the same point in space, because they are all together horizontally. However, they are clearly separated vertically because each quantum field is on a different energy scale. Therefore, it is much easier to excite an electron in the electron field than it is to excite a Higgs boson in the Higgs field, because it requires less energy or is at a smaller vertical distance from us. Even the vertical dimension could explain why renormalization (under dimensional regularization) works. A technique that makes the Standard Model of Particles (or part of it) useful by adding a fractal dimension to avoid infinities in its calculations. A mathematical trick that nobody knows why it works, and that Darwinian Physics would justify, since it would work through the vertical dimension. In fact, Darwinian Physics would explain the reason why the calculations diverge towards infinity at all, since by considering smaller and smaller scales, it simply approaches nothingness, the singularity, by descending in the vertical dimension.

Here we can see more clearly that Darwinian Physics is an authentic theory of quantum gravity, unifying these two pillars of physics in a single coherent cosmological model that conceives of the universe as a sphere that expands radially, just as a balloon is inflated. A vertical dimension that corresponds to the true dimension of time. It was only necessary to understand time in a presentist way, not in an eternalist way. In this way one can understand the reason why Einstein's gravity is a phenomenon belonging to the highest space-time holons (post-physical macro) and therefore with the highest degree of order in the cosmos, being the farthest from nothingness. Therefore, it is a fairly continuous space-time, due to the great accumulation of holons at these higher scales, and is so certain and predictable due to the high degree of causal inertia. Particles (quantum fields), on the other hand, are located in the lower space-time holons (post-physical micro) and are thus closer to nothingness. This explains the quantification and randomness characteristic of quantum mechanics. Therefore, the vertical dimension plays a key role in the unification, explaining the great differences between the two pillars of physics, which are due to different scales or vertical distance. All that was necessary to make this connection was to recognize the becoming of time and to conceive of it as a dimension truly distinct from the space dimensions.

In this sense, it can be said that if Einstein revolutionized Newton's gravity by replacing its Euclidean or flat geometry with Riemannian or curved geometry, the new paradigm would revolutionize Einstein's gravity by incorporating fractal geometry in the vertical dimension. A very special dimension which, to be rigorous, is semi-fractal because it cannot be perfectly fractal since nature is clearly not scale-invariant (conformal). The universe does not look or behave in the same way at the largest scales as it does at the smallest scales. This is the difference between general relativity and quantum mechanics. However, there is a fractal quality that is conserved throughout the vertical dimension because it is the product of the same underlying mechanism, holistic evolution. However, the fact that it is semi-fractal in no way diminishes its peculiarity and relevance, because nature has never had a perfectly non-fractal geometry, because perfect circles or triangles have never been observed. There are only approximations, because those perfect and idyllic geometric figures only exist in the imaginary world of mathematics. However, no one has ever complained, so no one should complain that the vertical dimension is not perfectly fractal. In the real world, nothing is perfect because it is not absolute.

But before we close this chapter, we cannot fail to mention two of the most important cosmological problems that are closely related to gravity: black holes and dark matter. Let's start with the first one, the astronomical object that remains after a very massive star, more than 4 times the mass of our Sun, runs out of fuel, having burned most of its hydrogen. This is where the gravitational force begins to win, which does not decrease with time, because now we know that it is time itself (inertia). And time never stops. Then, when the nuclear energy pushing the star outward stops, gravity finally takes over and collapses the star. In this way, the outer layers of the star shoot out in one of the most powerful explosions in existence, a supernova, whose brightness can eclipse an entire galaxy. The core, however, cannot escape gravity and rapidly contracts into a stellar black hole. However, it is a big mistake to think of these monsters as something physical or material, with a surface or something like that, because it is pure empty space-time, but extremely curved. This is because the core of the star has contracted to a singularity. This is because nothing physical can resist gravity, not even the self-repulsion of matter or Pauli's exclusion principle. The essential characteristic of fermions is that one particle of matter cannot occupy the same space as another. Therefore, the electrons in an atom must be in different orbits to create structure. Now, this repulsive force is not broken. The problem is that in these very large stars, this principle works against them because the repulsion is so strong that it generates even more gravity. Remember that all kinds of energy bend space-time. Therefore, to some extent, no matter how strong the force resisting gravity is, it will also generate even more gravity. In fact, the already formed black hole does not even need this stellar matter, because gravity itself is gravitational and is also energy. In other words, after a certain threshold, the curvature of space-time is self-sustaining. Therefore, black holes are not physical objects, but pure space-time phenomena. Just extremely curved empty space. Literally a hole in space-time. This is why Darwinian Physics can understand them, because it has the holarchy, the radial or vertical dimension mentioned above. In this sense, black holes are very similar to particle accelerators, only they are much more powerful and efficient drills, because they really manage to penetrate the whole of existence, layer by layer, holon by holon, all the way to its core. And since black holes have the maximum possible gravity, they also have the maximum possible collapse, piercing all of existence until they reach the nothingness itself. Such a deep perforation of the cosmos explains the most peculiar feature of black holes, the event horizon. The boundary that separates the interior of the black hole from the rest of the universe. It is a very important boundary, because the objects that cross it cannot get out again, because they would have to reach a speed higher than the speed of light. Which we know is impossible, but this time not because of dogma, but because you cannot go faster than time itself. That is why black holes are black, because not even light can escape from them. An essential characteristic that makes more sense in Darwinian Physics, since the most concrete explanation for the blackness of the interior of black holes is that simply from there on, physical reality no longer exists. In other words, beyond the event horizon, even pre-physical spacetime holons would be reached. That is why it is black, because light as a manifestation of time in physical reality can no longer exist there, because there is no physical reality.

Again, this is supported by general relativity, because in its mathematics the singularity of the black hole is not interpreted as a location in space, but as a moment in time, specifically in the future. Thus, when the event horizon is crossed, falling towards the singularity is as inevitable as tomorrow. In other words, it seems that space is transformed into time inside the black hole, which again makes sense in Darwinian Physics, because what makes physical reality special is movement, displacement through space. However, inside a black hole we find a pre-physical space-time reality, where existence is fundamentally temporal, not spatial, since there are only holons, not local existences. That is why the name Black Hole is very appropriate, since they would be authentic open wounds on the surface of the cosmos, revealing this stratification of existence in its radial dimension, penetrating not only to ever lower post-physical space-time holons, but even to the pre-physical spacetime holons, to nothingness itself. This is why current physics no longer works beyond the event horizon, because there is simply no physical reality beyond it.

All of this reinforces the idea in Darwinian Physics that the Big Bang is happening right now, because just as there was a singularity at the beginning of the universe, there would be a singularity in black holes. In fact, mathematically, they are the same singularity. This is why many physicists also call the big bang a white hole, because in general relativity the two are completely equivalent, but with the variable of time reversed. In a black hole, nothing can get out, while in a big bang, nothing can get in. A difference that is explained in Darwinian Physics, because in a Big Bang the cosmos expands, so time points or flows outward, upward in the vertical dimension. Whereas in a black hole, due to gravity or the extreme curvature of space-time, the same time that ascends from the vertical dimension now descends, flows into nothingness. In other words, time loops in black holes, so the big bang singularity, where time begins, and the black hole singularity, where time ends, would be exactly the same thing, nothingness. Closing the loop. So, in black hole mathematics the singularity is always in the future, whereas in big bang mathematics the singularity is always in the past. Two extremes meet because they would be the same in a presentist theory: nothingness. This simply explains the reason why billions of singularities exist today at the centers of black holes, which are mathematically identical to the singularity that must have existed at the Big Bang 13.8 billion years ago, because black holes are literally holes in the cosmos. Therefore, just as if we dig several very deep holes in the surface of the earth, they all end up in the same place, the core of the earth, so all black holes end up in the same place, the center of the cosmos, the nothingness. The origin and the end of time.

Meanwhile, the other great gravitational mystery of Parmenian physics is the enigmatic Dark Matter, a supposedly invisible matter that no experiment has been able to detect, although it has been intensively searched for with large detectors since the 1980s. Despite this, the majority of physicists and astronomers still believe in its existence. Only the theoretical candidates have become more and more exotic and convoluted to match the lack of experimental evidence. A classic modus operandi in today's theoretical physics. So, it would be an exotic and invisible matter that would account for no more and no less than 80% of all matter in the universe. The first hint of this suspicion came in the 1930s, when the Swiss astronomer and physicist of Bulgarian origin Fritz Zwicky (1898-1974), observing clusters of galaxies (groups of hundreds to thousands of galaxies), noticed that the galaxies were moving at a speed much greater than the gravity that could be generated by such galaxies to keep them together. Therefore, he suspected that there must be much more matter than is observed. He called this invisible matter dark matter. However, he believed that it was normal but opaque matter, such as dust, planets, or very faint stars.

Much later, in the late 1970s, the American astronomer Vera Rubin observed the same problem, but in the galaxies themselves, where the stars farthest from the center orbited at the same speed as the stars closer to the center. This is contrary to what is known about gravity, which, like light, weakens with distance. It is not the same to see the brightness of a flashlight in front of your face as it is 10 meters away. That is why, in the solar system, the planets farther from the sun orbit more slowly than those closer, but on a galactic scale this principle seems to be violated. And since in the Galaxy it is much more difficult to suppose that there is five times more dust and opaque objects that we do not see, because obviously we would have detected them with our high-tech telescopes that can see a wide electromagnetic spectrum (from gamma, X-rays, visible light, infrared, radio, and so on); it has been suggested that there must be a new type of matter that interacts very weakly with light (hence we do not see it) and even with itself, since there is no evidence that it forms structures, since no one has ever been killed by an invisible dark matter asteroid, let alone seen collisions of planets and stars with equivalent but invisible objects made of dark matter. So, this Dark Matter must form a gigantic bubble of dust that surrounds all galaxies, explaining the high velocity of the stars farthest from the galactic center, because in reality they would still be in the center, but this time of a gigantic bubble of Dark Matter. Something that seemed very crazy at first, but was reinforced as more evidence was added in favor of gravitational lensing (distortion of space-time) being stronger than the amount of normal matter observed. And even signs of this amount of extra matter have been found in the CMB itself, which was formed in the early universe.

This is the evidence for dark matter, but the evidence against it is almost completely omitted from the release. For example, dark matter predicts a maximum density in the nuclei of small galaxies, whereas observations show a rather flat distribution. Dark matter also predicts many more small satellite galaxies around large galaxies. For example, our Milky

Way galaxy should have hundreds, but it actually has only a few dozen. Moreover, these small satellite galaxies are usually aligned with the galactic disk. This does not explain the dark matter that is supposed to be distributed around the galaxy that surrounds it. We also know from observations that the mass of a galaxy is correlated with the fourth power of the rotational velocity of its outermost stars. This is known as the Tully-Fisher effect. An observational fact that cannot be explained by the dark matter hypothesis. And there are many more such things, like Renzo's rule; the correlation between luminosity and rotation curves; the collisions of clusters of galaxies at high speed, like the Bullet Cluster, and so on. Not to mention that no experiment has yet detected a dark matter particle, although it has been searched for decades.

In view of the above, many propose to modify gravity, but not Einstein's gravity, but Newton's gravity. An option called MOND (modified Newtonian dynamics), which does not fit with all observations either, especially in galaxy clusters and in the CMB. Something very curious, because the same situation was repeated in the 19th century, in an amazingly similar way with respect to the precession anomaly in the orbit of Mercury, which could not be explained by Newton's gravity either. This also led to the postulation of dark matter. In this case, a new dark planet, which they called Vulcan, which theoretically had not been observed because it would be very close to the Sun, but whose gravity would explain this precession in Mercury. However, since this supposed new planet had not yet been found, in spite of all the astronomical efforts made, it was postulated that Newton's equations would also have to be modified. For example, the "2" in r^2 of his formula was changed. Then, to match what was observed in Mercury's orbit, that "2" had to be "2.00000016," which solved the precession. The new problem is that precession also occurs in Venus, Earth, and Mars, but to a lesser extent. So, this little modification worked beautifully for Mercury, but failed miserably for the other planets. Something very similar to the current MOND, which works perfectly for galaxies, but fails miserably for clusters and other observations. This is the problem with ad hoc theories, which are tailor-made for a particular problem, but end up making it worse or creating more problems in general. The Vulcan hypothesis, on the other hand, fully respected Newton's theory of gravity, but the supposed new planet was still unobserved.

And as we all know, the real solution to the problem of mercury precession was not to add something new or modify Newton's theory, but to invent a new theory of gravity. In this case, general relativity. Perhaps we need to do the same with the problem of galaxy rotation, because just as it happened with Newton's gravity, which failed when gravity was very intense, as it was in the case of Mercury being very close to the Sun, Einstein's gravity may fail on larger cosmological scales, on the scales of galaxies and clusters of galaxies. This is exactly what Darwinian Physics postulates by reinventing gravity. Thus, by integrating the vertical dimension, talking about stars, galaxies, and galaxy clusters means talking about different scales, so it is logical in Darwinian Physics that gravitational behavior can change. Then, in the same way that new arrangements (holons) emerge to allow the formation of larger and larger physical structures through vertical holistic evolution, such as: strong nuclear interaction, which is the order that allows the formation of atomic nuclei, electromagnetic interaction, which is the order that allows the formation of atoms and molecules, and Einstein's gravity, which is the order that allows the formation of planets and stars, it is reasonable to think that on larger scales a new order, a stronger gravity, is needed to allow the formation of galaxies and galaxy clusters. In fact, Darwinian Physics would explain this difference or tension that exists between observations of galaxy rotations and galaxy clusters (why dark matter works better in clusters while MOND works better in galaxies) because they are simply different orders, because they are at different scales or vertical distances. Therefore, what physicists call dark matter would be just another gravitational (space-time) phenomenon that allows the formation of even larger structures (galaxies and galaxy clusters) to continue the VHE of physical reality. Explanation of why no dark matter particles have been detected.

CHAPTER XV

THE PROBLEM OF ORIGIN: BIG BANG AND COSMIC INFLATION

With an integrated cosmological model that unifies the two pillars of current physics, we can begin to answer the biggest and most important questions about the universe. And what better place to start than with the most difficult problem of current Parmenian physics, the origin of the universe. A problem that, ironically, has already been answered in the second part of the book, because Darwinian Physics starts from there, from nothingness. But now it will be much clearer how, especially after the previous chapter. Thus, the origin of the universe actually starts from a singularity, and from there it expands, with each new holon emerging thanks to holistic evolution, resulting in this cosmic globe we call the universe. In contrast, the current big bang theory says absolutely nothing about the bang. About what caused it. All it refers to is the observational fact that the Universe is expanding, due to the redshift of the light from galaxies (which reveals that they are moving away from each other), along with the discovery of the Cosmic Microwave Background or CMB (which proves that long ago the Universe was much denser and hotter than it is today). So, in the past, the universe was getting smaller and smaller. Something that follows from the equations of general relativity itself, but which Einstein, being a faithful Parmenian, refused to believe in the first place. So, both fact and theory support the central idea that the universe started at one point, at a singularity, and has been expanding ever since.

So far so good, except that there are a number of observations that do not fit with what the Big Bang theory predicts, the most serious flaws being the horizon and flatness problems. The first is the incredible homogeneity of the temperature of the CMB, which varies by only one part in 100,000 (0.00001°K) from an average of 2.7°K (-270°C). An extremely small difference, even though it is the temperature between very distant points, from one end of the observable universe to the other. And this is a big problem, literally, because our observable universe has a diameter of 93 billion light-years. But the entire universe is only 13.8 billion years old. So clearly there has not been enough time for heat or light to travel from one end to the other to equalize the temperature. In the same way, it takes a long time for a stove or heater to heat the entire room to an even temperature. You may think that this is not possible today, but the universe was much smaller in the past. True, but there was also less time. For example, if we go back to the time when the observable universe was only the size of the solar system, that was just a nanosecond after the big bang (10^{-9}s). The problem is that light takes 8 hours to travel that distance. Clearly much more than the available nanosecond. And if we go back even further, the universe will be even smaller, but there would be even less time, so again the light at the ends of the universe would not have had time to communicate. In fact, there would be more than 10^{40} regions of the observable universe that could not have been in contact with each other, yet they all turned out to be very similar in temperature. Quite a coincidence. In other words, the universe was born too big for light to travel through and equalize the temperature. And anyway, when we look at the early universe, it already has an incredibly homogeneous temperature, so how was that possible?

The other problem with the big bang theory is flatness, and this is due to the fact that matter and energy curve space-time, creating the phenomenon we know as gravity. So, one would expect that at the beginning of the universe, when all matter and energy was so concentrated, it would be very curved. However, as we saw in the previous chapter, cosmological measurements made by the Max Planck Space Telescope indicate an extremely flat universe. This is because if our universe were curved, the curvature would have to be less than 1%. This does not make sense according to current cosmology, because the curvature increases with time. This is very important because if the current curvature of the universe is only 0.4%, for example, it would mean that the curvature at the time of the Big Bang must have been 10^{62} times smaller. That is, a 1 followed by 62 zeros is less curved than 0.4%. This is the equivalent of throwing a bowling ball in a straight line and having it deviate only 0.4% from its direction after rolling a light-year distance. Obviously insane, so there must have been a mechanism that flattened the universe in such a brutal way. And all this without mentioning the biggest problem of the big bang theory, its origin: What or who caused the bang of the big bang, what was there before, why was the entropy so low, and so on.

To solve these problems, except for the third one of course, current physics proposes a patch to the big bang theory called cosmic inflation. A mechanism consisting of a scalar quantum field that came into action very soon, only about 10^{-35} seconds (a 0, followed by 34 zeros and one second) after the Big Bang. And it lasted a similar time, 10^{-30} seconds (one 0 followed by 29 zeros and one second). But despite its ridiculous brevity, an unimaginably short moment, it had equally unimaginable consequences, expanding the entire universe some 10^{30} times its size (a 1 followed by 30 zeros). The equivalent of expanding a proton to the size of our sun. With this monstrous and fleeting expansion, physicists believe they have solved the horizon problem, because the pre-inflation universe was small enough for light to have time to travel through it and homogenize the temperature before it stretched. And it solves the flatness problem, because by inflating the universe so gigantically, we now see it as flat, just as we see the Earth as flat only because it is too big a sphere to see its curvature from the surface.

Despite this supposed success in correcting the flaws of the Big Bang theory (except for how the universe began), cosmic inflation does not avoid bringing its own problems, the most important of which is that it has no hard observational evidence to support it. This is because such a bestial expansion should have left large scars in the form of primordial gravitational waves, which should be observable in the CMB through the detection of B-modes (polarization of light). However, this signature of inflation is completely absent in the most detailed observations of the CMB. This is very serious, because the theory of cosmic inflation is obviously an ad hoc theory, since it is tailor-made to solve very specific problems of the Big Bang theory. Therefore, it was very important that it predicted something that was not expected, that had not been observed before, and that existed only as a necessary product of that mechanism. And that was the primordial gravitational waves. But as soon as that signal was not observed, theoretical physicists began to design an avalanche of new versions that would produce weaker primordial gravitational waves to justify the fact that they had not been observed. In other words, the theory of cosmic inflation ceased to be scientific, at least in Popperian terms, because it is not falsifiable; because it is not even a theory, but a theoretical framework for building as many inflationary models as one wants. Therefore, like string theory, it is not surprising that it is not falsifiable, because there is no theory to falsify, but a set of models that are created and discarded as new data emerge. This is a very dangerous tendency because, impressively, there is a large majority of renowned and important astronomers, cosmologists and very influential theoretical physicists who are strongly pushing for the Nobel Prize in Physics to be awarded to this theory. This would be ominous because it would be the officialization of the decline of theoretical physics as a serious and reliable science.

In view of the above, if we cannot falsify cosmic inflation by observation (empirical consistency), we will have to do so according to the new definition of science that we have given in this book, evaluating it according to the criteria of scientificity of internal and holistic consistency. In this sense, the biggest problem of the inflationary model would be its incredible instability and convenience, since it emerged and lasted for ridiculously short times, as we have recently seen. So why did the universe expand more than a million trillion trillion trillion times in less than a trillionth of a trillionth of a trillionth of a trillionth of a second, why so briefly and intensely, why did it disappear immediately afterwards, why was it so timely? The conditions for such a cosmic inflation to occur are so special that the probability of it happening as needed is incredibly low. Just to consider one parameter of all those that must be provided in inflationary models, the one that reproduces the distribution of matter in the universe on a large scale, requires a fine-tuning with an error of less than 15 decimal places (10^{-15}), as the famous British physicist and Nobel Prize winner Roger Penrose demonstrated in the late 1980s. However, the requirements do not end here, because cosmic inflation cannot account for all the energy in the universe, since the inflationary field requires some energy to activate and generate the expansion of a space-time bubble. And how much energy is required to create an entire universe? 10 kilos. Right, 10 kilos to create all the galaxies, stars and planets that exist. It may not seem like much, especially compared to the entire contents of the universe, but in the LHC, the heaviest particle created from the vacuum weighs only 10^{-20} grams. So, for 10 kilograms (a million trillion times more) of energy to spontaneously appear is still literally an act of God. For all these requirements and more, the chances of inflation having occurred as necessary to give us the universe we observe are close to one in a googol (10^{100}).

Faced with this absurd probability, the inflationists resorted to something equally absurd to justify it, the multiverse. Because it is a quantum mechanism, it is impossible to stop, because it is a spontaneous phenomenon, just like radioactivity. So, if it happened once, it must happen again and again, with the same probability. So, it makes no sense that it happened only once, about 10^{-35} seconds after the big bang, and then mysteriously disappeared forever in such a timely manner. Then the inflationary physicists explained that it is indeed so, that the inflationary mechanism acts in a constant and random way like any other quantum phenomenon, but all those countless times it acts again and again, we do not see them in this universe, because they occur in other universes. This is how the modern versions of these models, called chaotic or eternal cosmic inflation, were born, by turning a gigantic flaw into a supposed virtue, because they justify this negligible probability of having exactly the cosmic inflation, we need by appealing to the anthropic principle, which (applied to cosmology) says that: any valid description of our universe must be consistent with our existence. In other words, any cosmological theory must allow for the existence of intelligent beings who wonder why the universe is the way it is; otherwise, there would be no one to ask such a question. So, using this principle, inflationary physicists say that cosmic inflation was so special in our universe because there are infinite cosmic inflations that create infinite parallel universes, the vast majority of which are inhospitable or incredibly short universes that collapse immediately afterward. In this way, it was statistically inevitable that a universe with the conditions and values of the constants we observe would appear; because only in these universes could life and intelligent beings have evolved, wondering about the reason for these constants and initial conditions so particular. Similarly, it was by chance that life evolved on this planet, thanks to the existence of a large number of planetary systems. Not because this planet was special, but because it is one of so many billions of planets. So, we are simply the winners of the multiverse lottery. We only have to make the small concession of accepting the existence of infinite parallel universes.

As discussed in the first part of the book, this wild card of the anthropic principle has its limits because it cannot be used indefinitely in a legitimate way. For example, it is true that the search for a fundamental explanation of why the planet Earth is special makes no sense, since it is explained statistically. There are simply so many planets that one of them had to be at the ideal distance from its star, have the necessary size to maintain an atmosphere, have a Jupiter to protect it from comets, have a moon to stabilize the planet's rotation, and so on and so forth. However, to continue this argument ad infinitum is completely absurd, because in the same way that it is legitimate to say that our universe is not special because there is a multiverse, it would be equally legitimate to say that our multiverse should not be special because there is a multi-multiverse. And in the same way, it would also be valid to argue that the multi-multiverse would not be special because there is a multi-multi-multiverse, and so on. Especially since physicists have already invented many kinds of multiverses. For example, there is the multiverse where the laws are the same, but the initial conditions change. Another multiverse where the values of the constants also change. But we can even follow Max Tegmark's multi-multi-multi-multiverse, where every conceivable mathematical structure is a universe. Therefore, if we validate the use of the anthropic principle in the case of inflation, then we must also accept the existence of Max Tegmark's mathematical multiverse and all conceivable multiverses to infinity. A situation not unlike the multiverse of Rick and Morty (an animated series). So, we have two choices: accept that there are multiple types of infinite multiverses; or apply Ockham's razor and recognize that it is only legitimate to apply the anthropic principle to what we can observe and be sure exists, the universe.

Many problems and assumptions could be forgiven, if at least the eternal cosmic inflation would say something about the real problem of the Big Bang, its origin, what caused it. However, even its infinite multiverse cannot solve this problem, because its eternity only applies to the future. This is because the inflationary field can expand small regions of space-time by inflating infinite universes from now on, but it cannot do the same indefinitely into the past. This is because it cannot violate the postulates of general relativity, which require the existence of a singularity in the past. This conclusion is based on a theorem formulated by the original author of the theory of cosmic inflation himself, the American physicist and cosmologist Alan Harvey Guth (born in 1947), together with two other colleagues: Arvind Borde and Alex Vilenkin (hence this theorem is known as BGV, after the initials of their last names). In this theorem, they show that, according to the principles of general relativity, any universe that has been expanding throughout its history cannot be infinite in the past, but must have a space-time boundary in that direction. In other words, eternal inflation is actually not eternal, because it cannot be infinite in the past, because it cannot eliminate the singularity that must have existed before the inflation acted, and therefore cannot avoid the question of what caused the bang in the big bang theory. A question it obviously cannot answer. So, they wanted to eliminate the question by eliminating the singularity in the past, by extending the laws eternally. However, they themselves have shown that the problem of the horizontal origin is unavoidable.

In fact, unbelievably, cosmic inflation exacerbates this problem because, due to its quantum nature, it would have wiped out all pre-inflation information, leaving only its fluctuations. So, we cannot know what happened before the inflation. A period of time censored by physics, which becomes longer and more important precisely because of the absence of these primordial gravitational waves. The excuse given to justify the fact that these gravitational waves have not yet been observed is that they must have been weaker than expected. This means that the universe was colder when inflation was at work. So, the longer the detection of these primordial gravitational waves is delayed, the weaker they must have been, and the colder the universe must have been when inflation occurred. This means that there was more time between the Big Bang and the inflationary phase. Thus, according to the latest observations, which have not found the B modes predicted by inflation, we know that the universe could not have been hotter than 10^{15} GeV, which means that we cannot go back more than 10^{-35} seconds. A time equivalent to the entire observable universe being 1.5 meters across. This is the maximum we can go back with respect to the Big Bang, because the longer the detection of these primordial gravitational waves is delayed, the larger the universe would have been when cosmic inflation began. Therefore, the inflation theory would be further and further away from the origin, from the singularity (10^{-43}s and 10^{-35}m), and therefore further and further away from explaining the cause of the Big Bang.

In conclusion, the theory of cosmic inflation is just a set of phenomenological models about the conditions moments after the Big Bang, introduced ad hoc to explain the flatness and homogeneity of the universe, which the Big Bang theory cannot explain. Therefore, it is not an explanation of the beginning of the universe, on the contrary, it makes it impossible to go back to the moment of the big bang. So obviously it cannot explain where space-time comes from, because it needs it, like any quantum field. Let us remember that inflation is also a quantum field, and as such it is located in space-time. A space-time that has already integrated all the other quantum fields, because it does not create them, of course, since that would be heresy, since it would be equivalent to creating the laws of physics. Therefore, inflation does not explain the origin of the universe, nor the origin of space-time, nor the origin of quantum fields, nor the origin of laws, nor does it require infinite parallel universes. Therefore, cosmic inflation is just a very unsatisfactory and very convenient conjecture that, far from improving the Big Bang theory, makes it worse, because not only does it not address the incredibly special conditions of low entropy that must have existed at the beginning, at "t = 0;" but it adds its own fine adjustments, making something already very improbable even more improbable.

This is a radically different situation from that of Darwinian Physics, because it is not a simple patch to the Big Bang theory, but a reinvention of the theory that directly addresses the problem of the origin of the universe. It does not avoid it. In fact, this is how the development of the theory itself began. A holistic genesis, already described in chapters VIII and IX, which takes seriously the only possible beginning of everything, nothingness. Something unimaginable for today's physics. But although it is a very different theory, it is very consistent with the evidence available about the first stages of the universe; and especially about its singular beginning, which is a result of Einstein's equations of gravity, which physicists consider a flaw in the theory, but which in Darwinian Physics would have a real meaning, being nothingness itself. So, far from being a problem or a drawback, the BGV theorem would be another support for this proposal, since it would justify the nihilistic origin of the cosmos. A space-time point that is inevitable in Einstein's theory, where everything would emerge and expand through the emergence of ever larger concentric holons through holistic evolution. A birth and evolution of the universe that is not only much more economical and satisfactory, since everything is created from nothingness, but also more organic than the mechanical expansion described in the current Big Bang theory, proposing a much more complex and progressive growth, like all Darwinian evolution.

This is how one begins to solve the third question that current physics can only ignore and deny: what caused the bang in the big bang? Without resorting to nonsensical assumptions like the multiverse. And all this without neglecting the other two problems that cosmic inflation addresses. For example, the flatness problem in Darwinian Physics is solved by integrating holistic evolution into general relativity. This reverses the evolution of curvature because, contrary to what is expected in cosmology, existence does not become more curved with time. On the contrary. It becomes flatter. Just as when we blow up a balloon, its curvature does not increase but decreases as it gets bigger and bigger. Again, just as we see the earth as flat compared to, say, an apple. And what is inflating the universe is just dark energy, the kind of energy that is becoming more and more abundant in the universe, and that in general relativity subtracts the positive curvature that creates matter, which becomes more and more dilute as the universe gets bigger and bigger. In addition, it must be remembered that the curvature of the entire universe occurs with respect to the vertical dimension, so the calculation could be different. In any case, the universe should be tending toward a flat or Euclidean geometry, which is fully consistent with the increase in the degree of order due to holistic evolution. That is why the measurements of the Max Planck Space Telescope show that our universe is very flat, because it confirms that it is very old, or ordered, because it has evolved holistically enough to be so. Something that is reinforced by what was said in the previous chapter, the fact that the outermost layers of existence or the largest spacetime holons are being observed by measuring the farthest cosmological distances possible. So, it makes sense that because of their size, they would be the flattest or most ordered holons, especially compared to our reality, which is much smaller.

So we have an incredibly flat universe due to its high degree of holistic evolution; however, it cannot be perfectly flat because the universe cannot be infinite, as many physicists would like it to be, in order not to have to deal with a nothingness beyond it. Ignoring that an infinite universe would be much worse, because of all the absurd consequences we saw in Chapter V, the most reasonable thing to do is to simply assume that our universe is an unimaginably large sphere. A scenario that is not only more reasonable, but also much more likely, since perfect flatness would have to be with infinite decimal zeros. An absolute zero. Instead, no matter how negligible the curvature, the universe would eventually have to close like any circle or sphere. This means that our universe would be circumnavigable. Therefore, if we traveled through it in a straight line in any direction (horizontally), after enough time we would end up at the same place we started the journey. Similar to what would happen if we traveled around the planet in the same direction. An obviously hypothetical situation, since it would be impossible, since the maximum speed "c" does not allow us to circumnavigate such a vast and, above all, expanding universe. However, it is relaxing to think that in this paradigm there cannot be an "I Hitler" beyond the one who won the Second World War, since the universe is not infinite.

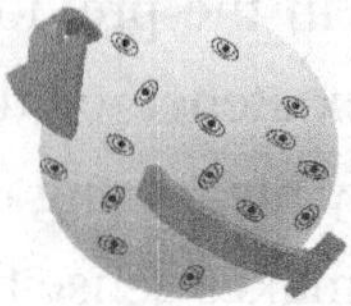

But the question remains. If the universe is finite in that it has a positive curvature, then what is beyond it, where does the universe expand to? The answer given by Darwinian Physics is literally nothing. Recall from the previous chapter that wherever we look, we are pointing toward nothingness. If we look at the largest scales of the universe, as far out as we can, we are going back in time, looking at the first galaxies, the first stars, to the first light, the CMB. Something that happened 380 thousand years ago after the big bang. As far back as we can see, with light. But if we could see even further, we would still be colliding with the big bang, which is nothingness. And vice versa, if we look at smaller and smaller scales, through particle accelerators, or even better, through black holes, exactly the same thing happens, we end up back in the nothingness, where new big bangs are created. So, whether we look farther away or closer, we run into nothingness. The only difference is that upward the journey is longer, the geodesic spiral winds around the nothingness several times as it points toward the first Big Bang.

On the other hand, the journey down is more direct, towards the last big bang, the most recent present.

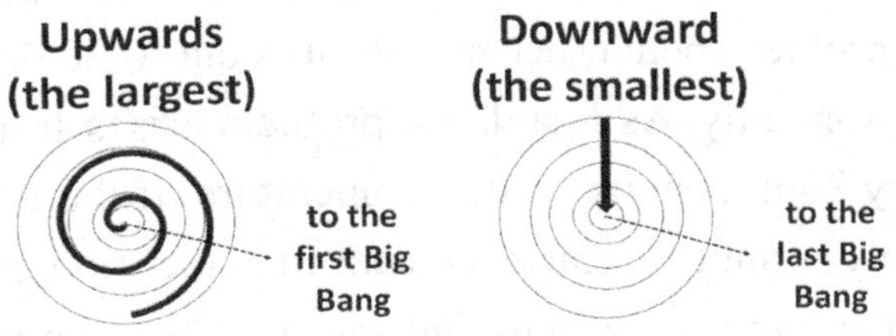

In this way, it is affirmed that the form of everything that exists is closed, since there is nothing beyond. There is only the universe and nothingness, literally. Again, this saves us an infinite amount of space if the geometry were flat or open, in addition to the infinite moment of eternity and the infinite parallel universes. So, in Darwinian Physics, there is nothingness beyond the cosmos, because there cannot be anything smaller than nothingness itself, the spacetime point or singularity; nor can there be anything larger than the universe itself, because there is no future, only the present. This means that there is only what there is, a cosmic globe of a certain size in the now, and it will continue to expand outward, toward a future that has not yet materialized into a new present. Therefore, since there is no future, there is nothing beyond. So, you cannot go beyond the universe, because there is no place beyond, because there is no more space; there is no more time. So, it is wrong to think of the universe as a bubble in a vast background, because there is no such background. There is just nothing beyond everything. The original definition of the word universe as the totality of all that exists has been rescued.

As for the second problem of the Big Bang theory, that of the horizon, Darwinian Physics would solve it by integrating vertical causality. Exactly, the radial dimension of the cosmos. To understand how this solution works, one must also understand the problem better, because it turns out to be deeper than it seems, ironically. As I said, the problem arises from the extraordinary homogeneity of the temperature of the CMB. This is extremely strange, because it means that, thermodynamically, the universe had a very high entropy from the beginning, because it was in thermal equilibrium. Something that makes no sense, since it means that in the early infancy of the universe, it already seemed extremely old. It would be the cosmological equivalent of The Curious Case of Benjamin Button (2008 American film). So, what physicists would expect to see is very large temperature differences, that corners of the universe are much hotter than others. This would be a state of low entropy, like the example of the freshly made latte machiato, where the coffee and milk are clearly separated. In contrast, what we see at the beginning of the universe is the opposite, almost total uniformity. Therefore, what the CMB reveals to us, in entropic terms, would be to see the future instead of the past. This thermodynamic paradox is resolved again in Darwinian Physics by adding time, because what is really strange to physicists is how such distant parts of the universe can be at the same temperature. It seems that they are communicating, although this is not possible because the speed of light forbids such a distant connection, being much slower than the expansion and size of the universe. However, physicists forget that there is an instantaneous natural phenomenon at any distance, quantum entanglement. Therefore, in Darwinian Physics, the horizon problem is solved by showing that every point in space, no matter how far apart they are, is entangled. In other words, the homogeneity of the temperature of the universe is actually a vertically caused phenomenon. This is confirmed by the fact that the variations in temperature are completely random. The letter of the representation of vertical causality.

So the problem was very well named because, ironically, it arises from thinking purely horizontally. If the only causality considered is horizontal, then the homogeneity of the CMB is obviously meaningless, because you would be using presents to connect very distant points in space. A distance that far exceeds the speed of light, the shortest or fastest possible present. On the other hand, if we include vertical causality and realize that time comes from below, from nothing, it should not be surprising that it manifests simultaneously throughout the cosmos in a single instant. In fact, this was the norm in the holistic evolution of the primitive universe, in its pre-physical stage, since it functioned practically with pure vertical causality, since there were no local existences (physical reality), but only holons. Therefore, the present manifested itself globally in the whole holon, through those quantum fluctuations (randomness), as any spontaneous phenomenon, because it came directly from nothingness.

Moreover, vertical causality allows us to do something impossible for cosmic inflation, to go back to the Big Bang. Let us recall that, due to the absence of primordial gravitational waves, inflation cannot go back further than 10^{-35}s, a time that increases as the detection of these gravitational waves in the CMB is delayed. Therefore, from this time to the Big Bang singularity (10^{-43}s) is, by definition, unknowable to current physics. An impossibility that does not apply to Darwinian Physics, because it starts from the singularity itself, from nothingness. This means that it can explain the initial conditions of the universe without the need for inflation, because although it seems to be an unimaginably fleeting stage compared to our scale (holons and much larger presents), in that period of time many things happened because many space-time holons emerged. Therefore, there was a great evolution of the pre-physical spacetime reality, which is the reality responsible for generating the cosmological conditions that inflation supposedly solves. This flatness and homogeneity, in addition to these random fluctuations in the CMB. Even holistic presentism and its vertical causality is so effective in solving the horizon problem that it reverses the question: why should it be different, why should there be large temperature differences at all, who created them, where did they come from? Questions that Darwinian Physics avoids when explaining the CMB, because it would be the proof of the nihilistic origin of the cosmos. That is why the primordial B-mode or gravitational waves predicted by cosmic inflation are not observed. In fact, Darwinian Physics predicts just the opposite, that there is nothing else to be found, no additional sign, pattern or regularity that physicists predict and desperately search for: magnetic monopoles (Grand Unification Theory or GUT), cosmic strings, Hawking's rings (Roger Penrose's Conformal Cyclic Cosmology or CCC), and so on. A total lack of evidence that frustrates physicists a lot, but that would be the proof of the creatio ex nihilo of the universe predicted by Darwinian Physics. So, in this case, the absence of evidence is proof of nothingness, ironically because, first, no one expected it, and second, as we saw above, it was certain even today that much more would be found in the CMB. But despite their great hopes and efforts, observations with the most precise instruments have yielded nothingness.

CHAPTER XVI

The Problem at the End: The Dark Energy

It is not only important to know how it all began, but also how it will end. This question arose with the big bang theory, since it was previously believed that the universe was eternal, like God. However, since it was discovered that the universe had a beginning, it is now relevant to ask if there will be an end. Thus, after Edwin Hubble's observational confirmation of the expansion of the universe in 1929, the most natural thing to do was to believe that this expansion must be slowing down due to the gravitational pull. It is as if the Big Bang were the equivalent of throwing a ball into the air. The normal thing is to expect the ball to slow down as it goes up, due to the action of the Earth's gravity, until the ball slows down completely and then falls toward the Earth. The same thing should happen to the universe, it would slow down its expansion after the initial impulse of the big bang, due to the gravity of all the galaxies, until a moment comes when it will contract and collapse back to the same starting point. In the hope that a rebound will be generated that will produce a new big bang, and so on. Thus, the first cosmological model of modern physics was cyclical, where there would be an eternal series of Big Bangnes, followed by a Big Crunch, and then a Big Bounce. A very simple and reasonable model, similar to the swing of a pendulum, which would be in eternal motion if there were no friction, as would be the case with the universe itself, which is all there is. The only theoretical problem was figuring out how to generate this bounce after the whole universe collapsed into a singularity. But except for that detail, which would have to be solved, many physicists were convinced that it had to be this way. That's why all the astronomical effort was to measure how much the expansion was slowing down. When the Big Crunch would happen.

So now you can understand the great shock in the physics community when they discovered in 1998 that the expansion of the universe was actually accelerating. That there was something pushing harder and harder on the universe, an unknown kind of energy, the famous dark energy. Since then, the most likely final scenario of the cosmos has been the Big Freeze, assuming that this dark energy is constant, the famous cosmological constant (Λ), as specified in the current Standard Model of cosmology, or ΛCDM. Under these circumstances, the death of the universe would be thermal, in the sense that it would be a very, very slow end, in fact eternal, until it cools down infinitely. Like a campfire where only ashes are left to cool slowly through the night. Another scenario is now falling apart, because the latest observations show that the cosmological constant is most likely not constant, since it would increase with time. Therefore, the most likely end of the universe so far would be the Big Rip, where the expansion of the universe would accelerate faster and faster, not only pushing the galaxies apart, but also eventually pulling the stars within the galaxies apart, as the dark energy in the void between the stars would now be strong enough to overcome the gravity between them. But as dark energy continued to increase, it would eventually overcome the gravity between the planets themselves, the electromagnetism that binds molecules and atoms, and the strong nuclear force that binds the quarks in the nucleus of the atom. In other words, dark energy will become so intense that it will rip all matter apart, leaving behind an increasingly dilute soup of fundamental particles.

So, everything indicates that the ΛCDM cosmological model faces many challenges, either because of the cosmological constant (Λ), the cold dark matter (CDM), the flat universe it assumes, and not to forget that it needs the inflation mechanism to solve the problems of the Big Bang. A mechanism that we have already seen in the previous chapter and that is not without problems and controversies. However, a characteristic of not only this cosmological model, but any other model based on general relativity, is that they are not really eternalistic models, as we would expect from a Parmenian paradigm with absolute laws, but they are actually sempiternal. That is, they are eternal in the future, but finite in the past. A mathematical truth that is supported by the BGV theorem we saw in the previous chapter, and that not even chaotic or eternal cosmic inflation can escape. And this is very strange, because modern cosmology puts us at an incredibly special time; for we would be only 13.8 billion years after the Big Bang. That sounds like a lot, but it is only about three times as old as our planet is now. An insignificant amount of time compared to the eternity that will be in the future if the universe continues to expand indefinitely. And although our existence in this time would be justified by the anthropic principle, since we are in the time when most stars are born and therefore there is more probability for life to emerge, this does not change the fact that we are in an incredibly special time, for why was there a big bang only 13.8 billion years ago, if the future is eternal, why has the universe recently begun to expand if it will expand forever? One could say that the cosmological or Copernican principle is violated, but not spatially as traditionally used in current cosmology; but this time temporally, because we would be in an incredibly privileged moment, very close to the beginning of time, even though there is an eternity in the future. The temporal equivalent of being very close to the center of the universe. So, according to current cosmology, we are not special in space, but we are special in time; because the universe has just awakened from a sleep that was supposed to be eternal. All because of the ever-insoluble problem of the origin of the current paradigm.

Obviously, Darwinian Physics proposes a different cosmology and a different ending because, as always, it directly addresses the mystery or the problem. It does not avoid it. And in this case, the mystery of dark energy. And as it is already a tradition in the book, what solves it is the ingredient that modern physics lacks, time. Thus, once again, an extravagant hypothesis is abandoned, such as a negative energy that has anti-gravitational effects and is intrinsic to the vacuum for its own sake. Instead, dark energy is interpreted in Darwinian Physics as traditional energy because it is caused by the same thing, time. However, its consequence is quite different from what we are used to in physical reality because it is how time manifests in our mother reality, the spacetime holon. So, it is not that it is essentially different because it has the same cause, time; but its effect is quite different because it corresponds to a different main reality. An expansion of the universe, which can be well understood by the analogy of the cosmic globe used in the chapter on gravity, and the Big Bangnes, which pump out nothingness in the form of new presents, serve not only to make our clocks tick-tock, but also to expand the universe itself. In other words, they allow us to continue the holistic evolution by accumulating more information that makes the whole cosmos expand from its vertical radial dimension by adding new layers or holons to existence.

Remember that the latter tries to become more and more existential, to distance itself vertically more and more from the nothingness; and for this it has to age ($+\Delta$ dt), but also to grow, to become bigger ($+\Delta$ dr). So, it makes a lot of sense in Darwinian Physics that on the largest scales, at the distances separating the most distant galaxies, more space is being created. Which is exactly what dark energy would do. An expansion that is homogeneous and isotropic because everything is moving away from everything else. Again, like the balloon analogy, where each point on its surface moves away from the others as the balloon inflates, creating the illusion that each point perceives itself to be the center of the expansion. In this way, the Copernican Principle is fulfilled in Darwinian Physics, but applied to a finite universe; because the only privileged point of reference would be nothingness, the nucleus of the cosmos. In the horizontal spatial dimensions in which we live, there would be no special places, just as there are no special points on the surface of the globe. This new worldview explains the expansion of the universe much better, because it reveals its origin, which is time itself. A very weak expansion by our standards, since every second one kilometer of space expands by the equivalent of one proton. However, this is to be expected at these scales, since dark energy corresponds to the growth or holistic evolution of larger, more ordered spacetime holons. Therefore, it is normal for these changes to occur at this rate, just as we find the changes that occur at geological scales on this planet to be very slow.

So far from being a great mystery, the expansion of the universe is something that follows naturally in Darwinian Physics. In fact, this great tension between the measurements of the cosmological constant (Λ) made in the CMB (early universe), which gives 67.4 km/s/Mpc, and the Λ calculated by measuring the galaxies (recent universe) of 73.04 km/s/Mpc; would also be another confirmation of the integral growth of the cosmos, since a dark energy that becomes stronger and stronger is the most expected in Darwinian Physics. Let us remember that in this paradigm there was not a single impulse, a single big bang, but many. One at a time. Therefore, the universe is constantly being pushed, because what is pushing is time itself, due to holistic presentism. Moreover, exponential expansion is quite natural in any Darwinian evolution. For example, biological evolution also increases rapidly, since more than half of the entire history of life on Earth was single-celled, mainly bacteria and archaea. But in the last 500 million years, in the so-called Cambrian explosion, virtually all the complex multicellular life that has ever existed arose. The same is true of cultural evolution, because for the last 300,000 years of human history, more than 95% of us have been simple hunter-gatherers. But the really interesting stuff started in the last 15,000 years, with the invention of agriculture and writing. And from there, the most interesting thing has happened in the last 200 years, with the industrial revolutions; where humanity went from 1.5 billion people to almost 9 billion people. So Darwinian evolutions have this exponential behavior. So, it is not surprising that the expansion of the universe is accelerating.

But what happens next, will it expand forever like in the Big Rip? Does this mean that the cosmological model of Darwinian Physics is also eternal? To find out, we need to know the continuation of the history of the cosmos, how the universe continues to evolve. To do this, we must return to the subject of information in Chapter XII, because we said there that information is related to the formation of increasingly complex structures. Which in turn is related to the second law of thermodynamics and its constant increase of entropy. Which, as we know, is equivalent to the degree of order or stability of the universe for Darwinian Physics. Then, as we saw in that chapter, complexity, or the production of new information, will have a maximum and then decay as long as entropy (degree of order) continues to increase. Remember that complexity lies in the middle between low and high entropy, since at these extremes there is only simplicity. Then, just as there is a mechanism that creates new information, in this case the present itself, through the collapse of Ψ (creation of factual quantum information), there would also be a mechanism that destroys information (from now on we will simply speak of information, as a synonym for factual quantum information), for example through Hawking radiation and the evaporation of black holes. This is a result that current theoretical physics tries to reverse at all costs, but which would be most natural in Darwinian Physics.

For in the same way that complexity (new properties) emerges and is created in nature, it is also true that things tend to decay (lose properties). In fact, this is the most common thing, due to the second law of thermodynamics. This is why we age and die, while things decay and deteriorate. Therefore, under the Darwinian definition of information as a simple property, it makes sense for information to be destroyed because it is normal for things to lose their properties. A new car will function and have a well-defined color and shape, while the old car is the one that has significantly lost its function, color, and shape. Therefore, the loss of properties is also the loss of information. A very normal process, which physicists accept quite well, but only when it corresponds to the information of very high or macroscopic physical things; but which they religiously deny that it happens in the case of the properties of things they consider fundamental, absolute and eternal, such as the particles or quantum fields of the Standard Model.

In Darwinian Physics, on the other hand, everything can decay, not only things composed of fundamental particles, but also the fundamental particles themselves. If everything came from nothing, then everything can return to nothingness. Thus, in a very, very distant future, not only will black holes destroy the information of macroscopic physical things, but the very atoms and particles that are thought to be eternal, such as the electron, will also decay and become photons. In this way, just as the information of the higher physical holons (galaxies, stars, planets, mountains, a television, and so on) can disappear and be destroyed, so too can the information or properties of the lower physical holons be destroyed because they are still higher realities of other more fundamental realities. In this case, space-time holons. In other words, all physical reality is metastable and therefore subject to decay. A possibility that is even considered by current physics itself, since in quantum field mechanics there is the possibility that the Higgs field, which is responsible for giving mass to particles (forming the first photon bubbles) and thus allowing structures and things to form, decays to a lower energy level. The so called "true vacuum." All physics as we know it disappears in the process. This is why the stability of the Higgs field is one of the most important problems in theoretical physics.

However, the end of physical reality is not a problem for Darwinian Physics. In fact, it is a prediction, because in this new paradigm it allows not only the creation of new information, new properties, but also its destruction, the extinction of information. We already know the process that creates the former, called Vertical Holistic Evolution (VHE); but the process that follows is exactly the opposite and is called Vertical Holistic Devolution (VHD). This is a very late stage in the life cycle of the universe, as more information is being created than is being destroyed. However, this process of constantly creating new realities and information cannot go on forever, because there is no such thing as infinity or eternity when there is nothingness. So, the VHE cannot go on forever and reach infinite complexity. At some point, the balance must change, with more information being destroyed than created. This second stage of the cosmos would consist of the extinction or decay of holons. And it occurs in the reverse order of their creation. In other words, it begins with the elimination of the last realities to emerge. From the higher and more complex holons to the lower and simpler ones.

This, which seems to contradict everything said in the book, is actually an inevitable consequence of the ordering process itself. The difference is that in the VHE stage, existence tolerated the emergence of more complex realities because it helped it to gain more order, to grow rapidly. Especially in the beginning, when existence had a very high level of chaos, having just emerged from nothingness. However, at a much more mature stage of its evolution, much more orderly and stable, existence can do without these higher holons. In fact, not only does it not need them, but it needs to get rid of them in order to further increase its level of order. This can only be achieved by eliminating the higher holons, since they contain high degrees of chaos, since they are created by means of holistic equilibrium. In this way, by eliminating the more complex realities, the Cosmos can continue to increase its degree of purity to get even closer to that 100% order ($P(H) = 1$). This is why the higher realities must eventually disappear, because existence must rid itself of the degree of chaos they contain. Therefore, if the Cosmos wants to continue to increase its level of order, to become more existential until it reaches total uniformity, it has to get rid of these complex realities. This is relatively easy, since the highest holons are also the most unstable, precisely because they are the most complex and therefore have a greater accumulation of holistic equilibria (combination of chaos and order).

Then, much later in the history of the universe, after a long period of VHE, an even longer period will begin, the VHD, in which more information will begin to be erased than is created. Thus, if we consider the classification of holons described in Chapter X, the first reality to disappear would be the main symbolic holon, civilizations, or the world of ideas. This should not surprise us, since we would not expect our civilization, with its particular way of life and its constant political and religious conflicts, to be very sustainable. In fact, we will be lucky if the current level of civilization does not decline by the next century. Then this main reality would sooner or later die out (hopefully later), leaving only its mother reality, or the wild life forms, the main living holon; which is a much more stable or ordered reality in comparison. Not for nothing has it existed on the planet for almost 4 billion years, compared to 5 thousand years of human civilization. However, from this main reality, the higher holons will also become extinct. The more complex species would follow, from large mammals to amphibians, reptiles, jellyfish, insects, and plants, until only bacteria and other simpler forms of life would remain. But even these forms of life will become extinct as our sun expands and becomes a red giant. Bringing biochemistry back to chemistry down to the most basic realities in the main physical holon. A physical reality that will become more and more simple, dull and orderly as the stars exhaust their fuel by exploding and/or dying out until everything is swallowed by black holes. Something that will happen in at least one googol (10^{100}) years. A dark epoch where there will be only black holes, orbiting and merging, evaporating very slowly for the next trillions of trillions of trillions of trillions of years due to Hawking radiation. But because black holes radiate more and more intensely as they lose mass and size, they will eventually shine like a star and explode as a supernova. This means that they will produce enough energy to create massive particles and even atoms, thanks to Einstein's equation ($E = mc^2$). Therefore, after a long time of being swallowed up by black holes, there will be physical things again. So far, Darwinian Physics agrees with current physics, but the former predicts that this decay process will continue because all matter is unstable in the very long term, including protons and electrons. Therefore, they will also decay, but it has never been observed, and we will probably never observe it, because their half-life (their degree of order) is incredibly long. But they are not eternal. Therefore, far, far in the future, some $(10^{100})^{100}$ years, it is expected that even electrons will eventually decay into photons. A light that will also become more ordered, decaying into less and less energetic photons, as the universe continues its accelerated expansion, stretching the wavelengths of photons, as it is doing now. Therefore, light will become colder and fainter as it travels through an increasingly empty and vast universe, until existence reaches its desired absolute darkness. The end of physical reality.

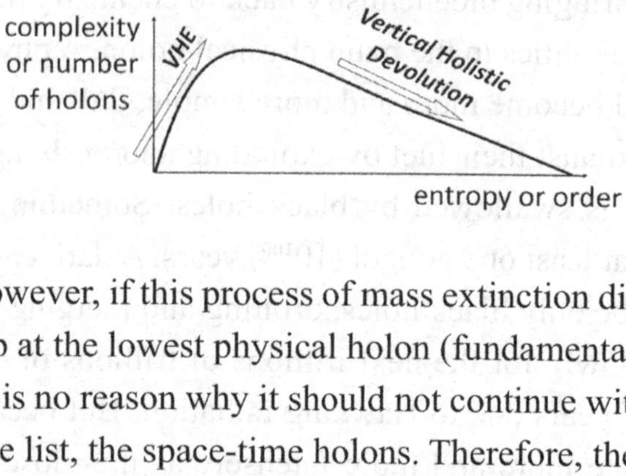

However, if this process of mass extinction did not arbitrarily stop at the lowest physical holon (fundamental particle), then there is no reason why it should not continue with the next ones on the list, the space-time holons. Therefore, these holons will also eventually decay, disintegrating layer by layer until they reach the first holon where it all began. By then, all higher realities will have disappeared to allow H_1 (first holon) to reach its goal. Pure or absolute order, a $P(H) = 1$. This simply means that the space-time point will repeat the same state it has been in since the beginning of existence. Only this time it will do so with perfect certainty, because although its probability of repeating the same state has gradually increased from the 50% it had at the beginning, it has never reached 100%. It had only approached it, always with a 0.99999...9. A behavior that may sound strange, but is quite normal in physics. In fact, it happens in both directions, for example with 0°K (absolute zero), which cannot be reached, but can always be approached, such as the 0.000000000038°K achieved with Bose-Einstein condensates by a team of German researchers in 2021. Another example is the speed of mass particles achieved at the LHC, which is 99.99999999% of the speed of light. Speeds that are even higher in cosmic rays, but none reach the speed of light. They are only infinitesimally close to "c." Well, in the very, very distant future of the universe, it finally reaches P(H) = 1 (100%); because in an anti-absolute physics, it is also an anti-infinite physics, so the sequence of nine decimal places cannot go on forever. It has to stop at some point, hence the need for this constant g in the pilot model. Moreover, as has happened in the interpretation of many worlds, it makes no sense to work with probabilities with infinite possibilities.

Contrary to what one might expect, however, this absolute certainty will not last forever. On the contrary. It will be incredibly ephemeral, because it will last only a moment, the shortest possible moment, a fundamental present or Planck time. This is because immediately after that moment in which one first has the absolute certainty that the same state will be repeated again, in the next present the certainty will begin to diminish again, after a whole existence in which the certainty and the degree of order have only increased. Having again a P(H) = 0.999999999...9. In other words, after a completely ordered existence, with exactly 0% probability of being nothing, the probability of the other state of nothingness appearing in the immediately following instant increases again. Something that will happen again in the next and the next moment. This reappearance of chaos after absolute order has been achieved may sound completely contradictory, but it is an inevitable consequence of the presentism of Darwinian Physics. Remember that presences are real and fundamentally independent, arising from nothingness. The same is true of flipping a coin several times. The fact that it has come up heads 10 or 20 times in a row does not make it more likely that the next flip will come up tails, because it still has a 50% chance of doing so. The same thing happens, but this time in a fundamental way, with the quantum properties of particles, such as the spin of the electron being 50% up or down. Then, the absolute certainty obtained when "t = g" refers exclusively to this moment, to this present; because the probabilities are independent, due to presentism. Therefore, at that moment, the probability of becoming nothingness was exactly 0%; however, after that moment, the probability of the sequence of successive repetitions being interrupted increases again. Then, after the cosmos has reached the supposed eternity (t = g), it simply goes to the other side, where this time the degree of nothingness increases again, instead of increasing the degree of order, as it had been up to "g."

This is the result that can be deduced from the pilot model of Darwinian Physics presented in the second part of the book, because the only constraint we had placed on causal inertia was that time be positive, because a negative time makes no sense, just as a negative temperature, speed, or distance makes no sense. Then "g" is just the radius of the circumference formula used to represent this causal inertia. Therefore, the true limit of time would be the diameter, i.e. "2g."

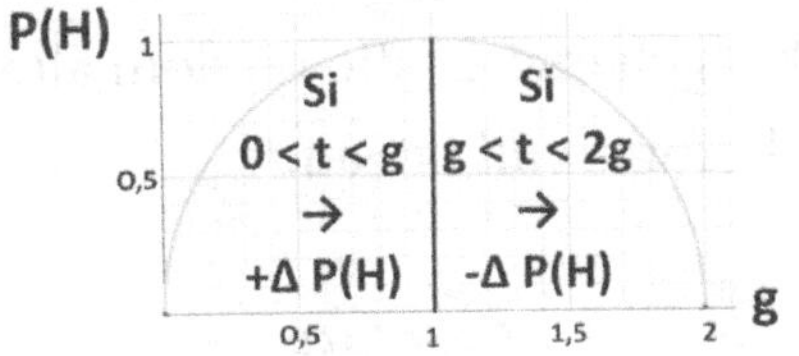

In the previous figure we can see that when all higher holons disappear, at "t = g," this is the time when H_1 finally reaches its pure order, an absolute certainty. However, there is still the other half of the time "g < t < 2g," where P(H), after having reached its maximum, goes down again, as it happened in the first stages of existence. This means that immediately after existence has reached absolute order, the furthest one can be from nothingness, the degree of chaos increases again, the cosmos approaches nothingness again.

An increase in the degree of chaos that occurs very gradually, in a very subtle way. So much so that it would be imperceptible for many, many billions of years. Something that is not very relevant as there will be nothing physical that can be there to observe it, of course, as all higher realities are extinguished. However, despite its slowness, time begins to play against existence, this time creating a negative feedback loop instead of a positive one. Therefore, eventually the level of uncertainty will be relevant enough for the second state of nothingness to re-emerge after an entire existence based on the other state. It will take its time, a significant part of "g," but after many presents in which nothingness pulsed with the same state, it will be inevitable that the probability of the second state will end up definitively interrupting the long sequence of consecutive repetitions that we have called existence. And with this interruption will come the final death of the cosmos, returning to pure chaos, nothingness, with its characteristic random alternation between its two states. As before it all began.

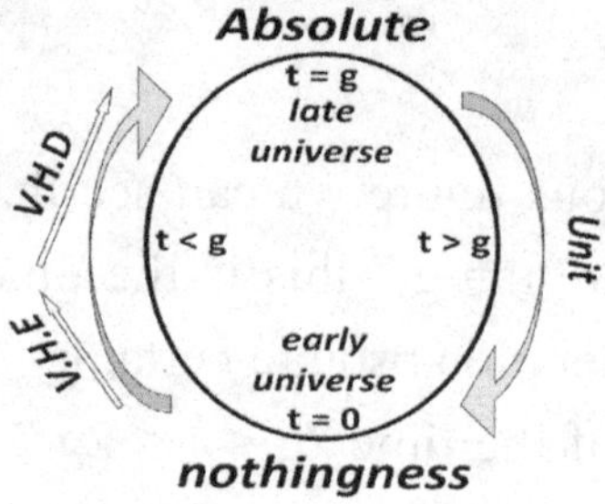

This last stage of the cosmos when "t > g" is called Unity because it contrasts with the stage of multiplicity when "t < g" where there were not only multiple holons but also multiple local existences in physical reality. This underscores the fact that existence does not fundamentally seek complexity, but that this was only a by-product of its true end, unity. Let us remember that being means not changing, being unchanging. And for that there can be only one state. The opposite of nothingness, which is pure change, is therefore in constant quantum superposition between two states, fluctuating wildly. That is why causal inertia seeks to subtract this base 2 from the formulation of the pilot model of Darwinian Physics, to reduce it to 1, precisely to be the opposite of nothingness, the "one" that Parmenides defended so much. But despite having reached absoluteness, the most natural or stable state is nothingness, not being. So, this pure being will eventually disintegrate back into nothingness.

This turns the cosmological model of Darwinian Physics into a cyclical one, because if the cosmos returns to the nothingness from which it originated, nothing would prevent a new existence from emerging from it again. Repeating the same process over and over again in an endless cycle of being and nothingness. So, the end of one universe would only be the preparation for a new one. Re-justifying the inherently dual nature through the life and death of the cosmos itself. So, although the first model of Darwinian Physics is only a toy model that still needs a lot of development, it would be revealing that the universe works similarly to a simple harmonic oscillator. This is because, like the latter, we have two extremes that constantly oscillate, only instead of oscillating between potential and kinetic energy, we would now have the constant oscillation between everything and nothing. A process in which pure potentiality is transformed into definite certainty and then disappears in total chaos, and then begins all over again. This means that there will have been infinite universes before ours, and there will be infinite universes after we are gone, because nothingness has always been and always will be. This may sound similar to the multiverse, but it is actually quite different because these infinite universes are not parallel. That is, there are no universes that coexist with ours at the same time. There is only one universe at a time, within an infinite sequence of universes, each of which exists completely separate from the rest, spatially and causally separated by nothingness. This allows Darwinian Physics to satisfy the two Copernican principles, both spatially and temporally. It fulfills the first because, as stated in the previous chapter, there is nothing physical that is privileged with respect to the expansion of the universe, because it is the nothingness that is the center that inflates existence in all directions, like a balloon. But it also fulfills the Copernican principle of time because, being a cyclic model, it also does not place us in a privileged moment, as happens in the sempiternal cosmological models of current physics, thanks to the fact that there have been infinite universes before and there will be infinite universes after ours.

Perhaps some might think that one universe at a time still makes us too special, because we would be the only ones in existence at any given time; therefore, a multiverse is still necessary to fulfill the Copernican principle. However, this is completely far-fetched and also unnecessary, because we already have a multiverse within our own universe, since each galaxy is technically already a universe in itself. In fact, until the 20th century, it was thought that the entire universe was just our galaxy, because the other galaxies that could be observed were thought to be clouds of dust orbiting around our galaxy. It was only with the advent of a new generation of more powerful telescopes that we realized that those dust clouds were actually other galaxies, as large or larger than ours, with hundreds of billions of stars, millions of light-years away. In fact, the nearest neighboring galaxy, Andromeda, which is similar in size to ours, is 2.5 million years away. This means that even if we had a phone there right now, our Hello! would not be answered for another 5 million years. Quite a slow conversation. So, pragmatically speaking, galaxies are completely unconnected places. This is why, in the early 20th century, they were not called galaxies, but rather island universes. A very apt name. And considering that there are an estimated 2 trillion galaxies with 100 billion stars each, separated by several million light-years, we can conclude that in our universe we already technically have a multiverse. But if you are not convinced by the technical separation, there are also many zones of the universe with billions of galaxies that are causally disconnected from us. That is, we will never see them, and they will never see us. This is because the estimate of 2 trillion galaxies mentioned above is only in our observable universe; however, the entire cosmos is at least 250 times larger than our observable universe due to that small curvature that we have not yet been able to measure in the CMB. A large cosmos that continues to expand. Therefore, we already have a causally (horizontally) disconnected multiverse, as postulated in cosmic inflation or string theory, because those zones will never observe or communicate with each other, since they are moving away faster than the light coming from those regions. Only these parallel universes are not in an extra space-time called multiverse with infinite bubble universes, but are very distant zones within the same space-time we call universe.

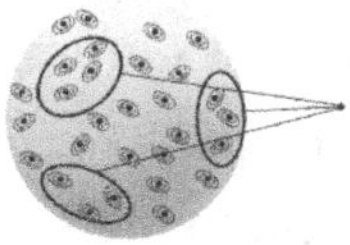

As seen in the figure above, not only can we technically consider each galaxy as a universe, but there are also several zones on the surface of this cosmos that are completely disconnected from each other, so that they function as true parallel universes, since nothing that happens in those zones can affect the others, since their light will never reach those other places. Therefore, to continue to argue for the multiverse through the Copernican principle is not only absurd, but completely redundant, since these other unobservable universes are already here, in this one universe. As the monk William of Ockham said: Pluralitas non est ponenda sine necessitate.

In spite of everything, one could still criticize that the cyclic cosmology of Darwinian Physics is equivalent to the infinite multiverse so much criticized in this book, because the only difference would be that all those absurdities like "your Marie Curie" or "your Brad Pitt" would no longer be in these parallel universes, but in these infinite universes that existed before and that will exist after. So, it is an equally absurd and disturbing result. But it is not the same thing at all, because the serial universes of Darwinian Physics are separated by nothingness itself. So, there is no way for the universes to transmit, interact, or inherit any kind of information or regularity. They are as disconnected as possible as each one creates its own space-time and its own laws (regularities) in the now. Remember that in Darwinian Physics there is true novelty and strong emergence. In the Parmenian paradigm, on the other hand, you have to share space-time and laws among these infinite parallel or serial universes, since there is no creation of new things in absoluteness or determinism. This is why many physicists look for evidence of these other parallel universes, because there may be a causal relationship or interaction between these universes. Therefore, they hope to find evidence of collisions between universes that could have left a trace in the CMB or through gravitational waves, since we would be in the same space-time. Therefore, they theorize that it would be possible to send signals between parallel universes through gravitational waves, using black holes, for example, which would act as channels or bridges between universes. In fact, many believe that the Big Bang is a white hole on the other side of a black hole in a parent universe; or that time travel would be trips to these other universes, as happens in superhero movies. A radically different situation in Darwinian Physics, because in the serial universes it is completely impossible to make any kind of communication between them, because they are completely separate, so any interaction between them is simply impossible. Therefore, in practice and in principle, the fact that there have been infinite universes before and that there will be infinite universes after is completely irrelevant, because each universe is unique, because there will be nothing before and there will be nothing after, literally.

CHAPTER XVII

THE PROBLEM OF COMPLEXITY: FINE TUNING

The last major problem we will deal with in this book is fine-tuning. This problem is closely related to the topic of the previous chapter and is concerned with explaining the origin of the values of the constants or parameters used in the two pillars of physics. Values such as the coupling constants, the speed of light (c), particle masses and charges, the gravitational constant (G), Planck's constant ($\hbar$), the fine structure constant (α), the cosmological constant (Λ), the small mass of the Higgs boson (naturalness problem), and so on. In the Standard Model of Particles alone, more than twenty parameters are needed, which are not derived from any theory, but have to be entered by hand from experimental measurements. So, the question arises. If there is no reason to justify the values of these parameters, why do we have these values and not others? This question is not trivial, because it seems that any change in any of these values, no matter how small, would not result in the universe we live in. On the contrary, it would produce very different, inhospitable, matterless, starless, or incredibly short universes. For example, if "α" were 4% larger, and therefore the electromagnetic interaction were more intense, the nuclear reaction at the center of the stars, the one that generates light and heat, could not take place due to the electrical repulsion of the atomic nuclei; or if "Λ" were more intense, the universe would not have had time to form galaxies because it expands too fast. And conversely, if "G" were more intense, the universe would collapse immediately after the Big Bang; and so on for each parameter. So, it seems that all values of the constants are finely tuned for galaxies, stars, planets and us to exist. Something that again violates the Copernican principle by making our universe special again, but no longer because of place or time, but because it has a very particular configuration of values for the constants of nature.

Faced with this unknown, many theoretical physicists found in this problem the perfect argument to continue defending their inevitable multiverse. As we saw in the previous chapter, the hypothesis of parallel universes is not only absurd, incredibly uneconomical, but also useless, because it is completely unnecessary to fulfill the Copernican space principle. However, since the multiverse is an inevitable consequence in the Parmenian theories because it is a defect of their absolute laws, we try to turn this defect into a virtue by invoking the anthropic principle again. In this way, the fine-tuning problem would be solved, because we simply live in a universe with these values for the constants, because only in universes with similar values can intelligent beings evolve, wondering about the reason for such particular values of these constants. Again, it would be like asking why our planet is so finely tuned to have life. There are just so many planets in the universe that you had to win the lottery. Well, the same argument is used, but this time it is applied to the values of the constants. The problem, again, is that there is not just one type of multiverse, but several types, such as the multiverse of chaotic or eternal cosmic inflation, the multiverse of the string theory landscape, the multiverse of Hugh Everett's many-worlds interpretation, the multiverse of the flat universe, Max Tegmark's mathematical multiverse, and so on. Many types of multiverses that do not converge and often exclude each other. Then, as stated in Chapter XV, we have two choices: we accept that a multiverse exists; or the infinite parallel universes are simply a clear sign that the theory is wrong, in a way equivalent to when a zero appears in the divisor in physicists' calculations.

In fact, the repeated comparison made in this book between the multiverse and God is reinforced here because both hypotheses have been used as a solution to the fine-tuning problem. This is because many modern theologians argue that the one truly responsible for fine-tuning the values of the constants of nature so that we could exist was God, who designed the universe in six days for just that purpose. A parallelism that should come as no surprise, since invoking both an all-powerful being living in the clouds and infinite parallel universes requires the same leap of faith. Therefore, every time a physicist so vigorously defends the fine-tuning problem to justify his multiverse, he is providing a supposedly scientific argument in favor of God, since both hypotheses are equivalent. In fact, the two are so similar that a similar situation has already occurred, but centuries ago in another science. We are talking about biological fine-tuning. In the time of Charles Darwin, as you know, there was a great debate about the validity of evolution by Natural

Selection. In this context, many argued that, for example, since the eye is such a complex object, it would be impossible for it to have arisen by chance, as it would be like finding a sophisticated pocket watch in the forest. Therefore, given its complexity and fine structure, someone must have designed and built it. In the case of the watch, it was a watchmaker, while in the case of the eye, it had to be God.

The same situation is repeated again, but this time in physics. We would only have to replace the eye with the universe itself and God with the multiverse and we would have exactly the same situation. However, we already know how the problem of biological fine-tuning was solved without the need to invoke something immeasurable like God; thanks to the fact that Charles Darwin provided a natural mechanism that allows us to build biological structures as complex and sophisticated as an eye, through a process based on the inherent randomness of genetic mutations and the selection exerted by environmental conditions. A very progressive evolutionary process (over thousands or millions of years) based on trial and error (disease and extinction). An impressive and resounding success that Darwinian Physics tries to reproduce, but this time in physics. Again, without the need for something immeasurable like infinite parallel universes, by having a natural mechanism of evolution.

So the conclusion we can draw from all of this is that when we talk about the problem of fine-tuning, we are really talking about the problem of complexity. How something incredibly complex could have come about, seemingly out of nowhere, without the need for a designer or a creator. The only difference between these two types of fine-tuning would be the nature of the complex thing we are referring to. A few centuries ago that complexity was biological, whereas now it would be physical in nature. Therefore, since it is essentially the same problem, it should require the same solution. So, it is not surprising that holistic evolution is inspired by Darwinian evolution. Two types of evolution that, while seemingly very different, are based on the same principle: chance and balance. Darwinian evolution consists of an equilibrium between random genetic mutations (the equivalent of chaos), which produce new biological characteristics, and the environment, which selects them for adaptability (the equivalent of order). This is why Darwinian Physics can solve the problem of fine-tuning physics so well, because it uses the same formula that solved biological fine-tuning. This is how Darwinian Physics answers the question of why we have this particular complexity in the universe, with these particular values in the constants; because they simply evolved that way, thanks to holistic equilibrium. So, just like biological reality, physical reality naturally tends toward complexity, thanks to this vertical evolution.

Some will find this comparison between biology and physics very forced, but it is the most "natural" one. The two situations are very similar because, due to the randomness of Natural Selection, one could also say that there are infinite parallel trees of life; in the same way that the randomness of quantum particles justifies the invention of infinite parallel universes in Hugh Everett's interpretation, for example. Remember that every living thing is just a combination of four nucleotides (A, C, T, G); but any other combination would also be possible. It is not that you are the only possible genetic combination of your parents, because they could in principle produce millions of siblings just as healthy as you. Imagine, then, the number of alternative evolutionary biological trees that exist if each living being is only one possibility among several million. You would have to multiply all those millions by every living thing that has existed from LUCA to the present day. If there are millions of organisms in a single drop of water, imagine the number of alternative life forms that have existed on this planet in 4 billion years of evolution. A number that rivals any multiverse physicists have ever invented. And despite this enormous number of stories of parallel life on Earth, you won't meet a single biologist who advocates or has advocated the existence of a multi-biosphere, much less published studies in scientific journals. And that's because, unlike physicists, biologists have no problem accepting the existence of real randomness at the level of the genetic code. That is why all those countless alternative biological realities remain just that, alternatives, potential possibilities, thanks to the existence of ontological randomness in fundamental biology, at the level of genes. The same thing would happen in Darwinian Physics. Since it is a non-deterministic or non-absolutist paradigm, it recognizes the existence of ontological randomness at the level of the foundations of physics, which automatically makes the assumption of the parallel existence of all these potential possibilities in the form of a multiverse unnecessary.

Then, the recognition of nothingness not only allows us to solve the problem of measurement in quantum mechanics and the problem of time, but also solves the mystery of why the constants of physics have these values and not others. Something that can be understood very well, continuing with the parallelism with biology, because if what varies randomly in each generation in living reality is equivalent to what varies in each instant, in each random collapse of Ψ; then the properties that remain between generations, being the most stable characteristics that define a species, would be equivalent to the laws and constants of physics. This means that in the same way that Darwinian evolution can explain the origin of species, through the emergence of those properties that remain between generations, Darwinian Physics can explain the origin of the laws and constants of physics, through the disruptive emergence of new principal holons by VHE. In this way, the laws and constants of physics would be changes that come out of nothing, but remain in evolution by becoming basic realities that allow new, higher realities to emerge. To continue the analogy, although we are quite distant (evolutionarily speaking) from fish, we still retain the spine we inherited from them. The same is true of the limbs we inherited from amphibians, or the warm blood of mammals, or the opposite thumb of Homo habilis, and so on. In other words, evolution works with what it has, since nothingness can only utilize the existing potential by choosing among the possibilities that the now gives it. And it does this in a hierarchical way, building itself in layers (holoarchy). Thus, in the case of living reality, genetic mutations are much more recurrent in the most recent traits, while the oldest traits are much more stable, since they become the reservoirs of subsequent variations, having evolved first. The same happens with the great stability in the laws and constants of physics, since they correspond to the regularities of the lower holons of the cosmos and therefore to a higher HHE or order.

By better understanding how this evolutionary process actually works, one of the major criticisms of the evolving universe is automatically resolved, which is the observation of invariance in the laws and constants of physics throughout history. The most studied case is the fine structure constant or "α." For example, in a study published December 14, 2021 in Astronomy & Astrophysics, they measured the value of "α" when the universe was only 40% of its current age using the ESPRESSO high-resolution spectrograph installed on the VLT telescopes at the European Southern Observatory (ESO) in Chile, and found that its value has not changed since then. This means that the charge intensity of the electron has been the same for more than half the lifetime of the universe. With this and other data on other constants, physicists believe they have disproved the evolution of the universe because they have not changed over billions of years; however, such a result is perfectly compatible with Darwinian Physics, since the error lies in not understanding how Darwinian evolution works. In fact, expecting the fine structure constant to change over time would be the biological equivalent of expecting the number of cervical vertebrae to change during mammalian evolution. This is because you might think that because of the length of giraffes' necks, they have many more cervical vertebrae than you do, but giraffes still have the same number of vertebrae as your neck and all other mammals. That is, seven vertebrae, with the exception of the sloth and the manatee. The only difference is that the cervical vertebrae of the giraffe's neck are apparently more elongated. So, the observation that the number of neck vertebrae in mammal fossils has not changed since 50 million years ago is obviously not evidence that there has been no evolution in mammals during all that time. Darwinian evolution simply does not work that way, because Darwinian evolution focuses on the creation of new traits, not the modification of existing traits, since the former arise from the latter. That is why it needs them. That is why the main characteristics of mammals, such as having a spine, warm blood, hair, producing milk, and so on; have remained virtually unchanged over time. This does not mean that there has been no evolution, on the contrary. These main characteristics allow new characteristics to appear, and thus new species.

In the case of the evolution of physical reality, it is exactly the same. This is why the value of the fine-structure constant has not changed for much of the life of the universe, as have the other constants; because since they appeared, there has been no reason to change them, especially when many things have been built on top of them, such as electrons, atoms, molecules, stars, and us. Thus, the VHE essentially works in a bottom-up fashion; where new properties (regularities) emerge from pre-existing properties, which are much more stable as they become the main lower holons supporting the new realities that emerge and evolve on top of them. Thus, the evidence for the evolution of physical reality is fundamentally not to be found in the temporal variation (HHE) of the fine-structure constant; but it should be found in its variation with respect to energy or scale, since the VHE is a fundamentally vertical phenomenon, a product of holarchy. And surprise, surprise, that is exactly what physicists have discovered, that the fine-structure constant (and in general any coupling constant) is actually not constant, because it varies with the energy scale. Something that was confirmed by the unification of electromagnetism with the weak nuclear force, under the Higgs mechanism. Here it was found that the fine structure constant at 0°K is very close to 1/137 (1/137.035999206 or 0.0072973525682), but at 10^{15} °K it would be 1/127 or higher. This means that the fine structure constant increases with temperature. Therefore, in the first moments of the universe, we are talking about 10^{-11}s (almost a billionth of a second), the fine structure constant was larger. This is something that can be tested experimentally in the "now," because let's not forget that the LHC is a time machine, a drill that goes into the vertical dimension towards nothingness or the Big Bang. Thus, by observing these very small scales (very low holons), we can experimentally prove that the fine structure constant is not constant. However, since it evolved or appeared on these scales with the value 1/137,035999206..., there were no more reasons in Darwinian Physics for this value to change; just as there are no reasons for the number of cervical vertebrae to vary in the next mammals to appear through evolution. Therefore, this again confirms that the Darwinian evolution of physics is a presentist phenomenon, because its process or change we can observe not with a time machine to go into an eternalistic past, but by descending vertically into smaller scales of space with higher energy. So again, far from discrediting the Darwinian evolution of the universe, modern physics would prove it.

In conclusion, our universe was built layer by layer (holoarchy), in an indeterminate manner, just as the tree of life on Earth was built and is being built by evolution. This means that any other type of universe, with different laws and constants, was perfectly possible. In fact, it must have been already, because the cosmological model of Darwinian Physics is cyclical, as we saw in the previous chapter. Therefore, there must have been an infinite number of universes before ours, with very different laws and constants. Then, since we are in a universe with these laws and constants, we can also invoke the anthropic principle, because only in universes that have evolved in a similar way could intelligent beings arise to wonder why the universe is like this and not like some other. Just as the present tree of life was not the only possibility on Earth, nor was it the most complex or the simplest, so the universe in which we live is not the most surprising or the most disillusioning. It is only one among an infinite number of other possibilities. Again, using biology as a reference, it is entirely possible that other forms of life could have existed on this planet right now. The classic example, if the asteroid had fallen a few minutes earlier or later, it would have hit the ocean (remember, the Earth rotates). The most expected place, since these cover almost ¾ of the planet's surface, but also the least catastrophic place, because after falling on land it caused a huge amount of debris to rise into the atmosphere, which for so long blocked the light of the sun, completely stopping the food chain by stopping photosynthesis. But if the asteroid had fallen in the most likely place, the ocean, the dinosaurs would have had a better chance of survival. And most likely they would still be ruling the planet. But such possibilities, the result of small changes, are infinite. Any small change in the past would have had the potential to create a completely different ecosystem than we have today because of the butterfly effect. Insects, reptiles, birds, jellyfish, archaea, or other life forms that we cannot even imagine would dominate because they would be technically alien to us. So, there would have been biological complexity, yes or no, but to what extent and of what kind Natural Selection cannot tell us. It can only explain how life can evolve, not exactly into what.

It is the same with Darwinian Physics. It cannot predict exactly what kind of specific complexity we will have, the exact values of the constants of physics, or its laws, because it is simply the product of chance. But what Darwinian Physics can explain is why a universe emerges from nothing, and why it naturally tends toward greater complexity. Just as Darwinian Natural Selection explains why life evolves and becomes more complex. In both cases, however, the theory cannot predict exactly what the outcome will be, since in principle any other outcome could have been possible. So Darwinian Physics, like the multiverse, has to rely on the anthropic principle. Only this time it is used in a much more satisfactory way, unifying all physics by answering what time and quantum randomness are; and above all, it is much more economical, not requiring infinite eternal parallel universes.

That the universe is the product of evolution is not only the most reasonable explanation for the origin of the cosmos, the laws and constants of nature; it would also easily explain why physicists do not like the Standard Model of Particles at all. In fact, they find it terrible, as the famous string theorist and best-selling author Michio Kaku said in an interview with Michel Shermer on his YouTube channel Skeptic, where he was promoting his new book The God Equation: “The Standard Model of Particles is a theory that only its mother could love.” And the reason for this disdain is not only the aforementioned fine-tuning, but also the hierarchy problem, which refers to the large difference in the energy levels at which each fundamental interaction operates. This is surprising if theoretical physicists believe that all fundamental forces are manifestations of the same underlying phenomenon, the same superforce or theory of everything. Then there is the problem of flavor, or the reason for the existence of three generations of particles: why does the electron, for example, need two heavier versions of itself if they are not stable and therefore have no vital function in matter as we know it? Nor do we understand the reason for the gigantic difference in masses between particles, where the Higgs field is also involved. For example, the top quark has more than 180 times the mass of a proton, which is made up of three quarks. A weight equivalent to a tungsten atom. Or the electron (0.51 MeV) has a mass 2000 times less than the proton (938 MeV), while the neutrino weighs at least a million times less than the electron. A mass so small that it has not yet been possible to measure it. An abysmal difference in mass that physicists find very "unnatural." In addition, the model does not integrate gravity, matter or dark energy, along with several other anomalies such as the non-violation of CP symmetry in the strong interaction. In short, there are many reasons for physicists to want to make it "prettier" by adding more symmetries, more particles, more dimensions, and so on. Precisely the predictions that are conspicuous by their absence in the experiments.

For Darwinian Physics, on the other hand, it is the opposite, because the Standard Model of Particles as it stands is not only beautiful, but also incredibly natural, because its structure reveals an organic development, the product of evolution. That is why, to the displeasure of physicists, it is incredibly asymmetric, with violations of charge symmetry, parity, and time reversal. It also required the violation of baryonic symmetry for matter to exist in the universe and to avoid all its annihilation by antimatter. The weak nuclear interaction is also left-handed, since it does not interact with the right-handed component of particles. There are also no magnetic monopoles, as there are no single electrical charges. And let us not forget the total absence of the symmetry that hurts physicists the most, supersymmetry.

So, the Standard Model of Particles is far from the canons of beauty of the current foundations of physics, because it turns out to be much more complex and asymmetric than they want it to be. By contrast, by the standards of beauty of Darwinian Physics, the Standard Model of Particles is simply sublime. In this sense, today's theoretical physicists again resemble the physicists who opposed Darwin's Natural Selection in the second half of the 19th century, because they still have the same aesthetic sense. Only the form has changed. Instead of God, it is now only his word, the absolute laws. A mathematical God who created the universe flat, symmetrical, with values close to one (natural). But unfortunately for them, the physics of the Standard Model of Particles reveals a universe with a more varied and asymmetric structure. Something they obviously find unattractive and even grotesque. All because they are looking for a simplistic, artificial beauty, the fruit of a creator, an intelligent architect. For Darwinian Physics, on the other hand, the Standard Model of Particles, without speculative elements, seems precious because it reveals a truly natural construction, the product of complexity.

Thus, the Darwinian solution is truly elegant because it is integral and consistent not only with the whole of physics, but also with the rest of the sciences. As noted in the first part of the book, this new solution scores very high on the scientific criterion of holistic consistency. A quality that is particularly evident here, since the two evolutionary mechanisms, those of biology and of physics, complement each other perfectly. In fact, evolution by Natural Selection would be only one mechanism derived from holistic evolution. Its manifestation in the main living reality. Just as memetic evolution is a mechanism derived from evolution by Natural Selection. The manifestation of Darwinian evolution in the symbolic or cultural core reality.

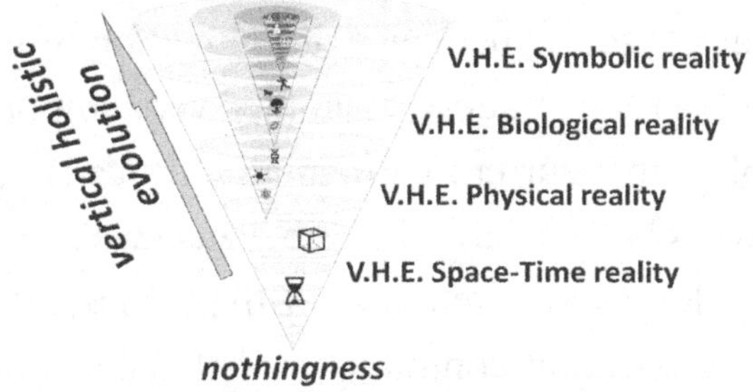

Then, by explaining emergence and complexity, Darwinian Physics would be transformed into a true theory of everything, since it would finally unify biology, psychology, and sociology with physics through the same fundamental natural mechanism. It would bring all the sciences into the same paradigm or theoretical framework, since we could now derive the emergence of new ideas and their tendency to be preserved in the cultural ecosystem through memetic evolution. An evolution that is supported by another subordinate evolution (holoarchical), the biological evolution (us). The same that also seeks to replicate and perpetuate itself in another ecosystem, this time biological, thanks to the fact that there is a lower main reality that sustains it, which also seeks to maintain and perpetuate its own patterns and regularities (laws and constants) through the evolution of physical reality. A main holon that in turn owes its existence to another even lower master reality that seeks the very same thing, space-time reality. And so on, until we arrive at the origin of all, nothingness. That which sustains the cosmos itself through time. So, in the same way that living beings try to pass on their genes to the next generation, and in the same way that politicians and religious people try to pass on their ideas (ideologies) to the next generations, the universe itself tries to extend its order (laws) to the next present.

Thanks to this consilience or unification of the sciences, Darwinian Physics can address another very important fundamental problem that current Parmenian physics is again unable to address, namely free will. This is because we can now finally connect neurology and psychology with physics, allowing us to resolve a long-standing contradiction between these sciences: the obvious experience of agency and will, and deterministic and absolute physics. It is a paradox very similar to that seen earlier in this book with chance and time. It is not surprising, then, that current physics has provided the same solution by denying its existence; for, like chance and becoming, free will is incompatible with absolute laws. So free will must be an illusion, something that exists only in our minds, just like time and chance. An even more absurd situation, since there is nothing more ironic than seeing physicists defend the non-existence of free will with such determination and will, as if their actions would change anything if they were right. Far from being anecdotal, however, the denial of free will can have far more serious consequences than the denial of time and chance, for free will is the foundation that supports our entire civilization, which is based on the fundamental assumption that we have a real ability to make our own choices. Thus, there could be no true democracy without the freedom to vote, nor could there be a free market without the freedom to buy and sell (if we have enough money, of course), nor could there be true justice without the freedom to choose between right and wrong. In fact, on this last point, there have already been experiments in psychology that show that believing in the non-existence of free will affects morality. For example, psychologists Kathleen Vohs of the University of Minnesota and Jonathan Schooler of the University of California at Santa Barbara conducted experiments in which one group of people was given a scientific text stating that free will is an illusion, and another group was given nothing. Then they played a series of games where you could cheat to get more money. The result was that the group that read the scientific text against free will cheated significantly more than the group that did not read the text. And this can be very serious, especially in these troubled times of great political, economic, and social conflict, when it is vital to believe that we are in control and that we have the will to make things better. Here again is the parallel between religion and Parmenian physics, for in the former there has also long been a great conflict over the compatibility of free will and God's omniscience. If he knows everything, including one's destiny, how is it possible for us to have true freedom; isn't everything already determined if God already knows? Exactly the same dilemma since the laws of physics are supposed to be the word of God.

Fortunately, Darwinian Physics is not so paradoxical, since it is fully compatible with this neurological-psychological phenomenon, as well as with time and chance. In fact, these three phenomena would be closely related, which explains why current Parmenian physics is so hostile to all three. This is because for there to be true freedom, there cannot be determinism, because the future would already exist. So, although free will is obviously a very complex phenomenon, whose direct explanation lies in living reality, through biology and neurology; in any case, physics has to allow it in principle, being its mother reality (holoarchy). Darwinian Physics, therefore, recognizes and explains this phenomenon, without falling into the extreme, naive and even dangerous simplism of saying that there is not a shred of freedom, responsibility or control, that everything is an illusion, because everything that has happened, is happening and will happen is already written in stone in the block universe. But neither does it go to the other extreme, that of complete freedom, denying any kind of genetic, social, economic, cultural, family predisposition, and the laws of physics themselves. Much as we would like to, we will not fly away like Neo in the Matrix. What Darwinian Physics really does is simply recognize free will for what it is, a complex phenomenon within a holoarchy that must be respected. Therefore, holistic presentism allows Darwinian Physics to define free will in a realistic way, recognizing the great weight of horizontal causality that determines much of our lives, such as: our culture, family, socioeconomic status, experiences, genetics, geography, and so on; but it also recognizes the existence of a small but important area of freedom, of chance, due to the vertical causality of the "now."

So, there is room for authentic agency, since, as the book points out several times, the future does not exist. And although the information in the "now" (past) has a great influence on how the next present will manifest (causal inertia), there is still real freedom in how it might manifest. Something that is proven in the true indeterminacy of quantum systems (Heisenberg's uncertainty principle and measurement problem). So, this very small freedom may mean nothing for many simple macroscopic systems, such as a rock, but for the most complex structure known in the whole universe, our brain, these small quantum fluctuations at the level of neurotransmitters traveling in the synapses between neurons could be significant because they can be greatly amplified by chaotic processes. In the same way, more and more quantum effects are being discovered that explain biological phenomena such as: the great efficiency of photosynthesis in plants, which could be due to quantum superposition; or the navigation of birds due to their great sensitivity to the Earth's magnetic field, which could be due to quantum entanglement. Then, thanks to vertical causality, this very small freedom of the now can be used to build our free will, amplified by a positive feedback loop. Exactly what the great complexity of our brain would allow us to do. A topic that will be explored and developed in depth in the next book, which will continue to explore the implications of this new scientific paradigm, but this time more focused on biology, neurology, psychology and sociology. So, if you are interested in these topics as well, if you are a true generalist, then watch for these next books to be published.

But to return to this book, not only would freedom not be an illusion, as physicists currently postulate, but it would be omnipresent, since it is inherent in time itself. Therefore, although the cosmological model of Darwinian Physics is cyclical, each universe that emerges is completely unique and free, since it is constructed in the "now," in an authentically spontaneous way. And as we already know, the "now" has its origin in the most uncertain and indeterminate place possible, in nothingness. Therefore, although there have been infinite universes before, our universe "now" is still authentically free, only it is less and less so because of its own history that it has built (causal inertia). A true independence that becomes even more apparent when we consider that this series of cyclic universes is actually separated by nothing. Therefore, there is no connection, no continuation from one universe to another, because there is no transfer of information. That is why each universe is really free, because even each present is free, at least in principle. So even if there is a mechanism that naturally generates complexity, predicting exactly what kind of complexity will emerge remains a matter of luck. Not because there is a flaw in the theory, but because nature itself does not know, due to ontological randomness. Thus, to know the laws and constants, one needs traditional physics, which can discover them through its reductionist-analytical approach, together with precise observations and exact mathematical models.

In view of the above, the reader is reminded that this book has neither the intention nor the purpose of discrediting or taking away any merit from current theoretical physics, since its incredible success is obvious. It is only a reminder that this alone is not enough, that a second type of physics is also necessary. A more philosophical and holistic one (synthesis), but equally rigorous, that complements and supports traditional effective physics, so that the latter can continue with its extraordinary success and efficiency, as already stated in the first part of the book. Because a new paradigm does not consist in denying the previous paradigm; on the contrary, it consists in expanding it, in surpassing it. Therefore, it is ridiculous to reject current physics, since it obviously works wonderfully in experiments. The true goal of Darwinian Physics is to serve as a support for the systematic and consistent integration of all exact and reductionist physical theories, to give them meaning and coherence by unifying them with the other scientific theories, in order to return to a single scientific worldview, because we have always had a single and integral universe. A consilience or unification of the sciences that would bring great benefits not only to science, but also to society. And not only by promoting the development of new technologies, but also by helping to fill the existential void that so many people are currently trying to fill by returning to religion (or worse, to new religions and sects), to extremist political ideologies, or to pseudo-scientific New Age beliefs. This is thanks to the provision of a clearly better alternative. The most complete and integral worldview possible, built on the most reliable source of knowledge that can exist, science. A scientific unification that would promote not only a much more harmonious integration with nature, but above all, among ourselves, by discouraging the fanaticism and tribalism characteristic of biased and partial worldviews that omit or deny the rest of the knowledge, and with it, the people who believe what they do not. This is precisely what holistic consistency seeks to prevent, by avoiding this process of systematic mutilation not only of science, but also of our vision of nature as a whole and of ourselves.

unification of the sciences that would bring great benefits not only to science, but also to society. And not only by promoting the development of new technologies, but also by helping to fill the existential void that so many people are currently trying to fill by returning to religion (or worse, to new religions and sects), to extremist political ideologies, or to pseudo-scientific New Age beliefs. This is thanks to the provision of a clearly better alternative: the most complete and integral worldview possible, built on the most reliable source of knowledge that can exist: science. A scientific unification that would promote not only a greater and more harmonious integration of [illegible], but above all among ourselves, by discouraging the fanaticism and tribalism characteristic of biased and partial worldviews that only deny the rest of the knowledge and with it the people who hold what they deny. This is precisely what [illegible] [illegible] seeks to prevent, by [illegible] the [illegible] of [illegible] not only of science but also of our vision of nature as a whole and of ourselves.

INVITATION

This has been a long and exhausting lonely adventure, not because I wanted it to be that way, quite the contrary. For more than ten years I have visited and written to several hundred physicists, yet only a few have been kind enough to reply, expressing that they are not even interested in knowing what it is all about.

I understand that there are many people who, due to the same thing I criticize academia for, extreme specialization or simply out of ignorance (Dunning-kruger effect), believe they know more than physicists, and create their own theories as to why Einstein was wrong. However, if you have read this far, you will know that I am not one of those people. Unfortunately such people abound, and they annoy and harass many physicists and scientists daily with their crazy and ridiculous speculations. So, I understand the great skepticism and even scorn I have received. That is why I ended up writing this book, being a rather desperate measure, practically the only way to be heard.

For this reason, if you are a physicist and you consider that the ideas presented here have value, please write to the mail that appears in the first pages of the book. This attitude of not wanting to understand quantum mechanics is really starting to be very harmful. And this proposal not only makes sense of quantum mechanics, but also does so by finally explaining what time is, unifying it with Einstein's theory of gravity. For all these reasons it is a proposal that has every merit to be considered. Its only sin is that it does not arise within academia. But if the latter is the problem, because of the extreme specialization, and if scientific knowledge is truly universal, truly democratic, then that is not a reason to dismiss this book a priori.

Made in the USA
Monee, IL
22 February 2025

12738729R00341